STUDENT'S SOLUTIONS MANUAL

David R. Lund
University of Wisconsin – Eau Claire

WITH

CHRIS FRANKLIN
University of Georgia

BERNARD J. MORZUCH
University of Massachusetts – Amherst

to accompany

INTRODUCTORY STATISTICS

FIFTH EDITION

NEIL A. WEISS

 **ADDISON-WESLEY**

An imprint of Addison Wesley Longman, Inc.

Reading, Massachusetts • Menlo Park, California • New York • Harlow, England
Don Mills, Ontario • Sydney • Mexico City • Madrid • Amsterdam

Reproduced by Addison-Wesley from camera-ready copy supplied by the author.

Copyright © 1999 Addison Wesley Longman.

All rights reserved. No part of this publication may be reproduced, stored in a retrieval system, or transmitted, in any form or by any means, electronic, mechanical, photocopying, recording, or otherwise, without the prior written permission of the publisher. Printed in the United States of America.

ISBN   0-201-88322-8

2 3 4 5 6 7 8 9 10 VG 0100

PREFACE

Student's Solutions Manual is designed to be used with the text *Introductory Statistics, Fifth Edition* by Neil A. Weiss. It provides solutions to every odd-numbered exercise in the text and solutions for all of the review problems at the ends of each of Chapters 1-16 of the text. The solutions are more than answers--intermediary steps in the process of solving the exercises are also provided. Throughout the text, the author has provided many similar exercises in pairs, one of them with an odd number and one of them with an even number. Thus, this solutions manual should enable you, the student, to complete almost all of the exercises in the text. In addition, you will find that when the solution of one of the odd-numbered exercises in this manual requires prior work from another exercise, that exercise is also odd-numbered and its solution is also in this manual.

The Review Problems at the end of each chapter will enable you to determine whether you have accomplished the goals listed in the Chapter Review. To maximize the benefit from your review activities, you should solve these problems prior to looking at the solutions in this manual.

D.R.L.
Eau Claire, WI

CONTENTS

CHAPTER 1 ANSWERS

Exercises 1.1

1.1 (a) The *population* is the collection of all individuals, items, or data under consideration in a statistical study.

(b) A *sample* is that part of the population from which information is collected.

1.3 Descriptive methods are methods for organizing and summarizing information and include graphs, charts, tables, averages, measures of variation, and percentiles.

Exercises 1.2

1.5 This study is inferential. Data from a sample of Americans are used to make an estimate of (or an inference about) average TV viewing time.

1.7 This study is inferential. Data from a 10% sampling of all 1990 U.S. death certificates are used to make an estimate of (or an inference about) leading causes of death.

1.9 This study is descriptive. It is a summary of attendance figures for selected sports.

1.11 This study is descriptive. It is a summary of the annual averages of the Dow Jones Industrial Averages.

1.13 (a) Since an estimate of the average life of a tire is sought, this study is inferential.

(b) Since winning times are simply listed, this study is descriptive.

(c) Since the politician does nothing more than obtain the number of votes cast for her opponent, this study is descriptive.

(d) Since the study intends to generate information about the drug to be used in estimating its effectiveness, this study is inferential.

(e) Since the study will be used to estimate the percentage of voters that will vote for the candidate, it is inferential.

(f) Since the economist is going to estimate the average income of all California residents, the study is inferential.

(g) Since the owner is summarizing the salaries of her employees, the study is descriptive.

1.15 (a) This is a descriptive statement. It is a summary of those surveyed.

(b) This would be an inferential statement. Data from a study would be used to make an estimate of (or an inference about) the percentage of Americans who would choose organically grown produce.

Exercises 1.3

1.17 Spreadsheet software, graphing calculators, and dedicated statistical software can be used to conduct statistical and data analyses.

1.19 (a) To point is to move the mouse so that the mouse cursor is on the desired item.

(b) To click is to press and release the mouse button when the cursor is on the desired item.

(c) To double click is to rapidly press and release the mouse button twice when the cursor is on the desired item.

(d)   To drag is to press and hold down the mouse button at the desired start location, move the mouse so the mouse cursor is at a desired end location, and then release the button. The item at the start location will be 'dragged' to the new location.

1.21 The user is being asked to click on the Menu item **Calc**, then click on the sub-menu item **Random Data**, and finally click on the next sub-menu item **Normal**.

1.23 C5

1.25 (a) First we use the method explained in Example 1.5 of the text to store the ENERGY consumption values in C1 and name the column ENERGY. Click anywhere on the worksheet, then click on the column-name cell for C1, type ENERGY, and then press the ENTER key.

The cursor should now be in row 1 of Column C1. If not, click on that cell and click on the data-entry arrow in the upper left corner of the worksheet so that it is pointing down. Now type 130 and press the ENTER key. Type 55 in row 2 of C1 and press the ENTER key. Continue in this manner until all 50 data values have been entered.

(e) We use the method explained in Example 1.6 of the text to import the data file from the file *1-25.dat* found in the Exercise directory of *DataDisk*. Choose **File ▶ Other Files ▶ Import Special Text...**, then type ENERGY2 IN THE **Store Data in columns(s)** text box, and click **OK**. Then type A:\Exercise\1-25 in the **File name** text box and click **Open**.

1.27 (a) First we use the method explained in Example 1.5 of the text to store the data on heights in C1 and name the column HEIGHT1. Click anywhere on the worksheet, then click on the column-name cell for C1, type HEIGHT1, and then press the ENTER key.

The cursor should now be in row 1 of Column C1. If not, click on that cell and click on the data-entry arrow in the upper left corner of the worksheet so that it is pointing down. Now type 65 and press the ENTER key. Type 67 in row 2 of C1 and press the ENTER key. Continue in this manner until all 11 data values have been entered.

Then store the data on weights in C2 in the same manner. Name the column WEIGHT1, and enter the data in C2.

(b) We use the method explained in Example 1.6 of the text to import the data file from the file *1-27.dat* found in the Exercise directory of *DataDisk*. Choose **File ▶ Other Files ▶ Import Special Text...**, then type HEIGHTS2 WEIGHTS2 in the **Store Data in columns(s)** text box, and click **OK**. Then type A:\Exercise\1-27 in the **File name** text box and click **Open**.

1.29 (a) First we use the method explained in Example 1.5 of the text to store the data on public college number of volumes in C1 and name the column PUBLIC1. Click anywhere on the worksheet, then click on the column-name cell for C1, type PUBLIC1, and then press the ENTER key.

The cursor should now be in row 1 of Column C1. If not, click on that cell and click on the data-entry arrow in the upper left corner of the worksheet so that it is pointing down. Now type 79 and press the $\boxed{\text{ENTER}}$ key. Type <u>41</u> in row 2 of C1 and press the $\boxed{\text{ENTER}}$ key. Continue in this manner until all 7 data values have been entered.

Then store the data on private college number of volumes in C2 in the same manner. Name the column PRIVATE1, and enter the 6 data values in C2.

(b) We use the method explained in Example 1.6 of the text to import the data file from the file *1-29.dat* found in the Exercise

directory of *DataDisk*. Choose **File ▶ Other Files ▶ Import Special Text...**, then type <u>PUBLIC2 PRIVATE2</u> in the **Store Data in columns (s)** text box, and click **OK**. Then type <u>A:\Exercise\1-29</u> in the **File name** text box and click **Open**.

Exercise 1.6

1.31 A census is generally time consuming, costly, frequently impractical, and sometimes impossible.

1.33 The sample should be representative so that is reflects as closely as possible the relevant characteristics of the population under consideration.

1.35 Dentists form a high-income group whose incomes are not representative of the incomes of Seattle residents in general.

1.37 (a) Probability sampling uses a randomizing device such as tossing a coin or die, or consulting a random number table to decide which members of the population will constitute the sample.

(b) False. It is possible for the randomizing device to randomly produce a sample which is not representative.

(c) Probability sampling eliminates unintentional bias, permits the researcher to control the chance of obtaining a nonrepresentative sample, and guarantees that the techniques of inferential statistics can be applied.

1.39 Simple random sampling.

1.41 (a) GLS, GLA, GLT, GSA, GST, GAT, LSA, LST, LAT, SAT.

(b) There are 10 samples, each of size three. Each sample has a one in 10 chance of being selected. Thus, the probability that a sample of three salaries is the first, second, or tenth sample on the list presented in part (a) is 1/10.

1.43 (a) MS, FS, LG, BI   MS, FS, BI, JC   MS, LG, JS, JC   FS, LG, JS, JC

MS, FS, LG, JS   MS, FS, JS, JC   MS, BI, JS, JC   FS, BI, JS, JC

MS, FS, LG, JC   MS, LG, BI, JS   FS, LG, BI, JS   LG, BI, JS, JC

MS, FS, BI, JS   MS, LG, BI, JC   FS, LG, BI, JC

(b) One procedure for taking a random sample of four representatives from the six is to write the initials of the representatives on six separate pieces of paper, place the six slips of paper into a box, and then, while blindfolded, pick four of the slips of paper. Or, number the representatives 1-6, and use a table of random numbers or a random-number generator to select four different numbers between 1 and 6.

(c) P(MS, BI, JS, and JC) = 1/15; P(FS, LG, BI, and JC) = 1/15

1.45   I am using Table IX to obtain a list of 25 random numbers between 1 and 685 as follows. First, I pick a random starting point by closing my eyes and putting my finger down on the table.

My finger falls on the three digits located at the intersection of line number 16 with columns 07, 08, and 09. (Notice that the first column of digits is labeled "00" rather than "01"). The selected digits are 552. This is my starting point.

I now go down the table and record the three-digit numbers appearing directly beneath 552. Since I want numbers between 1 and 685 only, I throw out numbers between 686 and 999, inclusive. I also discard the number 000.

After 552, I record 155 and 008 but skip 765.

Now that I've reached the bottom of the table, I move directly rightward to the adjacent column of three-digit numbers and go up.

I record 016, 534, 593, skip 964, record 667, 452, 432, 594, skip 950, record 670, 001, 581, 577, 408, skip 948, 807, 862, record 407, 047, and skip 977.

Now that I've reached the top of the table, I move directly rightward to the adjacent column of three-digit numbers and go down.

I record 422, 293, 182, skip 745, 911, 729, 815, 847, 892, record 163, skip 890, record 378, 596, skip 794, 852, record 428, skip 755, 748, and record 242.

I've finished recording the 25 random numbers. In summary, these are:

552	593	670	407	163
155	667	001	047	378
008	452	581	422	596
016	432	577	293	428
534	594	408	182	242

1.47   I am using Table IX to obtain a list of 10 random numbers between 1 and 500 as follows. First, I pick a random starting point by closing my eyes and putting my finger down on the table.

My finger falls on the three digits located at the intersection of line number 00 with columns 34, 35, and 36. The selected digits are 356. This is my starting point.

I now go down the table and record the three-digit numbers appearing directly beneath 356. Since I want numbers between 1 and 500 only, I throw out numbers between 501 and 999, inclusive. I also discard the number 000.

After 356, I skip 876, record 351, skip 717, record 239, 455, 431, 008, skip 900, 721, record 259, 068, 156, skip 570, 540, 937, 989, and record 047.

I've finished recording the 10 random numbers. In summary, these are:

356	239	431	259	156
351	455	008	068	047

1.49   (a)   The possible samples of size one are G   L   S   A   T

(b)   There is no difference between obtaining a sample of size one and selecting one official at random.

1.51   (a)   Set a range of 1 to 685 integers and have a random-number generator generate 25 numbers in this range. Match these numbers to a numbered list of the 685 employees.

(b)   Exercise for the student.

1.53 First, we store the numbers 1-500 in a column named INTER500 as follows:
Click on the column-name cell for C1 and type <u>INTER500</u>. Similarly, type
<u>SRS</u> in the column-name cell for C2.

- Choose **Calc ▶ Make Patterned Data ▶ Simple Set of Numbers...**

- Type <u>INTER500</u> in the **Store patterned data in** text box.

- Select the **Patterned sequence** option button

- Click in the **From first value** text box and type <u>1</u>

- Click in the **To last value** text box and type <u>500</u>

- Click **OK**

- Now choose **Calc ▶ Random Data ▶ Sample From Columns...**

- Type <u>10</u> in the small text box after **Sample**

- Click in the **Sample ☐ rows from column(s)** text box and specify
NUMBERS

- Click in the **Store Samples in** text box and type <u>SRS</u>

- Click **OK**

To print the numbers now in the SRS column, we choose **Manip ▶ Display
Data...**, specify SRS in the **Columns, Constants, and matrices to display**
text box, and then click **OK**. The output is then displayed in the
Session window and can be printed from there.

Exercises 1.7

1.55 (a)   Answers will vary, but here is the procedure: (1) Divide the
population size, 685, by the sample size, 25, and round down to
the nearest whole number; this gives 27.   (2) Use a table of
random numbers (or a similar device) to select a number between 1
and 27, call it k.   (3) List every 27th number, starting with k,
until 25 numbers are obtained; thus the first number on the
required list of 25 numbers is k, the second is $k+27$, the third is
$k+54$, and so forth (e.g., if $k=6$, then the numbers on the list are
6, 33, 60, ...).

(b)   Systematic random sampling is easier.

(c)   Yes, unless there is some kind of cyclical pattern in the listing
of the employees.

1.57 Yes, since there is no cyclical pattern in the listing. In fact,
because the listing is by sales, one could argue that in this case
systematic random sampling is preferable to simple random sampling.

1.59 (a)   Number the suites from 1 to 48, use a table of random numbers to
randomly select three of the 48 suites, and take as the sample the
24 dormitory residents living in the three suites obtained.

(b)   Probably not, since friends are likely to have similar opinions.

(c)   Proportional allocation dictates that the number of freshmen,
sophomores, juniors, and seniors selected be, respectively, 8, 7,
6, and 3. Thus a stratified sample of 24 dormitory residents can
be obtained as follows: Number the freshmen dormitory residents

from 1 to 128 and use a table of random numbers to randomly select 8 of the 128 freshman dormitory residents; number the sophomore dormitory residents from 1 to 112 and use a table of random numbers to randomly select 7 of the 112 sophomore dormitory residents; and so forth.

1.61   From the information about the sample, we can conclude that the population of interest consists of all adults in the continental U.S. The sample size was 2010 except that for questions about politics, only registered voters were considered part of the sample. The sample size for those questions was 1,637.

The overall procedure for drawing the sample was multistage (actually, three stages were used) sampling: the first stage was to randomly select 520 geographic points in the continental U.S.; then proportional sampling was used to randomly sample a number of households with telephones from each of the 520 regions in proportion to its population; finally, once each household was selected, a procedure was used to ensure that the correct numbers of adult male and female respondents were included in the sample.

The last paragraph indicates the confidence that the poll-takers had in the results of the survey, that is, that there is a 95% chance that the sample results will not differ by more than 2.2 percentage points in either direction from the true percentage that would have been obtained by surveying all adults in the actual population, or by more than 2.5 percentage points in either direction from the true percentage that would have been obtained by surveying all registered voters in the population. The last sentence says that smaller samples have a larger "margin of error," an explanation for the difference in the maximum percentage points of error for all adults and for registered voters.

Exercises 1.8

1.63   Observational studies can reveal only association, whereas designed experiments can help establish <u>cause and effect</u>.

1.65   Here is one of several methods that could be used: Number the women from 1 to 4753; use a table of random numbers or a random-number generator to obtain 2376 different numbers between 1 and 4753; the 2376 women with those numbers are in one group, the remaining 2377 women are in the other group.

1.67   Designed experiment. The researchers did not simply observe the two groups of children, but instead randomly assigned one group to receive the Salk vaccine and the other to get a placebo.

1.69   Observational study. The researchers simply observed the two groups.

1.71   (a)   Experimental units are the individuals or items on which the experiment is performed.

(b)   When the experimental units are humans, we call them subjects.

1.73   (a)   Experimental units: the twenty flashlights

(b)   Response variable: battery lifetime in a flashlight

(c)   Factors: one factor - battery brand

(d)   Levels of each factor: four brands of batteries were tested

(e)   Treatments: four brands of batteries were tested

1.75   (a)   Experimental units: the cacti used in the study

(b)   Response variable: the total length of cuttings at the end of 16 months

(c)   Factors: two factors - Broadleaf P-4 polyacrylamide and irrigation scheme

(d)   Levels of each factor: two levels of P4 (used or not used) and five irrigation schemes

(e)   Treatments: the ten combinations of levels of P4 and irrigation schemes resulting from testing each level of use of P4 with each of the five irrigation schemes

1.77 (a)   Experimental units: the product being sold

(b)   Response variable: the number of units of the product sold

(c)   Factors: two factors - display type and price

(d)   Levels of each factor: three types of display of the product and three pricing schemes

(e)   Treatments: the nine different combinations of display type and price resulting from testing each of the three pricing schemes with each of the three display types

1.79 This is a completely randomized design since the flashlights were randomly assigned to the different battery brands.

1.81 Double-blinding guards against bias, both in the evaluations and in the responses. In the Salk-vaccine experiment, double-blinding prevented a doctor's evaluation from being influenced by knowing which treatment (vaccine or placebo) a patient received; it also prevented a patient's response to the treatment from being influenced by knowing which treatment he or she received.

1.83 The statement seems to indicate that Ross believes that if women stay at home with their children, they are, on average, likely to be more depressed (psychologically disadvantaged position) than those who have a job and no children. This would be incorrectly implying a cause and effect relationship (staying at home is more likely to cause depression) as a result of an observational study. There may be other equally plausible explanations for the relationship found in her research.

REVIEW TEST FOR CHAPTER 1

1. Student exercise.

2. Descriptive statistics are used to display and summarize the data to be used in an inferential study. Preliminary descriptive analysis of a sample often reveals features of the data that lead to the choice or reconsideration of the choice of the appropriate inferential analysis procedure.

3. Descriptive study. The scores are merely reported.

4. Inferential study. A sample of thousands is used to make an inference about the mental health of millions.

5. Inferential study. A conclusion about all Americans is based on the information obtained from a sample of 20,000 households.

6. Descriptive study. The statistics reported are facts, not estimates made from a sample.

7. Inferential study. The table contains estimates of the victimization rates based on samples of 60,000 households per month.

8. A literature search should be made before planning and conducting a study.

9. (a)   A representative sample is one which reflects as closely as possible the relevant characteristics of the population under consideration.

(b)   Probability sampling involves the use of a randomizing device such as tossing a coin or die, using a random number table, or using computer software which generates random numbers to determine which members of the population will make up the sample.

(c)   A sample is a simple random sample if all possible samples of a given size are equally likely to be the actual sample selected.

10.   Since Yale is a highly selective and expensive institution, it is very unlikely that the students at Yale are representative of the population of all college students or that their parents are representative of the population of parents of all college students.

11.   (a)   This method does not involve probability sampling. No randomizing device is being used and people who do not visit the campus have no chance of being included in the sample.

(b)   The dart throwing is a randomizing device which makes all samples of size 20 equally likely. This is probability sampling.

12.   (a)   WA,OR,CA   WA,OR,AK   WA,OR,HI   WA,CA,AK   WA,CA,HI

WA,AK,HI   OR,CA,AK   OR,CA,HI   OR,AK,HI   CA,AK,HI

(b)   Each of the 10 samples in part (a) has a 1/10 chance of being chosen. Thus, the first sample in the list has a 1/10 chance of being chosen, the second sample in the list has a 1/10 chance of being chosen, and the tenth sample in the list has a 1/10 chance of being chosen.

13.   (a)   Table IX can be employed to obtain a sample of 50 random numbers between 1 and 7246 as follows. First, I pick a random starting point by closing my eyes and putting my finger down on the table.

My finger falls on four digits located at the intersection of some line number 16 with four columns. (Notice that the first column of digits is labeled "00" rather than "01"). This is my starting point.

I now go down the table and record the all four-digit numbers appearing directly beneath the first four-digit number which are between 0001 and 7246 inclusive. I throw out numbers between 7247 and 9999, inclusive. I also discard the number 0000. When the bottom of the column is reached, we move over to the next sequence of four digits and work our way back up the table. Continue in this manner. When 50 distinct four-digit numbers have been recorded, the sample is complete.

(b)   Starting in row 14, columns 16-19, we record 6293, 3331, 6151, 0940, 1928, skip 7328, move to the right and record 4124, 0219, 0341, 0208, 5470, skip 7830, 8363, 9865, 7604, record 7075, 6679, 4106, 6899, skip 9781, record 5236, 2293, 6984, skip 7485, 7315, 9708, move to the right, record 1424, 0315, skip 8645, record 4198, 4096, 7132, 0189, skip 7722, record 3094, 0529, 3909, 1966, 4204, skip 8054, record 5802, 1065, skip 9337, record 1946, skip 8612, record 3335, move to the right, record 6417, 0102, 5669, 4865, skip 7398, record 5203, 2200, 0788, 6875, 5945, 6562, 5109, 2930, 0353, 0538, 3098, 5306, 1793, 4830.

The final list of numbers is

6293, 3331, 6151, 0940, 1928, 4124, 0219, 0341, 0208, 5470,

7075, 6679, 4106, 6899, 5236, 2293, 6984, 1424, 0315, 4198,

4096, 7132, 0189, 3094, 0529, 3909, 1966, 4204, 5802, 1065,

1946, 3335, 6417, 0102, 5669, 4865, 5203, 2200, 0788, 6875,

5945, 6562, 5109, 2930, 0353, 0538, 3098, 5306, 1793, 4830.

14.   (a)   Systematic random sampling is done by first dividing the population size by the sampling size and round the result down to

the next integer, say k. Then we select one random number, say r, between 1 and k. That number will be the first member of the sample. The remaining members of sample will be those numbered r+k, r+2k, r+3k, ... until a sample of size n has been chosen. Systematic sampling will yield results similar to simple random sampling as long as there is nothing systematic about the way the members of the population were assigned their numbers.

(b)   In cluster sampling, clusters of the population (such as blocks, precincts, wards, etc.) are chosen at random from all such possible clusters. Then every member of the population lying within the chosen clusters is sampled. This method of sampling is particularly convenient when members of the population are widely scattered and is appropriate when the members of each cluster are representative of the entire population.

(c)   In stratified random sampling with proportional allocation, the population is first divided into subpopulations, called strata, and simple random sampling is done within each stratum. Proportional allocation means that the size of the sample from each stratum is proportional to the size of the population in that stratum.

15.   (a)   Answers will vary, but here is the procedure: (1) Divide the population size, 7246, by the sample size 50, and round down to the nearest whole number; this gives 144. (2) Use a table of random numbers (or a similar device) to select a number between 1 and 144, call it k. (3) List every 144th number, starting with k, until 50 numbers are obtained; thus the first number on the required list of 50 numbers is k, the second is $k+144$, the third is $k+288$, and so forth (e.g., if $k=86$, then the numbers on the list are 86, 230, 374, ...).

(b)   Yes, unless for some reason there is a cyclical pattern in the listing of registered voters.

16.   (a)   Proportional allocation dictates that 10 full professors, 16 associate professors, 12 assistant professors, and 2 instructors be selected.

(b)   The procedure is as follows: Number the full professors from 1 to 205, and use Table IX to randomly select 10 of the 205 full professors; number the associate professors from 1 to 328, and use Table IX to randomly select 16 of the 328 associate professors; and so on.

17.   (a)   In an observational study, researchers simply observe characteristics and take measurements. In a designed experiment, researchers impose treatments and controls and *then* observe characteristics and take measurements.

(b)   An observational study can only reveal associations between variables, whereas a designed experiment can help to establish cause and effect relationships.

18.   This is an observational study. The researchers at the University of Michigan simply observed the poverty status and IQs of the children.

19.   (a)   This is a designed experiment.

(b)   The treatment group consists of the 158 patients who took AVONEX. The control group consists of the 143 patients who were given a placebo. The treatments were the AVONEX and the placebo.

20.   The three basic principles of experimental design are control, randomization, and replication. Control refers to methods for controlling factors other than those of primary interest. Randomization means randomly dividing the subjects into groups in order to avoid

unintentional selection bias in constituting the groups. Replication means using enough experimental units or subjects so that groups resemble each other closely and so that there is a good chance of detecting differences among the treatments when such differences actually exist.

21. (a) Experiment units: the doughnuts

(b) Response variable: amount of fat absorbed

(c) Factor(s): fat type

(d) Levels of each factor: four types of fat

(e) Treatments: four types of fat

22. (a) Experiment units: tomato plants

(b) Response variable: yield of tomatoes

(c) Factor(s): tomato variety and density of plants

(d) Levels of each factor: These are not given, but tomato varieties tested would be the levels of variety and the different densities of plants would be the levels of density.

(e) Treatments: Each treatment would be one of the combinations of a variety planted at a given plant density.

23. (a) This is a completely randomized design since the 24 cars were randomly assigned to the 4 brands of gasoline.

(b) This is a randomized block design. The four different gasoline brands are randomly assigned to the four cars in each of the six car model groups. The blocks are the six groups of four identical cars each.

(c) If the purpose is learn about the mileage rating of one particular car model with each of the four gasolines, then the completely randomized design is appropriate. But if the purpose is to learn about the performance of the gasoline across a variety of cars (and this seems more reasonable), then the randomized block design is more appropriate and will allow the researcher to determine the effect of car model as well as of gasoline type on the mileage obtained.

24. (a) First we use the method explained in Example 1.5 of the text to store the data on ages in C1 and name the column AGE1. Click anywhere on the data window, then click on the column-name cell for C1, type AGE1, and then press the ⎥ ENTER ⎥ key.

The cursor should now be in row 1 of Column C1. If not, click on that cell and click on the data-entry arrow in the upper left corner of the Data window so that it is pointing down. Now type 48 and press the ⎥ ENTER ⎥ key. Type 41 in row 2 of C1 and press the ⎥ ENTER ⎥ key. Continue in this manner until all 35 data values have been entered.

(b) We use the method explained in Example 1.6 of the text to import the data file from the file *r1-26b.dat* found in the Exercise

directory of *DataDisk*. Choose **File ▶ Other Files ▶ Import Special Text...**, then type AGE2 in the **Store Data in columns(s)** text box, and click **OK**. Then type A:\Exercise\r1-26b in the **File name** text box and click **Open**.

25. (a) First we use the method explained in Example 1.5 of the text to store the horsepower and mileage data in C1 and C2 respectively and name the columns HP1 and MPG1. Click anywhere on the data window, then click on the column-name cell for C1, type HP1, and then press the ‖ENTER‖ key.

The cursor should now be in row 1 of Column C1. If not, click on that cell and click on the data-entry arrow in the upper left corner of the Data window so that it is pointing down. Now type 155 and press the ‖ENTER‖ key. Type 68 in row 2 of C1 and press the ‖ENTER‖ key. Continue in this manner until all 12 data values have been entered.

Click on the column-name cell for C2, type MPG1, and press the ‖ENTER‖ key. The cursor should now be in row 1 of Column C2. Type 16.9 and press the ‖ENTER‖ key. Type 30.0 in row 2 of C2 and press the ‖ENTER‖ key. Continue until all 12 MILEAGE values have been entered.

(b) We use the method explained in Example 1.6 of the text to import the data file from the file *1r-25.dat* found in the Exercise directory of *DataDisk*. Both variables are in the same file.

Choose **File ▶ Other Files ▶ Import Special Text...**, then type HP2 MPG2 in the **Store Data in columns(s)** text box, and click **OK**. Type A:\Exercise\1r-25 in the **File name** text box and click **Open**.

26. (a) First we use the method explained in Example 1.5 of the text to store the male and female earnings data in C1 and C2 respectively and name the columns MEN1 and WOMEN1. Click anywhere on the data window, then click on the column-name cell for C1, type MEN1, and then press the ‖ENTER‖ key..

The cursor should now be in row 1 of Column C1. If not, click on that cell and click on the data-entry arrow in the upper left corner of the Data window so that it is pointing down. Now type 826 and press the ‖ENTER‖ key. Type 1790 in row 2 of C1 and press the ‖ENTER‖ key. Continue in this manner until all 8 data values have been entered.

Click on the column-name cell for C2, type WOMEN1, and press ‖ENTER‖. The cursor should now be in row 1 of Column C2. Type 1994 and press the ‖ENTER‖ key. Type 510 in row 2 of C2 and press the ‖ENTER‖ key. Continue until all 10 values have been entered.

(b) We use the method explained in Example 1.6 of the text to import the data file from the files *1r-26a.dat* and *1r-26b.dat* found in the Exercise directory of *DataDisk*. Each variable is in a

separate file this time. Choose **File ▶ Other Files ▶ Import Special Text...**, then type MEN2 in the **Store Data in columns(s)** text box, and click **OK**. Then type A:\Exercise\1r-26a in the **File**

name text box and click **Open**. Then choose **File ▶ Other Files ▶ Import Special Text...**, then type WOMEN2 IN THE **Store Data in columns(s)** text box, and click **OK**. Then type A:\Exercise\1r-26b in the **File name** text box and click **Open**.

27. First, we store the numbers 1-7246 in a column named VOTERS as follows: Click on the column-name cell for C1 and type VOTERS. Similarly, type SRS in the column-name cell for C2.

■   Choose **Calc ▶ Make Patterned Data ▶ Simple Set of Numbers**

- Type <u>VOTERS</u> in the **Store patterned data in** text box.
- Click in the **From first value** text box and type <u>1</u>
- Click in the **To last value** text box and type <u>7246</u>
- Click **OK**

- Now choose **Calc ▶ Random Data ▶ Sample From Columns...**
- Type <u>50</u> in the small text box after **Sample**

- Click in the **Sample** ☐ **rows from column(s)** text box and specify VOTERS
- Click in the **Store Samples in** text box and type <u>SRS</u>
- Click **OK**

To print the numbers now in the SRS column, we choose **Manip ▶ Display Data...**, specify SRS in the **Columns, Constants, and matrices** to display text box, and then click **OK**. The output is then displayed in the Session window and can be printed from there.

CHAPTER 2 ANSWERS

Exercises 2.1

2.1 (a) Eye color, model of car, and brand of popcorn are qualitative variables.

(b) Number of eggs in a nest, number of cases of flu, and number of employees are discrete, quantitative variables.

(c) Temperature, weight, and voltage are quantitative continuous variables.

2.3 (a) Qualitative data result from observing and recording the characteristics described by a qualitative variable such as color or shape.

(b) Discrete, quantitative data are numerical data that arise from a finite or countably infinite set of numbers. Usually they result from counting something.

(c) Continuous, quantitative data are numerical data that arise from the numbers in some interval. They are usually the result of measuring something such as temperature which can take any value in a given interval.

2.5 Of qualitative and quantitative (discrete and continuous) types of data, only qualitative involves non-numerical data.

2.7 The second column consists of *quantitative, continuous* data. This column provides measurements in terms of pounds of tobacco produced.

2.9 These data provide information on the number of employees by industry. Thus, they are *quantitative, discrete* data.

2.11 (a) The area figures are *quantitative, continuous* data. They are obtained from square mile measurements.

(b) Stating that "Africa is largest in area and second largest in population" provides information concerning rank. Thus, the type of data contained in this statement is *quantitative, discrete*.

(c) The population figures are *quantitative, discrete* data. They provide information on the number of people in each continent.

(d) The fact that Madame Curie was born in Europe provides non-numerical information. Data that give non-numerical information are *qualitative*.

2.13 (a) The second column ranks the commercial banks by amount of deposits. Thus, it consists of *quantitative, discrete* data.

(b) The third column provides measurements in terms of dollar deposits. Thus, it consists of *quantitative, continuous (or discrete)* data.

2.15 In problems 2.6, 2.10, 2.12 (a), and 2.13 (a), the data is ranked which is ordinal data.

Exercises 2.2

2.17 One of the main reasons for grouping data is that grouping often makes a rather complicated set of data easy to understand.

2.19 The three most important guidelines in choosing the classes for grouping a data set are: (1) the number of classes should be small enough to provide an effective summary but large enough to display the relevant characteristics of the data; (2) each piece of data must belong to one, and only one, class; and (3) whenever feasible, all classes should have the same width.

2.21   If the two data sets have the same number of data values, either a frequency distribution or a relative-frequency distribution is suitable. If, however, the two data sets have different numbers of data values, using relative-frequency distributions is more appropriate because the total of each set of relative frequencies is 1, putting both distributions on the same basis.

2.23   In the first method for depicting classes we used the notation **a≤b** to mean values that are greater than or equal to a and up to, but not including b, such as 30≤40 to mean a range of values greater than or equal to 30, but strictly less than 40. In the alternate method, we used the notation a-b to indicate a class that extends from a to b, including both. For example, 30-39 is a class that includes both 30 and 39. The alternate method is especially appropriate when all of the data values are integers. If the data include values like 39.7 or 39.93, the first method is more advantageous since the cutpoints remain integers whereas in the alternate method, the upper limits for each class would have to be expressed in decimal form such as 39.9 or 39.99.

2.25   When grouping data using classes that each represent a single possible numerical value, the midpoint of each class would be the same as the value in that class. Thus listing the midpoints would be redundant.

2.27   The first class to construct is 40≤50. Since all classes are to be of equal width, and the second class begins with 50, we know that the width of all classes is 50-40=10. All of the classes are presented in column 1. The last class to construct does not go beyond 150≤160, since the largest single data value is 155. Having established the classes, we tally the energy consumption figures into their respective classes. These results are presented in column 2, which lists the frequencies. Dividing each frequency by the total number of observations, which is 50, results in each class's relative frequency. The relative frequencies for all classes are presented in column 3. By averaging the lower and upper class cutpoints for each class, we arrive at the class midpoint for each class. The class midpoints for all classes are presented in column 4.

Consumption (mil. BTU)	Frequency	Relative frequency	Midpoint
40≤50	1	0.02	45
50≤60	7	0.14	55
60≤70	7	0.14	65
70≤80	3	0.06	75
80≤90	6	0.12	85
90≤100	10	0.20	95
100≤110	5	0.10	105
110≤120	4	0.08	115
120≤130	2	0.04	125
130≤140	3	0.06	135
140≤150	0	0.00	145
150≤160	2	0.04	155
	50	1.00	

2.29   Since the first cutpoint is 25 and the class width is 1, the first class is 25≤26, the second is 26≤27, etc. All of the classes are presented in column 1. The last class to construct does not go beyond 32≤33 since the largest single data value is 32.8. Having established the classes,

we tally the annual salaries into their respective classes. These
results are presented in column 2, which lists the frequencies.

Dividing each frequency by the total number of observations, which is
35, results in each class's relative frequency. The relative
frequencies for all classes are presented in column 3. By averaging the
numbers that appear as the lower and upper cutpoints for each class, we
arrive at the midpoint for each class. The midpoints for all classes
are presented in column 4.

Starting salary ($thousands)	Frequency	Relative frequency	Midpoint
25<26	3	0.086	25.5
26<27	3	0.086	26.5
27<28	5	0.143	27.5
28<29	9	0.257	28.5
29<30	9	0.257	29.5
30<31	4	0.114	30.5
31<32	1	0.029	31.5
32<33	1	0.029	32.5
	35	1.001	

2.31 The first class to construct is 40-49. Since all classes are to be of
equal width, and the second class begins with 50, we know that the width
of all classes is 50-40=10. All of the classes are presented in column
1. The last class to construct does not go beyond 150-159 since the
largest single data value is 155. Having established the classes, we
tally the energy consumption figures into their respective classes.
These results are presented in column 2, which lists the frequencies.
Dividing each frequency by the total number of observations, which is
50, results in each class's relative frequency. The relative
frequencies for all classes are presented in column 3. By averaging the
lower cutpoint for each class with the lower cutpoint of the next higher
class, we arrive at the class midpoint for each class. The class
midpoints for all classes are presented in column 4.

Consumption (mil. BTU)	Frequency	Relative frequency	Midpoint
40-49	1	0.02	45
50-59	7	0.14	55
60-69	7	0.14	65
70-79	3	0.06	75
80-89	6	0.12	85
90-99	10	0.20	95
100-109	5	0.10	105
110-119	4	0.08	115
120-129	2	0.04	125
130-139	3	0.06	135
140-149	0	0.00	145
150-159	2	0.04	155
	50	1.00	

2.33 Since the first lower cutpoint is 25 and the class width is 1, the next lower cutpoint is 26. Since the data are recorded to one decimal place, the first class is 25.0-25.9, the second is 26.0-26.9, etc. All of the classes are presented in column 1. The last class to construct does not go beyond 32.0-32.9 since the largest single data value is 32.8. Having established the classes, we tally the annual salaries into their respective classes. These results are presented in column 2, which lists the frequencies.

Dividing each frequency by the total number of observations, which is35, results in each class's relative frequency. The relative frequencies for all classes are presented in column 3. By averaging the lower cutpoint for each class with the lower cutpoint of the next higher class, we arrive at the midpoint for each class. The midpoints for all classes are presented in column 4.

Starting salary ($thousands)	Frequency	Relative frequency	Midpoint
25.0-25.9	3	0.086	25.5
26.0-26.9	3	0.086	26.5
27.0-27.9	5	0.143	27.5
28.0-28.9	9	0.257	28.5
29.0-29.9	9	0.257	29.5
30.0-30.9	4	0.114	30.5
31.0-31.9	1	0.029	31.5
32.0-32.9	1	0.029	32.5
	35	1.001	

2.35 Classes are to be based on a single value. Since each data value is one of the integers 0 through 6, inclusive, the classes will be 0 through 6, inclusive. These are presented in column 1. Having established the classes, we tally the number of sales into their respective classes. These results are presented in column 2, which lists the frequencies. Dividing each frequency by the total number of observations, which is 52, results in each class's relative frequency.

The relative frequencies for all classes are presented in column 3. Since each class is based on a single value, the midpoint is the class itself. Thus, a column of midpoints is not given, since this same information is presented in column 1.

Number of cars sold	Frequency	Relative frequency
0	7	0.135
1	15	0.288
2	12	0.231
3	9	0.173
4	5	0.096
5	3	0.058
6	1	0.019
	52	1.000

2.37 Each data value is one of the integers 0 through 7, inclusive. Appropriate classes are 0 through 7, inclusive. These are presented in column 1. Having established the classes, we tally the number of days missed into their respective classes. These results are presented in column 2, which lists the frequencies. Dividing each frequency by the total number of observations, which is 80, results in each class's relative frequency. The relative frequencies for all classes are presented in column 3. Since each class is based on a single value, the midpoint is the class itself. Thus, a column of midpoints is not given, since this same information is presented in column 1.

Number of days missed	Frequency	Relative frequency
0	4	0.050
1	2	0.025
2	14	0.175
3	10	0.125
4	16	0.200
5	18	0.225
6	10	0.125
7	6	0.075
	80	1.000

2.39 The classes are the six different NCAA wrestling champions. These are presented in column 1. The frequency distribution of wrestling champions is presented in column 2. Dividing each frequency by the total number of champions during the period 1968-1997, which is 30, results in each class's relative frequency. The relative frequency distribution is presented in column 3.

Champion	Frequency	Relative frequency
Oklahoma	1	0.033
Oklahoma State	5	0.167
Iowa State	6	0.200
Iowa	17	0.567
Arizona State	1	0.033
	30	1.000

2.41 (a) The classes are presented in column 1. With the classes established, we then tally the closing prices (Last) into their respective classes. These results are presented in column 2, which lists the frequencies. Dividing each frequency by the total number of Last values, which is 30, results in each class's relative frequency. The relative frequencies for all classes are presented in column 3. By averaging the lower and upper cutpoints for each class, we arrive at the midpoint for each class. The midpoints for all classes are presented in column 4.

Last (Dollars)	Frequency	Relative frequency	Midpoint
30<40	2	0.067	35
40<50	7	0.233	45
50<60	2	0.067	55
60<70	8	0.267	65
70<80	5	0.167	75
80<90	1	0.033	85
90<100	3	0.100	95
100&Over	2	0.067	
	30	1.001	

(b)   Since there is no upper limit or cutpoint for the values in the last class, it is not possible to calculate a midpoint.

2.43   (a)   The classes are presented in column 1. With the classes established, we then tally the exam scores into their respective classes. These results are presented in column 2, which lists the frequencies. Dividing each frequency by the total number of exam scores, which is 20, results in each class's relative frequency. The relative frequencies for all classes are presented in column 3. By averaging the lower and upper cutpoints for each class, we arrive at the midpoint for each class. The midpoints for all classes are presented in column 4.

Score	Frequency	Relative frequency	Midpoint
30-39	2	0.10	35
40-49	0	0.00	45
50-59	0	0.00	55
60-69	3	0.15	65
70-79	3	0.15	75
80-89	8	0.40	85
90-100	4	0.20	90
	20	1.00	

(b)   The first six classes have width 10; the seventh class had width 11.

(c)   Answers will vary, but one choice is to keep the first six classes the same and make the next two classes 90-99 and 100-109.

2.45   (a)   Tally marks for all 50 students, where each student is categorized by height and weight, are presented in the following contingency table.

Height (in)

		60-65	66-71	72-77	Total
Weight (lb)	90-129	||| |||		||||		
	130-169	||| ||	||| ||| |||			
	170-209		|||	||| |||	
	210-249						
	250-289					
	Total					

(b)   Tally marks in each box of the above table are counted. These counts, or frequencies, replace the tally marks in the contingency table. For each row and each column, the frequencies are added, and their sums are recorded in the proper "Total" box.

Height (in)

Weight (lb)	60-65	66-71	72-77	Total
90-129	11	4	0	15
130-169	7	13	1	21
170-209	0	3	8	11
210-249	0	1	1	2
250-289	0	0	1	1
Total	18	21	11	

(c)   The row and column totals represent the total number of students in each of the corresponding categories. For example, the column total of 18 indicates that 18 of the students in the class are between 60 and 65 inches tall, inclusive.

(d)   The sum of the row totals is 50, and the sum of the column totals is 50. The sums are equal because they both represent the total number of students in the class.

(e)   Dividing each frequency reported in part (b) by the grand total of 50 students results in a contingency table that gives relative frequencies.

Height (in)

Weight (lb)	60-65	66-71	72-77	Total
90-129	0.22	0.08	0.00	0.30
130-169	0.14	0.26	0.02	0.42
170-209	0.00	0.06	0.16	0.22
210-249	0.00	0.02	0.02	0.04
250-289	0.00	0.00	0.02	0.02
Total	0.36	0.42	0.22	

(f)   The 0.22 in the upper left-hand cell indicates that 22% of the students are between 60 and 65 inches tall, inclusive, *and* weigh between 90 and 129 lb, inclusive. The 0.30 in the upper right-hand cell indicates that 30% of all students weigh between 90 and 129 lb, inclusive. A similar interpretation holds for the remaining entries.

2.47   With the data in a column named BTUS and *DataDisk* in floppy drive a, we type in Minitab's Session window the command %a:\macro\group.mac 'BTUS' and press the ⌷ENTER⌷ key. We are given three options for specifying the classes. Since we want the first class to have cutpoints 40 and 50, we select the second option (2), press the ⌷ENTER⌷ key, and then type 40 50 when prompted to enter the lower and upper cutpoints of the first class. The resulting output is

Grouped-data table for BTUS   N = 50

Row	LowerCut	UpperCut	Freq	RelFreq	Midpoint
1	40	50	1	0.02	45
2	50	60	7	0.14	55
3	60	70	6	0.12	65
4	70	80	3	0.06	75
5	80	90	6	0.12	85
6	90	100	7	0.14	95
7	100	110	5	0.10	105
8	110	120	3	0.06	115
9	120	130	2	0.04	125
10	130	140	3	0.06	135
11	140	150	0	0.00	145
12	150	160	2	0.04	155

2.49 With the data in a column named NUMBER and *DataDisk* in floppy drive a, we type in Minitab's Session window the command %a:\macro\group.mac 'NUMBER' and press the ⌑ENTER⌑ key. We are given three options for specifying the classes. Since we want the first class to have midpoint 0 and a classwidth of 1, we select the first option (1), press the ⌑ENTER⌑ key, and then type 0 1 when prompted to enter the midpoint and classwidth of the first class. The resulting output is

Grouped-data table for CARS   N = 52

Row	LowerCut	UpperCut	Freq	RelFreq	Midpoint
1	-0.5	0.5	7	0.135	0
2	0.5	1.5	15	0.288	1
3	1.5	2.5	12	0.231	2
4	2.5	3.5	9	0.173	3
5	3.5	4.5	5	0.096	4
6	4.5	5.5	3	0.058	5
7	5.5	6.5	1	0.019	6

Note: This exercise may also be carried out by following the procedure in Example 2.10.

2.51 With the data from the Champion column in a column named CHAMP (for purposes of the output, the column entries should be limited to 8 letters, as Oklahoma, OklaSt, Iowa, IowaSt, ArizSt),

■   Choose **Stat ▶ Tables ▶ Tally...**

■   Click in the **Variables** text box and select CHAMP

■   Click in the **Counts** and **Percents** boxes under **Display**

■   Click **OK**. The results are

CHAMP	Count	Percent
ArizSt	1	3.33
Iowa	17	56.67
IowaSt	1	3.33
Oklahoma	1	3.33
OklaSt	10	33.33
N=	30	

Exercises 2.3

2.53 A frequency histogram shows the actual frequencies on the vertical axis whereas the relative frequency histogram always shows proportions (between 0 and 1) on the vertical axis.

2.55 Since a bar graph is used for qualitative data, we separate the bars from each other to emphasize that there is no special ordering of the classes; if the bars were to touch, some viewers might infer an ordering and common values for adjacent bars.

2.57 (a) Each rectangle in the frequency histogram would be of a height equal to the number of dots in the dot diagram.

(b) If the classes for the histogram were based on multiple values, there would not be one rectangle for each column of dots (there would be fewer rectangles than columns of dots). The height of each rectangle would be equal to the total number of dots between each adjacent pair of cutpoints. If the classes were constructed so that only a few columns of dots corresponded to each rectangle, the general impression of the shape of the distribution should remain the same even though the details did not appear to be the same.

2.59 (a) The frequency histogram in Figure (a) is constructed using the frequency distribution presented in this exercise; i.e., columns 1 and 2. The lower class limits of column 1 are used to label the horizontal axis of the frequency histogram. Suitable candidates for vertical-axis units in the frequency histogram are the integers 0 through 10, since these are representative of the magnitude and spread of the frequency presented in column 2. The height of each bar in the frequency histogram matches the respective frequency in column 2.

(b) The relative-frequency histogram in Figure (b) is constructed using the relative-frequency distribution presented in this exercise; i.e., columns 1 and 3. It has the same horizontal axis as the frequency histogram. We notice that the relative frequencies presented in column 3 range in size from 0.00 to 0.20. Thus, suitable candidates for vertical-axis units in the relative-frequency histogram are increments of 0.05, starting with zero and ending at 0.20. The height of each bar in the relative-frequency histogram matches the respective relative frequency in column 3.

(a)

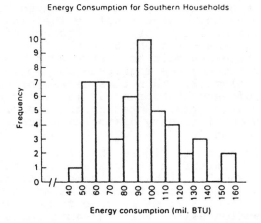

(b)

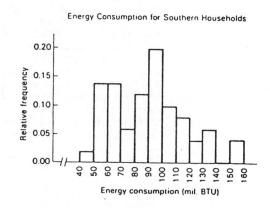

2.61 (a)   The frequency histogram in Figure (a) is constructed using the frequency distribution presented in this exercise; i.e., columns 1 and 2. The lower class limits of column 1 are used to label the horizontal axis of the frequency histogram. Suitable candidates for vertical-axis units in the frequency histogram are the integers 0 through 9, since these are representative of the magnitude and spread of the frequencies presented in column 2. The height of each bar in the frequency histogram matches the respective frequency in column 2.

(b)   The relative-frequency histogram in Figure (b) is constructed using the relative-frequency distribution presented in this exercise; i.e., columns 1 and 3. It has the same horizontal axis as the frequency histogram. We notice that the relative frequencies presented in column 3 range in size from 0.029 to 0.257. Thus, suitable candidates for vertical axis units in the relative-frequency histogram are increments of 0.05, starting with zero and ending at 0.30. The height of each bar in the relative-frequency histogram matches the respective relative frequency in column 3.

(a)   (b)

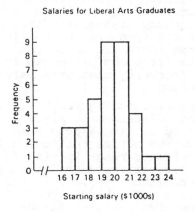

2.63 (a)   The frequency histogram in Figure (a) is constructed using the frequency distribution presented in this exercise; i.e., columns 1 and 2. Column 1 demonstrates that the data are grouped using classes based on a single value. These single values in column 1 are used to label the horizontal axis of the frequency histogram. Suitable candidates for vertical-axis units in the frequency histogram are the even integers within the range 0 through 16, since these are representative of the magnitude and spread of the frequencies presented in column 2. When classes are based on a single value, the middle of each histogram bar is placed directly over the single numerical value represented by the class. Also, the height of each bar in the frequency histogram matches the respective frequency in column 2.

(b)   The relative-frequency histogram in Figure (b) is constructed using the relative-frequency distribution presented in this exercise; i.e., columns 1 and 3. It has the same horizontal axis as the frequency histogram. We notice that the relative frequencies presented in column 3 range in size from 0.019 to 0.288. Thus, suitable candidates for vertical-axis units in the relative-frequency histogram are increments of 0.05, starting with zero and ending at 0.30. The middle of each histogram bar is placed directly over the single numerical value represented by the class. Also, the height of each bar in the relative-frequency histogram matches the respective relative frequency in column 3.

(a)   **(b)**

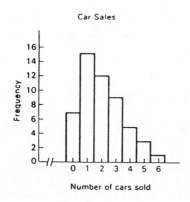

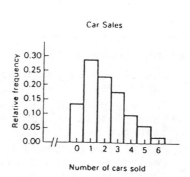

2.65   The horizontal axis of this dotplot displays a range of possible ages. To complete the dotplot, we go through the data set and record each age by placing a dot over the appropriate value on the horizontal axis.

Trucks in Use

2.67   (a)   The pie chart in Figure (a) is used to display the relative-frequency distribution given in columns 1 and 3 of this exercise. The pieces of the pie chart are proportional to the relative frequencies.

(b)   The bar graph in Figure (b) displays the same information about the relative frequencies. The height of each bar matches the respective relative frequency.

(a)   (b)

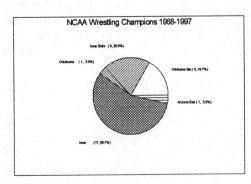

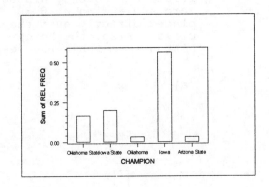

2.69 The graph indicates that:

(a)   20% of the patients have cholesterol levels between 205 and 209, inclusive.

(b)   20% are between 215 and 219; and 5% are between 220 and 224. Thus, 25% (i.e., 20% + 5%) have cholesterol levels of 215 or higher.

(c)   35% of the patients have cholesterol levels between 210 and 214, inclusive. With 20 patients in total, the number having cholesterol levels between 210 and 214 is 7 (i.e., 35% x 20).

2.71 Consider columns 1 and 3 of the energy-consumption data given in Exercise 2.59. Compute the midpoint for each class presented in column 1. Pair each midpoint with its corresponding relative frequency found in column 3. Construct a horizontal axis, where the units are in terms of midpoints and a vertical axis where the units are in terms of relative frequencies. For each midpoint on the horizontal axis, plot a point whose height is equal to the relative frequency of the class. Then join the points with connecting lines. The result is a relative-frequency polygon.

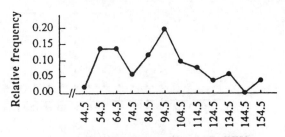

2.73 (a)   Consider all three columns of the energy-consumption data given in Exercise 2.59. Column 1 is now reworked to present just the lower cutpoint of each class. Column 2 is reworked to sum the frequencies of all classes representing values less than the specified lower cutpoint. These successive sums are the cumulative frequencies. Column 3 is reworked to sum the relative frequencies of all classes representing values less than the specified cutpoints. These successive sums are the cumulative relative frequencies. (Note: The cumulative relative frequencies can also be found by dividing the corresponding cumulative frequency by the total number of pieces of data.)

Less than	Cumulative frequency	Cumulative relative frequency
40	0	0.00
50	1	0.02
60	8	0.16
70	15	0.30
80	18	0.36
90	24	0.48
100	34	0.68
110	39	0.78
120	43	0.86
130	45	0.90
140	48	0.96
150	48	0.96
160	50	1.00

(b)   Pair each class limit in reworked column 1 with its corresponding cumulative relative frequency found in reworked column 3. Construct a horizontal axis, where the units are in terms of the cutpoints and a vertical axis where the units are in terms of cumulative relative frequencies. For each cutpoint on the horizontal axis, plot a point whose height is equal to the cumulative relative frequency. Then join the points with connecting lines. The result, presented in Figure (b), is an ogive based upon using cumulative relative frequencies. (Note: A similar procedure could be followed using cumulative frequencies.)

Energy Consumption for Southern Households

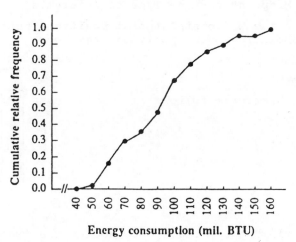

2.75 With the raw, ungrouped data from Exercise 2.27 in a column named BTUS,

■   Choose **Graph ▶ Histogram...**

■   Select BTUS for **Graph1** for the **X Variable**

- Click on the **Options** button and select **Frequency** for the **Type of histogram.**
- Select **Cutpoint** for the **Type of Intervals**
- Click on the **Midpoint/Cutpoint positions** button and type 1:71/7 in the **Midpoint/Cutpoint positions** text box
- Click **OK**
- Click **OK**

Then repeat the above process selecting **Percents** instead of **Frequency** for the **Type of Histogram.** The resulting histograms follow.

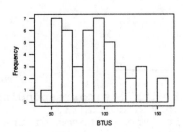 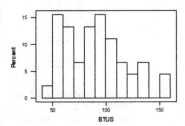

2.77 With the raw, ungrouped data from Exercise 2.35 in a column named CARS,

- Choose **Graph ▶ Histogram...**
- Select CARS for **Graph1** for the **X Variable**
- Click on the **Options** button and select **Frequency** for the **Type of histogram.**
- Select **Midpoint** for the **Type of Intervals**
- Click on the **Midpoint/Cutpoint positions** button and type 0:6/1 in the **Midpoint/Cutpoint positions** text box
- Click **OK**
- Click **OK**

The resulting histogram follows:

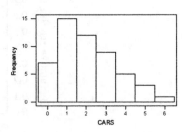

2.79 (a) The total must be obtained by adding the frequencies for each class. Thus, 1 + 7 + 7 + 3 + 6 + 10 + 5 + 4 + 2 + 3 + 2 = 50 households were sampled.

(b) The lower cutpoint for the sixth class is 90 and the upper cutpoint is 100. The midpoint for the sixth class is halfway between 90 and 100 and is therefore 95.

(c) The common class width is 10. This calculation is based upon subtracting the values reported for any successive pair of lower cutpoints. (This can also be verified by constructing each class interval around its respective midpoint and then subtracting the midpoints of any successive pair of intervals.)

(d)   The number consuming at least 100 million BTU and less than 109 million BTU is the frequency of the seventh class. Thus, five households had an energy consumption between 100 and 109 million BTU.

2.81   To begin, we store the exam score data in a column named SCORES. Then

- ■   Choose **Graph ▶ Character graphs ▶ Dotplot...**

- ■   Specify SCORES in the **Variables** text box.

- ■   Click **OK**

The computer output is:

We could also have used **Graph ▶ Dotplot...** in the above commands to get a high resolution dotplot.

2.83   (a)   There are 37 dots in the dotplot. Thus, 37 trucks were sampled.

(b)   Each dash (-) or tick mark (+) on the horizontal axis of this dotplot is equivalent to half of a year. Between five and eight years, inclusive, there are one tick mark and seven dashes. The total number of dots above this set of eight markings is 7. Thus, seven trucks were between five and eight years old, inclusive.

Exercises 2.4

2.85   For data sets with many values, a frequency histogram is more suitable for displaying the data since the vertical axis can be scaled appropriately for any number of data values. With very large sets of data, the stem-and-leaf plot would likely be much too large for display purposes unless the font size were made very small, rendering the plot nearly useless.

2.87   Depending on how 'compact' the data is, each of the original stems can be divided into either 2 or 5 stems to increase the number of stems and make the diagram more useful.

2.89   (a)   Construction of a stem-and-leaf diagram for the contents data begins with a vertical listing of the numbers comprising the stems. These numbers are 91, 92, ..., 106. To the right of this listing is a vertical line which serves as a demarcation between the stems and leaves that are about to be added. Each leaf will be the right-most digit -- the units digit -- of each number presented in the data set. The completed stem-and-leaf diagram is presented in Figure (a).

(b)   An ordered stem-and-leaf diagram consists of ordering numerically the leaves of each stem. The ordered stem-and-leaf diagram is presented in Figure (b).

	(a)			(b)
91	4		91	4
92			92	
93			93	
94	6		94	6
95	9 7		95	7 9
96	4		96	4
97	7 5 4 7		97	4 5 7 7
98	6 9 4 8 7		98	4 6 7 8 9
99	0 6 1 9 5 7		99	0 1 5 6 7 9
100	1		100	1
101	4 8 0 7		101	0 4 7 8
102	5 8		102	5 8
103	0 1		103	0 1
104			104	
105			105	
106	0		106	0

2.91 (a) Construction of a stem-and-leaf diagram for the energy consumption data begins with a vertical listing of the numbers comprising the stems. These numbers are 4, 5, 6, ..., 15. To the right of this listing is a vertical line which serves as a demarcation between the stems and the leaves that are about to be added.

Each leaf will be the right-most digit -- the units digit -- of each number presented in the data set. The completed stem-and-leaf diagram is presented in Figure (a).

(b) An ordered stem-and-leaf diagram consists of ordering numerically the leaves of each stem. The ordered stem-and-leaf diagram is presented in Figure (b).

	(a)			(b)
4	5		4	5
5	8 4 5 1 0 5 5		5	0 1 4 5 5 5 8
6	4 7 9 6 0 6 2		6	0 2 4 6 6 7 9
7	7 5 8		7	5 7 8
8	6 7 1 0 3 3		8	0 1 3 3 6 7
9	7 6 3 4 9 7 6 0 1 7		9	0 1 3 4 6 6 7 7 7 9
10	1 0 9 4 2		10	0 1 2 4 9
11	1 3 1 3		11	1 1 3 3
12	9 5		12	5 9
13	0 9 6		13	0 6 9
14			14	
15	5 1		15	1 5

2.93 (a) Using *one* line per stem in constructing the stem-and-leaf diagram means vertically listing the numbers comprising the stems *once*. The leaves are then placed with their respective stems in the usual fashion. The completed diagram is presented in Figure (a).

(b) Using *two* lines per stem in constructing the stem-and-leaf diagram means vertically listing the numbers comprising the stems *twice*. In turn, if the leaf is one of the digits 0 through 4, it is

placed with the first of the two stem lines. If the leaf is one
of the digits 5 through 9, it is placed with the second of the two
stem lines. The completed stem-and-leaf diagram using two lines
is presented in Figure (b).

(a)   (b)

```
                                   2 | 7 7 5
 2 | 7 7 5                         3 | 3 3 1 3
 3 | 7 5 3 3 1 3 9                 3 | 7 5 9
 4 | 9 8 5 1 1 6 9 4 3 8 4 7 5 1 0 3   4 | 1 1 4 3 4 1 0 3
 5 | 7 3 6 4 3 0 1 6 6 1 9 3         4 | 9 8 5 6 9 8 7 5
 6 | 2 0 7 7 1 7 2 3 0 0           5 | 3 4 3 0 1 1 3
 7 | 9                             5 | 7 6 6 6 9
 8 | 3                             6 | 2 0 1 2 3 0 0
                                   6 | 7 7 7
                                   7 |
                                   7 | 9
                                   8 | 3
```

2.95 (a)   Using *two* lines per stem in constructing the stem-and-leaf diagram
means vertically listing the numbers comprising the stems *twice*.
In turn, if the leaf is one of the digits 0 through 4, it is
placed with the first of the two stem lines. If the leaf is one
of the digits 5 through 9, it is placed with the second of the two
stem lines. The completed stem-and-leaf diagram using two lines
per stem is presented in Figure (a).

(b)   Using *five* lines per stem in constructing the stem-and-leaf
diagram means vertically listing each number comprising the stem
five times. In turn, if the leaf is the digit 0 or 1, it is
placed with the first of the five stems; a 2 or 3 leaf digit is
placed with the second stem; a 4 or 5 with the third stem; a 6 or
7 with the fourth; and an 8 or 9 with the fifth. The completed
stem-and-leaf diagram using five lines per stem is presented in
Figure (b).

(a)
```
2 | 9 9
3 | 3 1 3 2 3 4 3
3 | 6 9 5 9 5 7 9 9 5 7 8
4 | 0 0 1 2 0 4 4 0 3 2 1 2 4 1 1 4 4
4 | 8 8 5 9 5 6 5 5 7 9 8 5 9 5 9 9 7
5 | 1 1 3 1 2 1
5 | 8 7 5 7 9 5 7
6 |
6 | 9
7 | 0 3
7 |
```

(b)
```
2 |
2 | 9 9
3 | 1
3 | 3 3 2 3 3
3 | 5 5 4 5
3 | 6 7 7
3 | 9 9 9 9 8
4 | 0 0 1 0 0 1 1 1
4 | 2 3 2 2
4 | 5 5 5 4 4 5 5 4 5 4 4
4 | 6 7 7
4 | 8 8 9 9 8 9 9 9
5 | 1 1 1 1
5 | 3 2
5 | 5 5
5 | 7 7 7
5 | 8 9
6 |
6 |
6 |
6 |
6 | 9
7 | 0
7 | 3
```

2.97 (a) The data rounded to the nearest 10 ml with the terminal 0 dropped are (with data ending in '5' rounded to the nearest even 10 ml) shown at the left with a 5-stem stem-and-leaf plot at the right.

102	98	102	98	98
99	96	96	103	96
99	91	101	99	103
99	100	98	97	102
106	103	99	100	100
100	101	95	100	99

```
 9  | 1
 9  |
 9  | 5
 9  | 6 6 6 7
 9  | 8 8 8 9 9 9 9 8 9 9
10  | 1 0 0 0 0 1 0
10  | 2 2 3 3 2 3
10  |
10  | 6
```

(b)   The data with the units digits truncated are shown at the left and the stem-and-leaf is on the right.

102	97	101	97	97	9	1
99	95	95	103	96	9	
98	91	101	98	102	9	5 5 4
98	100	98	97	101	9	7 7 7 6 7
106	103	99	99	99	9	9 8 8 8 8 9 9 9 9 9 8
99	101	94	99	98	10	1 1 0 1 1
					10	2 3 3
					10	4
					10	6

(c)   While the values of a number of the leaves are different in the two diagrams and some of them change from one stem to another, the general picture of the data is virtually the same in both diagrams, with the diagram in (b) looking slightly more symmetric.

2.99   (a)   With the data in a column named DAYS,

■   Choose **Graph ▶ Stem and Leaf...**

■   Select DAYS in the **Variables** text box

■   Click on the **Increment** text box and type 10 to produce one line per stem.

■   Click **OK.**

The result is

```
Stem-and-leaf of BTUS       N = 50
Leaf Unit = 1.0
     4      0  6999
     5      1  1
     5      2
     5      3
     6      4  5
    13      5  0145558
    19      6  046679
    22      7  578
    (6)     8  013367
    22      9  3466779
    15     10  01249
    10     11  113
     7     12  59
     5     13  069
     2     14
     2     15  15
```

(a)   Follow the same procedure used in part (a), except type 5 for the interval. The result is

```
Stem-and-leaf of BTUS       N  = 50
Leaf Unit = 1.0

    4      0  6999
    5      1  1
    5      1
    5      2
    5      2
    5      3
    5      3
    5      4
    6      4  5
    9      5  014
   13      5  5558
   15      6  04
   19      6  6679
   19      7
   22      7  578
   (4)     8  0133
   24      8  67
   22      9  34
   20      9  66779
   15     10  0124
   11     10  9
   10     11  113
    7     11
    7     12
    7     12  59
    5     13  0
    4     13  69
    2     14
    2     14
    2     15  1
    1     15  5
```

2.101 (a)   The first line of the computer output states that N = 50. Thus, there are 50 pieces of data.

(b)   The computer output illustrates that there are two lines per stem.

(c)   Percentages that are 80% or greater are found in rows 5 and 6 of the stem-and-leaf diagram. Counting the percentages in rows 5 and 6 results in sixteen percentages in total. This cumulative frequency is also found at the intersection of the first column of the output with the fifth row. This cumulative frequency 16 is a depth, and it displays the number of observations that lie in the fifth and sixth rows or, equivalently, the number of percentages that are 80% or greater.

(d)   Percentages that are 69% or less are found in rows 1 and 2 of the stem-and-leaf diagram. Counting the percentages in rows 1 and 2 results in eight percentages in total. This cumulative frequency is also found at the intersection of the first column of the output with the second row. This cumulative frequency 8 is a depth, and it displays the number of observations that lie in the first and second rows or, equivalently, the number of percentages that are 69% or less.

(e)   The largest percentage is 87%. This is found in the last row.

(f)   The percentages that are in the 60s are 64%, 65%, 66%, 66%, 67%, 67%, 68%, and 68%. These are found in the first two rows.

Exercises 2.5

2.103 (a)   The distribution of a data set is a table, graph, or formula that gives the values of the observations and how often each one occurs.

(b)   Sample data is data obtained by observing the values of a variable for a sample of the population.

(c)   Population data is data obtained by observing the values of a

(c)   Population data is data obtained by observing the values of a variable for all of the members of a population.

(d)   Census data is the same as population data, a complete listing of all data values for the entire population.

(e)   A sample distribution is a distribution of sample data.

(f)   A population distribution is a distribution of population data.

(g)   A distribution of a variable is the same as a population distribution, a distribution of population data.

2.105 A large sample from a bell-shaped distribution would be expected to have roughly a bell shape.

2.107 Three distribution shapes that are symmetric are bell-shaped, triangular, and rectangular, shown in that order below. It should be noted that there are others as well.

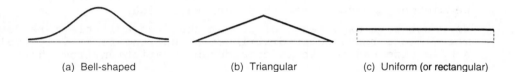

(a) Bell-shaped   (b) Triangular   (c) Uniform (or rectangular)

2.109 (a)   The distribution of the number of cars sold per week is *right skewed*.

(b)   The shape of the distribution of the number of cars sold per week is *right skewed*.

2.111 (a)   The distribution of the starting salaries from a sample of 35 liberal-arts graduates is *bell-shaped*.

(b)   The shape of the distribution of the starting salaries from a sample of 35 liberal-arts graduates is *symmetric*.

2.113 (a)   The distribution of cholesterol levels of 20 high-level patients is *left skewed*. *Note*: The answer *bell-shaped* is also acceptable.

(b)   The shape of the distribution of cholesterol levels of 20 high-level patients is *left skewed*. *Note*: The answer *symmetric* is also acceptable.

2.115 (a)   The distribution of the contents of a sample of 30 "one-liter" bottles of soft-drink is *bell-shaped*.

(b)   The shape of the distribution of the contents of a sample of 30 "one-liter" bottles of soft-drink is *symmetric*.

2.117 (a)   The distribution of the annual crime rates of the states in the United States is *right skewed*.

(b)   The shape of the distribution of the annual crime rates of the states in the United States is *right skewed*.

2.119 The precise answers to this exercise will vary from class to class or individual to individual.

(a)   We obtained a 50 random digits from a table of random numbers. The digits were

4 5 4 6 8 9 9 7 7 2 2 2 9 3 0 3 4 0 0 8 8 4 4 5 3

9 2 4 8 9 6 3 0 1 1 0 9 2 8 1 3 9 2 5 8 1 8 9 2 2

(b)   Since each digit is equally likely in the random number table, we expect that the distribution would look roughly rectangular.

(c)   Using single value classes, the frequency distribution is given by

The histogram is shown at the right.

Value	Frequency	Relative-Frequency
0	5	.10
1	4	.08
2	8	.16
3	5	.10
4	6	.12
5	3	.06
6	2	.04
7	2	.04
8	7	.14
9	8	.16

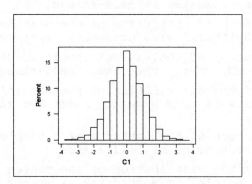

We did not expect to see this much variation.

(d)   We would have expected a histogram that was a little more 'even', more like a rectangular distribution, but the relatively small sample size can result in considerable variation from what is expected.

(e)   We should be able to get a more evenly distributed set of data if we choose a larger set of data.

(f)   Class project.

2.121 (a)   Following the procedure for Minitab given in the text, we obtained the relative frequency histogram for 3000 observations from a variable having the standard normal distribution.

(b)

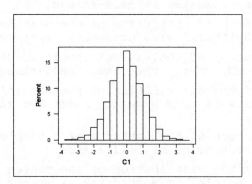

(c)   The histogram in part (b) has the shape of a standard normal distribution. The sample of 3000 is representative of the population from which the sample was taken.

2.123 (b)   Following the procedure for Minitab given in the text, we obtained the relative frequency histogram for 500 observations from a variable having a χ-square distribution with 5 degrees of freedom.

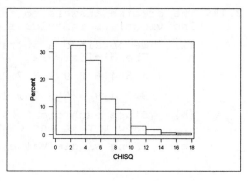

(c)   The χ-square distribution with five degrees of freedom is right skewed since the sample from that distribution exhibits that shape and should reflect the population shape.

Exercises 2.6

2.125 (a)   A truncated graph is one for which the vertical axis starts at a value other than its natural starting point, usually zero.

(b)   A legitimate motivation for truncating the axis of a graph is to place the emphasis on the ups and downs of the graph rather than on the actual height of the graph.

(c)   To truncate a graph and avoid the possibility of misinterpretation, one should start the axis at zero and put slashes in the axis to indicate that part of the axis is missing.

2.127 (a)   A good portion of the graph is eliminated. When this is done, differences between district and national averages appear greater than in the original figure.

(b)   Even more of the graph is eliminated. Differences between district and national averages appear even greater than in part (a).

(c)   The truncated graphs give the misleading impression that, in 1993, the district average is much greater relative to the national average than it actually is.

2.129 (a)   The problem with the bar graph is that it is truncated. That is, the vertical axis, which should start at $0 (in trillions), starts with $3.05 (in trillions) instead. The part of the graph from $0 (in trillions) to $3.05 (in trillions) has been cut off. This truncation causes the bars to be out of correct proportion and hence creates the misleading impression that the money supply is changing more than it actually is.

(b)   A version of the bar graph with an untruncated and unmodified vertical axis is presented in Figure (a). Notice that the vertical axis starts at $0.00 (in trillions). Increments are in halves of trillion dollars. In contrast to the original bar graph, this one illustrates that the changes in money supply from week to week are not that different. However, the "ups" and "downs" are not as easy to spot as in the original, truncated bar graph.

(c)   A version of the bar graph in which the vertical axis is modified in an acceptable manner is presented in Figure (b). Notice that the special symbol "//" is used near the base of the vertical axis to signify that the vertical axis has been modified. Thus, with this version of the bar graph, not only are the "ups" and "downs" easy to spot but the reader is also aptly warned by the slashes that part of the vertical axis between $0.00 (in trillions) and $3.05 (in trillions) has been removed.

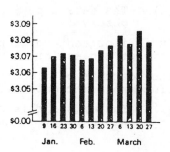

(a)   (b)

2.131 (a)   The brochure shows a "new" ball with twice the radius of the "old" ball. The intent is to give the impression that the "new" ball lasts roughly twice as long as the "old" ball. Pictorially, the "new" ball dwarfs the "old" ball in terms of size. From the perspective of measurement, if the "new" ball has twice the radius of the "old" ball, the "new" ball will have eight times the volume of the "old" ball (since the volume of a sphere is proportional to the cube of its radius, or the radius 2 to the third power equals 8). Thus, the scaling is improper because it gives the impression that the "new" ball lasts roughly eight times as long as the "old" ball rather than merely two times as long.

Old ball

New ball

(b)   One possible way for the manufacturer to illustrate the fact that the "new" ball lasts twice as long as the "old" ball is to present pictures of two balls, side by side, each of the same magnitude as the picture of the "old" ball and to label this set of two balls "new ball". (See below.) This will illustrate the point that a purchaser will be getting twice as much for the money.

Old ball

New ball

REVIEW TEST FOR CHAPTER 2

1.   (a)   A variable is a characteristic that varies from one person or thing to another.

  (b)   Variables are quantitative or qualitative.

  (c)   Quantitative variables can be discrete or continuous.

  (d)   Data is information obtained by observing values of a variable.

  (e)   The data type is determined by the type of variable being observed.

2.   It is important to group data in order to make large data sets more compact and easier to understand.

3.   The concepts of midpoints and cutpoints do not apply to qualitative data since no numerical values are involved in the data.

4.   (a)   The midpoint is halfway between the cutpoints. Since the class width is 8, 10 is halfway between 6 and 14.

(b)   The class width is also the distance between consecutive midpoints. Therefore the second midpoint is at 10 + 8 = 18.

(c)   The sequence of cutpoints is 6, 14, 22, 30, 38, ... Therefore the lower and upper cutpoints of the third class are 22 and 30.

(d)   An observation of 22 would go into the third class since that class contains data greater than or equal to 14 and strictly less than 22.

5.   (a)   The common class width is the distance between consecutive cutpoints, or 15 - 5 = 10.

(b)   The midpoint of the second class is halfway between the cutpoints 15 and 25, and is therefore 20.

(c)   The sequence of cutpoints is 5, 15, 25, 35, 45, ... Therefore the lower and upper cutpoints of the third class are 25 and 35.

6.   Single value grouping is appropriate when the data is discrete with only relatively few distinct observations.

7.   (a)   The vertical edges of the bars will be aligned with the cutpoints.

(b)   The horizontal centers of the bars will be aligned with the midpoints.

8.   The two main types of graphical displays for qualitative data are the bar chart and the pie chart.

9.   A histogram is better than a stem-and-leaf for displaying large quantitative data sets since it can always be scaled appropriately and the individual values are of less interest than the overall picture of the data.

10.   Bell-shaped   Right skewed   Reverse J shape   Uniform

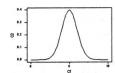  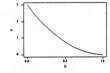 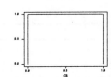

11.   (a)   Slightly skewed to the right. Assuming that the most typical heights are around 5'10", most heights below that figure would still be above 5'4", whereas heights above 5'10" extend to around 7".

(b)   Skewed to the right. High incomes extend much further above the mean income than low incomes extend below the mean.

(c)   Skewed to the right. While most full-time college students are in the 17-22 age range, there are very few below 17 while there are many above 22.

(d)   Skewed to the right. The main reason for the skewness to the right is that those students with GPAs below fixed cutoff points have been suspended by the time they would have been seniors.

12.   (a)   The distribution of the sample will reflect the distribution of the population, so it should be left-skewed as well.

(b)   No. The randomness in the samples will certainly produce different sets of observations resulting in shapes that are not identical.

(c)   Yes. We would expect both of the samples to reflect the shape of the population and be left-skewed.

13.   (a)   The first column ranks the hydroelectric plants. Thus, it

consists of *quantitative, discrete* data.

(b)   The fourth column provides measurements of capacity. Thus, it consists of *quantitative, continuous* data.

(c)   The third column provides non-numerical information. Thus, it consists of *qualitative* data.

14.   (a)   The first class to construct is 40-44. Since all classes are to be of equal width, and the second class begins with 45, we know that the width of all classes is 45 - 40 = 5. All of the classes are presented in column 1 of the grouped-data table in the figure below. The last class to construct does not go beyond 65-69, since the largest single data value is 69. Having established the classes, we tally the ages into their respective classes. These results are presented in column 2, which lists the frequencies. Dividing each frequency by the total number of observations, which is 42, results in each class's relative frequency. The relative frequencies for all classes are presented in column 3.

By averaging the lower and upper cutpoints for each class, we arrive at the midpoint for each class. The midpoints for all classes are presented in column 4.

Age at inauguration	Frequency	Relative frequency	Class midpoint
40<45	2	0.048	42.5
45<50	6	0.143	47.5
50<55	12	0.286	52.5
55<60	12	0.286	57.5
60<65	7	0.167	62.5
65<70	3	0.071	67.5
	42	1.001	

(b)   The lower cutpoint for the first class is 40. Since the data in the first class include values up to, but not including 45, the upper cutpoint for the first class is 45.

(c)   The common class width is 45 - 40 = 5.

(d)   The frequency histogram presented below is constructed using the frequency distribution presented above; i.e., columns 1 and 2. Notice that the lower cutpoints of column 1 are used to label the horizontal axis of the frequency histogram. Suitable candidates for vertical-axis units in the frequency histogram are the even integers within the range 0 through 12, since these are representative of the magnitude and spread of the frequencies presented in column 2. The height of each bar in the frequency histogram matches the respective frequency in column 2.

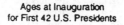

Ages at Inauguration
for First 42 U.S. Presidents

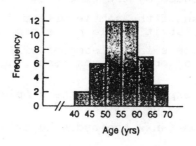

15.   The horizontal axis of this dotplot displays a range of possible ages
       for the 42 Presidents of the United States. To complete the dotplot, we
       go through the data set and record each age by placing a dot over the
       appropriate value on the horizontal axis.

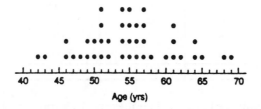

Ages at Inauguration
for First 42 U.S. Presidents

16.   (a)   Using *one* line per stem in constructing the ordered stem-and-leaf
             diagram means vertically listing the numbers comprising the stems
             once. The leaves are then placed with their respective stems in
             order. The ordered stem-and-leaf diagram using one line per stem
             is presented in Figure (a).

       (b)   Using *two* lines per stem in constructing the ordered stem-and-leaf
             diagram means vertically listing the numbers comprising the stems
             twice. In turn, if the leaf is one of the digits 0 through 4, it
             is ordered and placed with the first of the two stem lines. If
             the leaf is one of the digits 5 through 9, it is ordered and
             placed with the second of the two stem lines. The ordered stem-
             and-leaf diagram using two lines per stem is presented in Figure
             (b).

(a)

4	2 3 6 6 7 8 9 9
5	0 0 1 1 1 1 2 2 4 4 4 4 5 5 5 5 6 6 6 7 7 7 7 8
6	0 1 1 1 2 4 4 5 8 9

(b)

4	2 3
4	6 6 7 8 9 9
5	0 0 1 1 1 1 2 2 4 4 4 4
5	5 5 5 5 6 6 6 7 7 7 7 8
6	0 1 1 1 2 4 4
6	5 8 9

       (c)   The ordered stem-and-leaf diagram using two lines per stem
             corresponds to the frequency distribution of Problem 2(a).

17.   (a)   The following grouped-data table is constructed using classes
             based on a single value. Since each data value is one of the
             integers 0 through 6, inclusive, the classes will be 0 through 6,
             inclusive. These are presented in column 1. Having established
             the classes, we tally the number of busy tellers into their
             respective classes. These results are presented in column 2,
             which lists the frequencies. Dividing each frequency by the total
             number of observations, which is 25, results in each class's
             relative frequency. The relative frequencies for all classes are
             presented in column 3. Since each class is based on a single
             value, the class mark is the class itself. Thus, a column of
             midpoints is not given, since this same information is presented
             in column 1.

Number busy	Frequency	Relative frequency
0	1	0.04
1	2	0.08
2	2	0.08
3	4	0.16
4	5	0.20
5	7	0.28
6	4	0.16
	25	1.00

(b)   The following relative-frequency histogram is constructed using the relative-frequency distribution presented above in columns 1 and 3. Column 1 demonstrates that the data are grouped using classes based on a single value. These single values in column 1 are used to label the horizontal axis of the relative-frequency histogram. We notice that the relative frequencies presented in column 3 range in size from 0.04 to 0.28. Thus, suitable candidates for vertical axis units in the relative-frequency histogram are increments of 0.05, starting with zero and ending at 0.30. The middle of each histogram bar is placed directly over the single numerical value represented by the class. Also, the height of each bar in the relative-frequency histogram matches the respective relative frequency in column 3.

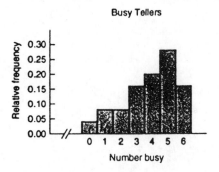

18.   (a)   The table below shows both the frequency distribution and the relative frequency distribution. If each frequency reported in the table is divided by the total number of students, which is 40, we arrive at the relative frequency (or percentage) of each class.

Class	Frequency	Relative Frequency
Fr	6	0.150
So	15	0.375
Jr	12	0.300
Sr	7	0.175

(b)   The following pie chart is used to display the percentage of accidental deaths in each of the four categories.

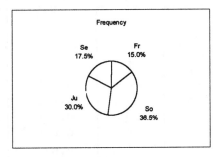

(c)   The following bar graph also displays the relative frequencies of each class.

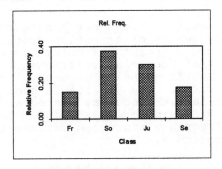

19.   (a)   The first class to construct is 400<1000. Since all classes are to be of equal width, we know that the width of all classes is 1000 - 400 = 600. All of the classes are presented in column 1 of the table below. The last class to construct is 6400<7000, since the largest single data value is 6560.91. Having established the classes, we tally the highs into their respective classes. These results are presented in column 2, which lists the frequencies. Dividing each frequency by the total number of observations, which is 36, results in each class's relative frequency. The relative frequencies for all classes are presented in column 3. By averaging the numbers that appear as the lower and upper cutpoints for each class, we arrive at the midpoint mark for each class. The midpoints for all classes are presented in column 4.

High	Freq.	Relative frequency	Midpoint
400<1000	16	0.444	700
1000<1600	9	0.250	1300
1600<2200	2	0.056	1900
2200<2800	2	0.056	2500
2800<3400	2	0.056	3100
3400<4000	3	0.083	3700
4000<4600	0	0.000	4300
4600<5200	0	0.000	4900
5200<5800	1	0.028	5500
5800<6400	0	0.000	6100
6400<7000	1	0.028	6700
	36	1.001	

(b)   The following relative-frequency histogram is constructed using the relative-frequency distribution presented above in columns 1 and 3. The lower cutpoints of column 1 are used to label the horizontal axis of the relative-frequency histogram. We notice that the relative frequencies presented in column 3 range in size from 0.000 to 0.444. Thus, suitable candidates for vertical axis units in the relative-frequency histogram are increments of 0.10, starting with 0.00 and ending at 0.50. The height of each bar in the relative-frequency histogram matches the respective relative frequency in column 3.

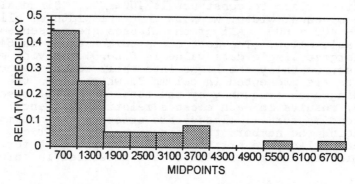

DJIA HIGHS

20.   (a)   The shape of the distribution of the inauguration ages of the first 42 presidents of the United States is *bell-shaped*.

(b)   The shape of the distribution of the number of tellers busy with customers at Prescott National Bank during 25 spot checks is *left skewed*.

21.   Answers will vary, but here is one possibility:

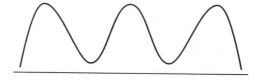

22. (a) Covering up the numbers on the vertical axis totally obscures the percentages.

     (b) Having followed the directions in part (a), we might conclude that the percentage of women in the labor force for 2000 is about three and one-third times that for 1960.

     (c) Not covering up the vertical axis, we find that the percentage of women in the labor force for 2000 is about 1.8 times that for 1960.

     (d) The graph is potentially misleading because it is truncated. Notice that vertical axis units begin at 30 rather than at zero.

     (e) To make the graph less potentially misleading, we can start it at zero instead of 30.

23. (a) First store the data in a column named AGES.   Then

     ■ Choose **Graph ▶ Histogram...**

     ■ Specify AGES in the **X** text box for **Graph 1**.

     ■ Click the **Options...** button

     ■ Select the **Cutpoint** option button from the **Type of Intervals** field

     ■ Select the **Midpoint/cutpoint positions** text box and type <u>40:70/5</u>

     ■ Click **OK**

     ■ Click **OK**

To print the result at the right from the Graph window, choose

File ▶ Print Window... .

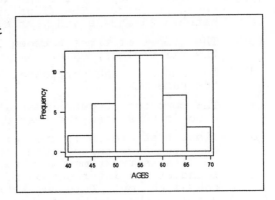

     (b) With the data already stored in the column name AGES,

     ■ Choose **Graph ▶ Character Graphs ▶ Dotplot...**

     ■ Specify AGES in the **Variables** text box.

     ■ Click **OK**

The computer output shown in the Session window is:

     Character Dotplot

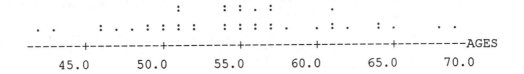

(c)   With the data already stored in the column named AGES,

- Choose **Graph** ▶ **Stem-and-Leaf...**

- Specify AGES in the **Variables** text box.

- Click on the **Intervals** text box and type <u>5</u>

- Click **OK**

The computer output is:

```
        Stem-and-leaf of AGES     N = 42
        Leaf Unit = 1.0
           2    4   23
           8    4   667899
          20    5   001111224444
         (12)   5   555566677778
          10    6   0111244
           3    6   589
```

24.   (a)   The number of times in the sample is found by summing the frequencies shown in the histogram. Therefore, there are $1 + 3 + 6 + 7 + 3 + 1 + 1 = 22$ times in the sample.

(b)   The lower cutpoint for the fourth class is 25 and the upper cutpoint is 30. The midpoint for the fourth class is halfway between 25 and 30 and is therefore 27.5.

(c)   The number of times between 30 and 34 minutes is the frequency of the fifth class, namely 3.

(d)   The label on the horizontal axis tells us that the data were given the name TIME.

25.   (a)   The computer output illustrates that there are two lines per stem used in this stem-and-leaf diagram.

(b)   Times that are 35 minutes or more are found in rows 6, 7, and 8 of the stem-and-leaf diagram. Counting the times in rows 6, 7, and 8 results in two times in total. This cumulative frequency is also found at the intersection of the first column of the printout with the sixth row. This cumulative frequency 2 is a depth, and it displays the number of observations that lie in the sixth, seventh, and eighth rows or, equivalently, the number of times that are 35 minutes or more.

(c)   Times that are less than 20 minutes are found in rows 1 and 2 of the stem-and-leaf diagram. Counting the times in rows 1 and 2 results in four times in total. This cumulative frequency is also found at the intersection of the first column of the printout with the second row. This cumulative frequency 4 is a depth, and it displays the number of observations that lie in the first and second rows or, equivalently, the number of times that are less than 20 minutes.

(d)   The longest time in the sample is 48 minutes. This is found in the last row.

(e)   The times that are in the 30s are 30, 31, 31, and 37 minutes. These are found in rows 5 and 6.

26.   (a)   Using Minitab to create the pie chart and bar chart, enter the type of accidental death in a column named TYPE and the frequency in a column named FREQ. Then

- ■   Choose **Graph ▶ Pie Chart...**

- ■   Click on **Chart Table,** enter CLASS in **Categories in** text box and enter FREQ in the **Frequencies in** text box. Enter 'Pie Chart of Classes in the **Titles** text box.

- ■   Click **OK.** The chart follows after part (b).

(b)   To create the bar chart, enter the class abbreviations in a column named CLASS and the relative frequencies from Problem 18 in a

column named REL FREQ, choose **Graph ▶ Chart...,** select REL FREQ for the **Y variable** in **Graph 1,** and CLASS for the X variable. Click **OK.**

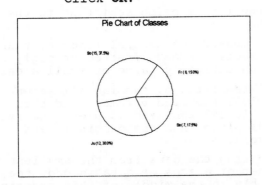

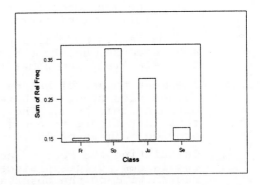

27.   (a)   Using the Minitab procedure described in the exercise, a sample of 1000 observations was obtained from a variable having an F-distribution with degrees of freedom (3,16) and stored in a column named FDIST. Your results will not be identical to these.

      (b)   ■   Choose **Graph ▶ Histogram...**

- ■   Specify FDIST in the **X** text box for **Graph 1.**

- ■   Click the **Options...** button

- ■   Choose **Percent** for type of histogram and **Cutpoints** for type of interval.

- ■   Click **OK.**

- ■   Click **OK.**

The relative-frequency histogram is shown at the right.

      (c)   Since the sample distribution reflects the distribution of the population, the F-distribution with degrees of freedom (3,16) is right-skewed.

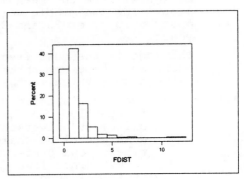

CHAPTER 3 ANSWERS

Exercises 3.1

3.1 The purpose of a measure of central tendency is to indicate where the center or most typical value of a data set lies.

3.3 Of the mean, median, and mode, only the mode is appropriate for use with qualitative data.

3.5 (a) The mean is the sum of the values (45) divided by n (9). The result is 5. The median is the middle value in the ordered list and thus is also 5.

(b) The mean is the sum of the values (135) divided by n (9). The result is 15. The median is the middle value in the ordered list and thus is 5, as before. The median is more typical of most of the data than is the mean and thus works better here.

(c) The mean does not have the property of being resistant to the influence of extreme observations.

3.7 The median is more appropriate as a measure of central tendency than the mean because, unlike the mean, the median is not affected strongly by the relatively few homes that have an extremely large or small areas.

3.9 The mean number of values is calculated first by summing all seven values presented, which results in 1351 thousand volumes, and then dividing this by 7. Thus, the mean is 193 thousand volumes. Carrying this last figure to one more decimal place than the original data provides a mean of 193.0 thousand volumes.

Calculating the median requires ordering the data from the smallest to the largest values. Since the number of pieces of data is odd, the median will be the observation exactly in the middle of this ordering, i.e., in the fourth ordered position. Thus, the median number of volumes is 79.0 thousand.

The mode is the most frequently occurring data value or values. In this exercise, there is no mode because none of the data values occurs more than once.

3.11 The mean age is calculated first by summing all 12 data values presented, which results in 402 years, and then dividing this sum by 12. Thus, the mean is 33.5 years. This figure is already rounded to one more decimal place than the original data.

Calculating the median requires ordering the data from the smallest to the largest values. Since the number of pieces of data is even, the median will be the mean of the two middle data values in the ordered list. The two middle values are the sixth and seventh ordered observations, or 33 and 36. Thus, the median age is (33 + 36)/2 = 34.5 years.

The mode is the most frequently occurring data value or values. In this exercise, there are three modes. They are 24, 28, and 37 years. Each value occurs twice in the data set.

3.13 (a) The mode is defined as the data value or values that occur most frequently. In this exercise, the data values are the six different universities. During the period 1968-1997, Iowa was the NCAA wrestling champion seventeen times, Oklahoma State five times, Iowa State six times, Oklahoma once, and Arizona State once. Since Iowa is the data value that occurs most frequently, it is the mode.

(b) Again, the data values are the five universities. There is no way to compute a mean or median for such data. Thus, it would not be

appropriate to use either the mean or the median here. In
general, the mode is the only measure of central tendency that can
be used for qualitative data.

3.15 (a)   The mean of the runner's places over the seven marathons that he
ran is 13.9. This is derived by summing the seven individual
places that he finished, which is 97, and dividing by 7. The
result is 13.857 which is rounded to 13.9.

(b)   The median of the data is the observation appearing in the middle
position after ordering the data from the smallest to the largest
values. The median is the fourth ordered observation, or 4.

(c)   The median provides a better descriptive summary of the data than
does the mean. Notice that six of the seven places finished by
the runner are more in line with the median 4 than with the mean
13.9. The median definitely illustrates where the majority of the
observations tend to cluster much better than does the mean in
this particular example. The reason that the mean performs so
poorly is that its value is greatly influenced by the magnitude of
the extreme observation 72. A median is not influenced by extreme
observations, since it is concerned with the position of an
observation rather than its magnitude. This is why statisticians
recommend using the median for ordinal data.

3.17 (a)   The low values of 2 and 4 look like outliers.

(b)   The mean based upon all 20 observations is computed to be 385/10 =
19.25 or 19.2.

(c)   The values 2 and 28, which are the individual low and high
observations, are deleted. The mean of the remaining data is
calculated to be 355/18 = 19.72, or 19.7. This is the 5% trimmed
mean for the data.

(d)   The values 2, 4, 27, and 28, which are the two low and two high
observations, are deleted. The mean of the remaining data is
calculated to be 324/16 = 20.25 or 20.2. This is the 10% trimmed
mean for the data.

(e)   The 10% trimmed mean provides the best measure of central tendency
for the data. Although two observations are eliminated that are
not outliers, the two that are obvious outliers are eliminated.

3.19 As a measure of central tendency, the midrange has the advantage of
being relatively easy to compute; it has the disadvantage of ignoring a
great deal of pertinent information--only the largest and smallest data
values are considered, whereas the remainder of the data is disregarded.

3.21 (a)   The modal class is defined as the class(es) having the largest
frequency. The modal class for the cholesterol-level data is the
class 210-214. It has the largest frequency, which is 7.

(b)   The mode of the raw data is 210. It appears four times in the
data set.

(c)   The mode 210 is contained in the modal class 210-214.

3.23 To begin, we store the age data in a column named AGES. Then, we

■   Choose **Calc ▶ Column Statistics...**

■   Select the **Mean** option button from the **Statistics** field

■   Click in the **Input variable** text box and specify AGES

■   Click **OK**

The output shown in the Sessions Window is

        Mean of AGES   =   33.500

- Choose **Calc ▶ Column Statistics...**

- Select the **Median** option button from the **Statistics** field
- Click **OK**

    Median of AGES   =   34.500

- Choose **Stat ▶ Tables ▶ Tally...**

- Specify AGES in the **Variables** text box
- Select the **Counts** check box
- Click **OK**

The output shown in the Sessions Window is

```
AGES   Count
 24      2
 27      1
 28      2
 33      1
 36      1
 37      2
 41      1
 43      1
 44      1

N=      12
```

There are three modes in this data (24, 28, and 37), all of which occur twice.

3.25 Using Minitab, we begin by entering the data in a column named CHAMP.

Then choose **Stat ▶ Tables ▶ Tally...**, specify CHAMP in the **Variables** text box, select the **Counts** check box, and click **OK**. The results are

```
CHAMP   Count
  AS      1
   I      17
  IS      6
   O      1
  OS      5
  N=      30
```

We see that Iowa (I) was the champion 17 times. Thus Iowa is the mode.

3.27 The mean price is calculated first by summing all 15 data values presented, which results in 726¢, and then dividing this sum by 15. Thus, the mean is 48.4¢.

Calculating the median requires ordering the data from the smallest to the largest values. Since the number of pieces of data is odd, the median will be the observation exactly in the middle of this ordering, i.e., in the eighth ordered position. Thus, the median price is 49¢.

The mode is the most frequently occurring data value or values. In this exercise, 52¢ is the mode because it occurs three times and none of the other data values occurs more than twice.

3.29 The mean height is calculated first by summing all 20 heights presented, which results in 897, and then dividing this sum by 20. Thus, the mean is 44.85 inches. Rounding this to one more decimal place than the original data, the mean is 44.8 inches.

Calculating the median requires ordering the data from the smallest to the largest values. Since the number of pieces of data is even, the median will be the mean of the two middle observations in this ordering, i.e., in the tenth and eleventh ordered positions. Thus, the median is (45 + 46)/2 = 45.5 inches.

The mode is the most frequently occurring data value or values. In this exercise, there are two modes (46 and 47), each of which occurs three times.

Exercises 3.2

3.31 Mathematical notation reduces the need to write out long descriptions and formulas.

3.33 For a given population, the population mean is not a variable; it is a parameter, a constant. The sample mean will vary from sample to sample and thus is considered to be a variable.

3.35 (a) $\Sigma x = 12 + 8 + 9 + 17 = 46$

(b) n = number of pieces of data = 4

(c) $\bar{x} = \Sigma x/n = 46/4 = 11.5$

3.37 (a) $\Sigma x = 12971$

(b) n = number of pieces of data = 8

(c) $\bar{x} = \Sigma x/n = 12971/8 = 1621.375$, which rounded to one decimal place is 1621.4

3.39 (a) $\Sigma x = 23.3$

(b) n = number of pieces of data = 10

(c) $\bar{x} = \Sigma x/n = 23.3/10 = 2.33$, which is given to two decimal places.

3.41 (a) $\bar{x} = \Sigma x/n = 588/6 = \$98.0$

(b)

x	x^2	$x-\bar{x}$	$(x-\bar{x})^2$
75	5,625	-23	529
98	9,604	0	0
130	16,900	32	1024
63	3,969	-35	1225
115	13,225	17	289
107	11,449	9	81
585	60,772	0	3148

From the bottom row, $\Sigma x^2 = 60,772$, $\Sigma(x-\bar{x}) = 0$, and $\Sigma(x-\bar{x})^2 = 3148$.

3.43 The expression Σxy represents the sum of products of the pairs of data; that is,

$$\Sigma xy = x_1y_1 + x_2y_2 + \ldots + x_ny_n.$$

The expression $\Sigma x\Sigma y$ represents the product of the sums of the two data sets; that is,

$$\Sigma x\Sigma y = (x_1 + x_2 + \ldots + x_n)(y_1 + y_2 + \ldots + y_n).$$

To show that these two quantities are generally unequal, consider three observations on x and three observations on y as presented in the first two columns of the following table:

x	y	xy
1	2	2
3	4	12
5	6	30
9	12	44

Consider also a third column which is the product of x and y, appropriately labeled xy. From the bottom row, we see that $\Sigma x = 9$, $\Sigma y = 12$, and $\Sigma xy = 44$.

From this simple example, it is obvious that Σxy does not equal $\Sigma x \Sigma y$, since $\Sigma xy = 44$ and $\Sigma x \Sigma y = (9)(12) = 108$.

Exercises 3.3

3.45   The purpose of a measure of dispersion is to show the amount of variation in a data set. By itself, any value of central tendency does not adequately characterize the elements of a data set. Central tendency merely describes the center of the observations but does not show how observations differ from each other. A measure of dispersion is intended to capture the degree to which observations differ among themselves.

3.47   When we use the standard deviation as a measure of variation, the reference point is zero since if all of the data values were equal, there would be no variation and the standard deviation is zero.

3.49   (a)   The mean of this data set is 5.   Thus

$\Sigma(x-\overline{x})^2 = (-4)^2 + (-3)^2 + (-2)^2 + (-1)^2 + 0^2 + 1^2 + 2^2 + 3^2 + 4^2 = 60$

Dividing by $n - 1 = 8$, $s^2 = 60/8 = 7.5$. The standard deviation, s, is the square root of 7.5, or 2.739.

(b)   The mean of this data set is 15.   Thus

$\Sigma(x-\overline{x})^2 = (-14)^2 + (-13)^2 + (-12)^2 + (-11)^2 + 10^2 + (-9)^2 + (-8)^2 + (-7)^2 + 84^2 = 7980$

Dividing by $n - 1 = 8$, $s^2 = 7980/8 = 997.5$. The standard deviation, s, is the square root of 997.5, or 31.583.

(c)   Changing the 9 to 99 greatly increases the standard deviation, illustrating that it lacks the property of resistance to the influence of extreme data values.

3.51   (a)   The defining formula for *s* is:

$$s = \sqrt{\frac{\Sigma(x-\overline{x})^2}{n-1}} \ .$$

The first three columns of the following table present the calculations that are needed to compute *s* by the defining formula:

x	x-$\bar{x}$	$(x-\bar{x})^2$	x^2
110	-4.2	17.64	12,100
122	7.8	60.84	14,884
132	17.8	316.84	17,424
107	-7.2	51.84	11,449
101	-13.2	174.24	10,201
97	-17.2	295.84	9,409
115	0.8	0.64	13,225
91	-23.2	538.24	8,281
125	10.8	116.64	15,625
142	27.8	772.84	20,164
1,142	0	2,345.60	132,762

With n = 10 and using the bottom figure of column 1, we find that $\bar{x} = \Sigma x/n = 1,142/10 = 114.2$. This value is needed to construct the differences presented in column 2. Column 3 squares these differences. The figure presented at the bottom of column 3 is the computation needed for the numerator of the defining formula for s. Thus,

$$s = \sqrt{\frac{2,345.6}{10-1}} = 16.1.$$

(b)   The shortcut formula for s with figures used from the bottom of Columns 1 and 4 of the previous table yields

$$s = \sqrt{\frac{132,762 - (1,142)^2/10}{(10-1)}} = 16.1.$$

(c)   The shortcut formula was a time-saver.

3.53 (a)   The range is 516 - 15 = 501.

(b)   The defining formula for s is:

$$s = \sqrt{\frac{\Sigma(x-\bar{x})^2}{n-1}}.$$

The first three columns of the following table present the calculations that are needed to compute s by the defining formula:

x	x-$\bar{x}$	$(x-\bar{x})^2$	x^2
79	-114	12,996	6,241
516	323	104,329	266,256
24	-169	28,561	576
265	72	5,184	70,225
41	-172	23,104	1,681
15	-178	31,684	225
411	218	47,524	168,921
1,351	0	253,382	514,125

With n = 7 and using the bottom figure of column 1, we find that $\bar{x}$ = $\Sigma x/n$ = 1351/7 = 193. This value is needed to construct the differences presented in column 2. Column 3 squares these differences. The figure presented at the bottom of column 3 is the computation needed for the numerator of the defining formula for s. Thus,

$$s = \sqrt{\frac{253,382}{7-1}} = 205.5 \ .$$

(c)   The shortcut formula for s is:

$$s = \sqrt{\frac{(\Sigma x^2) - (\Sigma x)^2/n}{(n-1)}} \ .$$

Columns 1 and 4 of the previous table present the calculations needed to compute s by the shortcut formula. Using the figures presented at the bottom of each of these columns and substituting into the formula itself, we get:

$$s = \sqrt{\frac{514,125 - (1,351)^2/7}{(7-1)}} = 205.5 \ .$$

(d)   The shortcut formula was easier to use. It required both fewer and easier column manipulations.

3.55 (a)   The range of a data set is defined to be the difference between the largest and smallest data values in the data set. The largest age is 44 years. The smallest age is 24 years. The range is 44 – 24 = 20 years.

(b)   The defining formula for s is:

$$s = \sqrt{\frac{\Sigma(x-\bar{x})^2}{n-1}}$$

The first three columns of the following table present the calculations that are needed to compute s by the defining formula:

x	x−$\bar{x}$	(x−$\bar{x}$)^2	x^2
37	3.5	12.25	1,369
28	− 5.5	30.25	784
36	2.5	6.25	1,296
33	− 0.5	0.25	1,089
37	3.5	12.25	1,369
43	9.5	90.25	1,849
41	7.5	56.25	1,681
28	− 5.5	30.25	784
24	− 9.5	90.25	576
44	10.5	110.25	1,936
27	− 6.5	42.25	729
24	− 9.5	90.25	576
402	0	571	14,038

With n = 12 and using the bottom figure of column 1, we find that $\bar{x}$ = Σx/n = 402/12 = 33.5 years. This value is needed to construct the differences in column 2. Column 3 squares the differences presented in column 2. The figure presented at the bottom of column 3 is the computation needed for the numerator of the defining formula s. Thus,

$$s = \sqrt{\frac{571}{12-1}} = 7.2 \text{ years.}$$

(c)   The shortcut formula for s is:

$$s = \sqrt{\frac{(\Sigma x^2) - (\Sigma x)^2/n}{(n-1)}} \ .$$

Columns 1 and 4 of the previous table present the calculations needed to compute s by the shortcut formula. Using the figures presented at the bottom of each of these columns and substituting into the formula itself, we get:

$$s = \sqrt{\frac{14,038 - (402)^2/12}{(12-1)}} = 7.2 \text{ years.}$$

(d)   The shortcut formula was easier to use. It required both fewer and easier column manipulations.

3.57   (a)   We will compute s for each data set using the shortcut formula. The shortcut formula requires the following column manipulations.

Data Set I			Data Set II	
x	x^2		x	x^2
0	0		10	100
0	0		12	144
10	100		14	196
12	144		14	196
14	196		14	196
14	196		15	225
14	196		15	225
15	225		15	225
15	225		16	256
15	225		17	289
16	256		142	2,052
17	289			
23	529			
24	576			
189	3,157			

For each data set, the calculations for s are presented in column 2 of the following table.

Data Set	s	Range
I	$\sqrt{\dfrac{3,157 - (189)^2/14}{(14-1)}} = 6.8$	24
II	$\sqrt{\dfrac{2,052 - (142)^2/10}{(10-1)}} = 2.0$	7

(b)   The range for Data Set I is 24 - 0 = 24. The range for Data Set II is 17 - 10 = 7. These are recorded in column 3 of the previous table.

(c)   Outliers increase the variation in a data set; in other words, removing the outliers from a data set results in a decrease in the variation.

3.59   If k=2, then $100(1 - 1/k^2)\% = 100(1 - 1/2^2)\% = 100(3/4)\% = 75\%$

If k=3, then $100(1 - 1/k^2)\% = 100(1 - 1/3^2)\% = 100(8/9)\% = 89\%$

3.61   (a)   At least 75% of the data must lie between 85 - 2(16.1) = 52.8 and 85 + 2(16.1) = 117.2 according to Chebychev's Rule.

(b)   In fact, twenty-nine of the thirty , or 97%, of the observations lie within two standard deviations of the mean. Since 97% is more than 75%, Chebychev's rule is verified for this data.

(c)   At least 89% of the data must lie between 85 - 3(16.1) = 36.7 and 85 + 3(16.1) = 133.3 according to Chebychev's Rule.

(d)   In fact, all thirty, or 10%, of the observations lie within three standard deviations of the mean. Since 100% is more than 89%, Chebychev's rule is verified for this data.

3.63 (a)   Chebychev's rule guarantees that *at least* 75% of the observations from any distribution lie within two standard deviations of the mean. It therefore applies to bell-shaped distributions for which the Empirical Rule says that *about* 95% of the data will lie within two standard deviations. Since 95% is greater than 75%, Chebychev's rule is verified for bell-shaped distributions. If a distribution is known to be bell-shaped, it is more appropriate to use the estimates provided by the Empirical Rule since those are estimates, not just lower limits.

(b)   Chebychev's rule guarantees that *at least* 89% of the observations from any distribution lie within two standard deviations of the mean. It therefore applies to bell-shaped distributions for which the Empirical Rule says that *about* 99.7% of the data will lie within two standard deviations. Since 99.7% is greater than 89%, Chebychev's rule is verified for bell-shaped distributions. If a distribution is known to be bell-shaped, it is more appropriate to use the estimates provided by the Empirical Rule since those are estimates, not just lower limits.

3.65 (a)   According to the empirical rule, approximately 68% of the observations should lie between 68.9 and 101.1, approximately 95% between 52.8 and 117.2, and approximately 99.7% between 36.7 and 133.3.

(b)   Twenty-six of the 30 (86.7%) lie within one standard deviation of the mean, 28 of the 30 (93.3%) lie within two standard deviations of the mean and 29 of the 30 (96.7%) lie within three standard deviations of the mean.

(c)   Only the percentage within two standard deviations of the mean is close to that provided by the empirical rule.

(d)   A stem-and-leaf diagram of the data is given by

```
 2 | 8
 3 |
 4 |
 5 | 7 8
 6 | 4 9
 7 | 4 9
 8 | 0 3 5 5 7 7 9 9
 9 | 0 2 3 4 4 5 6 6 7 7 7 8
10 | 0 0
```

This data is considerably left-skewed, explaining why the actual percentages and the empirical rule percentages are not consistent with each other.

(e)   Since the empirical rule is only appropriate for data that is roughly bell-shaped, it is not appropriate for data is considerably left-skewed.

3.67 (a)   The defining formula for s is:

$$s = \sqrt{\frac{\sum(x-\overline{x})^2}{n-1}} \; .$$

The first three columns of the following table present the calculations that are needed to compute s by the defining formula:

x	$x-\overline{x}$	$(x-\overline{x})^2$
300	−174	30,276
300	−174	30,276
940	456	217,156
450	− 24	576
400	− 74	5,476
400	− 74	5,476
300	−174	30,276
300	−174	30,276
1050	566	331,776
300	−174	30,276
4,740	0	711,840

With n = 10 and using the bottom figure of column 1, we find that $\overline{x} = \Sigma x/n = 4,740/10 = 474.0$. This value is needed to construct the differences presented in column 2. Column 3 squares these differences. The figure presented at the bottom of column 3 is the computation needed for the numerator of the defining formula for s. Thus,

$$s = \sqrt{\frac{711,840}{10-1}} = 281.24.$$

(b)

Salary x	Frequency f	Salary · Frequency xf
300	5	1500
400	2	800
450	1	450
940	1	940
1050	1	1050
	10	4740

The grouped data formula for the mean yields $\overline{x} = (\Sigma xf)/n = 4740/10 = 474$.

(c)   The answers for the sample mean are the same in parts (a) and (b).
      When data are grouped in classes each based on a single value, the
      term xf represents the total of f x's in each class, so the result
      is the same as it would be if the f x's were added together. Thus
      adding up the terms xf is the same as totaling all of the
      individual numbers.

(d)   The grouped-data formula for *s* is

$$s = \sqrt{\frac{\sum(x-\overline{x})^2 f}{n-1}} \ .$$

The columns of the table below give the necessary information for
computing s.

Salary x	Frequency f	$(x-\overline{x})$	$(x-\overline{x})^2$	$(x-\overline{x})^2 f$
300	5	-174	30,276	151,380
400	2	- 74	5,476	10,952
450	1	- 24	576	576
940	1	466	217,156	217,156
1050	1	576	331,776	331,776
	10			711,840

We have already found that $\overline{x} = 474$. This value is needed to construct
the differences presented in column 3. Column 4 squares these
differences and column 5 multiplies each squared difference by the
number of times f that it occurs. The figure presented at the bottom of
column 5 is the computation needed for the numerator of the grouped-data
formula for *s*. Thus,

$$s = \sqrt{\frac{711,840}{10-1}} = 281.24$$

(e)   The sample standard deviations are the same in parts (a) and (d).
      When the data are grouped in classes each based on a single value,
      the class midpoint is the same as the observation in each class.
      Thus the term $(x-\overline{x})^2 f$ counts each squared difference $(x-\overline{x})^2$ as
      many times f as it occurs, yielding the same result in each row as
      would have been obtained by adding up the $(x-\overline{x})^2$ quantities for
      each identical individual observation contributing to that row.
      The total of the $(x-\overline{x})^2 f$ terms in the grouped-data table is then
      the same as the total of the $(x-\overline{x})^2$ term in the ungrouped table.

3.69 Using Minitab, we enter the data in a column named AGES.   Then

■   Choose **Calc ▶ Column Statistics...**

■   Select the **Range** option button from the **Statistic** field

■   Click in the **Input variable** text box and specify AGES

■   Click **OK**

■   Choose **Calc ▶ Column Statistics...**

- Select the **Standard Deviation** option button from the **Statistic** field
- Click in the **Input variable** text box and specify AGES
- Click **OK**

The resulting output in the Session Window is

Range of AGES = 20.000

Standard deviation of AGES = 7.2048

3.71 Using Minitab, we enter the data in a column named PRICES.   Then

- Choose **Calc ▶ Column Statistics...**

- Select the **Range** option button from the **Statistic** field
- Click in the **Input variable** text box and specify PRICES
- Click **OK**

- Choose **Calc ▶ Column Statistics...**

- Select the **Standard Deviation** option button from the **Statistic** field
- Click in the **Input variable** text box and specify PRICES
- Click **OK**

The resulting output in the Session Window is

Range of PRICES = 11.000

Standard deviation of PRICES = 3.5010

Alternatively, one could enter the 15 data values in rows 1-15 of Column A of a spreadsheet like Excel, Lotus 1-2-3, or Quattro Pro.

Using Excel, in cell A17, enter the formula **=MAX(A1:A15)-MIN(A1:A15)**. In Lotus or Quattro, enter the formula **@MAX(A1.A15)-@MIN(A1.A15)**.   The result shown in the cell will be 11.00, the range.

In cell A18, for Excel, enter the formula, **=STDEV(A1:A15)**.   For Lotus or Quattro, enter the formula **@STDS(A1.A15)**.   The result shown in the cell will be 3.50, the sample standard deviation.

Finally, one could use a brute force method for finding the standard deviation with a hand held calculator, filling in a table as we have on several previous exercises.

3.73 Using Minitab, we enter the data in a column named HEIGHTS.   Then

- Choose **Calc ▶ Column Statistics...**

- Select the **Range** option button from the **Statistic** field
- Click in the **Input variable** text box and specify HEIGHTS
- Click **OK**

- Choose **Calc ▶ Column Statistics...**

- Select the **Standard Deviation** option button from the **Statistic** field
- Click in the **Input variable** text box and specify HEIGHTS
- Click **OK**

The resulting output in the Session Window is

> Range of HEIGHTS = 13.000

> Standard deviation of HEIGHTS = 3.3916

Alternatively, one could enter the 20 data values in rows 1-20 of Column A of a spreadsheet like Excel, Lotus, or Quattro.

Using Excel, in cell A22, enter the formula **=MAX(A1:A20)-MIN(A1:A20)**. The result shown in the cell will be 13.00, the range.

In cell A23, enter the formula, **=STDEV(A1:A20)**. The result shown in the cell will be 3.391553, the sample standard deviation.

Exercises 3.4

3.75 The median and interquartile range have the advantage over the mean and standard deviation that they are not sensitive to a few extreme values; they are said to be resistant.

3.77 An extreme observation may be an outlier, but it may also be an indication of skewness.

3.79 (a) The interquartile range is a descriptive measure of variation.

(b) It measures the spread of the middle two quarters (50%) of the data.

3.81 A modified boxplot will be the same as an ordinary boxplot when the maximum and minimum values lie inside the upper and lower limits, respectively. In other words, the maximum and minimum values are also the adjacent values.

3.83 First arrange the data in increasing order:

34   39   63   64   67   70   75   76   81   82

84   85   86   88   89   90   90   96   96   100

The position of the first quartile is $(n + 1)/4 = (20 + 1)/4 = 5.25$. Thus Q_1 is .25 of the distance from the fifth data value (67) to the sixth (70) and therefore $Q_1 = 67 + .25(70 - 3) = 67 + .75 = 67.75$.

The second quartile is the median of the entire data set. The number of pieces of data is 20, and so the position of the median is $(20 + 1)/2 = 10.5$, halfway between the tenth and eleventh data values. Thus the median of the entire data set is $(82 + 84)/2 = 83.0$. Thus $Q_2 = 83.0$.

The position of the third quartile $3(n + 1)/4 = 3(20 + 1)/4 = 15.75$. Thus Q_3 is .75 of the distance from the 15th data value (89) to the 16th data value (90) and therefore $Q_3 = 89 + .75(90 - 89) = 89.75$.

Interpreting our results, we conclude that 25% of the exam scores are below 67.75; 25% are between 67.75 and 83.0; 25% are between 83.0 and 89.75; and 25% are above 89.75.

3.85 First arrange the data in increasing order:

1   1   3   3   4   4   5   6   6   7   7   9   9   10   12   12   13   15   18   23   55

The position of the first quartile is $(n + 1)/4 = (21 + 1)/4 = 5.5$. Thus Q_1 is .5 of the distance from the fifth data value (4) to the sixth (4) and therefore $Q_1 = 4 + .5(4 - 4) = 4.0$.

The second quartile is the median of the entire data set. The number of pieces of data is 21, and so the position of the median is $(21 + 1)/2 = 11$. Thus the median of the entire data set is 7. Thus $Q_2 = 7.0$.

The position of the third quartile is $3(n + 1)/4 = 3(21 + 1)/4 = 16.5$. Thus Q_3 is .5 of the distance from the 16th data value (12) to the 17th data value (13) and therefore $Q_3 = 12 + .5(13 - 12) = 12.5$.

Interpreting our results, we conclude that 25% of the number of days spent in the hospital is less than 4; 25% are between 4 and 7.0; 25% are between 7.0 and 12.5; and 25% are above 12.5.

3.87 First arrange the data in increasing order:

3.3	5.7	6.6	7.7	8.3	8.6	8.9	9.2
10.2	10.3	10.6	11.8	12.0	12.7	13.7	

The position of the first quartile is $(n + 1)/4 = (15 + 1)/4 = 4$. Thus Q_1 is the fourth data value (7.7).

The second quartile is the median of the entire data set. The number of pieces of data is 15, and so the position of the median is $(15 + 1)/2 = 8.0$. Thus the median of the entire data set is 9.2. Thus $Q_2 = 9.2$.

The position of the third quartile is $3(n + 1)/4 = 3(15 + 1)/4 = 12$. Thus Q_3 is the 12th data value (11.8).

Interpreting our results, we conclude that 25% of the miles driven are below 7700; 25% are between 7700 and 9200; 25% are between 9200 and 11800; and 25% are above 11800.

3.89 (a)   From Exercise 3.83, recall $Q_1 = 67.75$ and $Q_3 = 89.75$. So the inner quartile range (IQR) is given by: $IQR = Q_3 - Q_1 = 89.75 - 67.75 = 22$. Thus, the middle 50% of the exam scores has a range of 22 points.

(b)   Min = 34, $Q_1 = 67.75$, $Q_2 = 83$, $Q_3 = 89.75$, Max = 100.

(c)   By constructing the lower and upper limits, you can determine if there are any potential outliers.
Lower limit = $Q_1 - 1.5(IQR) = 67.75 - 1.5(22) = 34.75$
Upper limit = $Q_3 + 1.5(IQR) = 89.75 + 1.5(22) = 122.75$
Since 34 is below the lower limit, it is a potential outlier.

(d)   (i)   (ii)

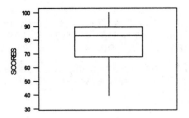   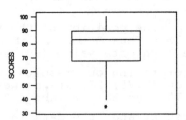

The figure above is the (i) boxplot and (ii) modified boxplot for the exam scores for the 20 students in an introductory statistics class. Note that the two boxes in the boxplot indicate the spread of the two middle quarters of the data and that the two whiskers indicate the spread of the first and fourth quarters. Thus, we see that there is less variation in the top two quarters of the exam score data than in the bottom two, and that the first quarter has the greatest variation of all. The (ii) modified boxplot indicates the potential outlier of 34.

3.91 (a) From Exercise 3.85 recall $Q_1 = 4$ and $Q_3 = 12.5$. So the inner quartile range (IQR) is given by: IQR $= Q_3 - Q_1 = 12.5 - 4 = 8.5$. Thus, the middle 50% of the days spent in short-term hospitals has a range of 8.5 days.

 (b) Min $= 1$, $Q_1 = 4$, $Q_2 = 7$, $Q_3 = 12.5$, Max $= 55$.

 (c) By constructing the lower and upper limits, you can determine if there are any potential outliers.

Lower limit $= Q_1 - 1.5(IQR) = 4 - 1.5(8.5) = -8.75$

Upper limit $= Q_3 + 1.5(IQR) = 12.5 + 1.5(8.5) = 25.25$

Since 55 lies outside the limits, it is a potential outlier.

 (d)   (i)   (ii)

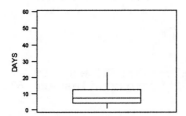
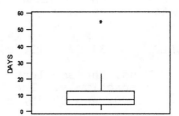

The figures above are the (i) boxplot and (ii) modified boxplot for the lengths of stay in short-term hospitals by 21 random patients. Note that the potential outlier of 55 is denoted in the modified boxplot.

3.93 Generally speaking, the accounting graduates have higher starting salaries than the liberal-arts graduates. In fact, more than half of the accounting graduates have higher starting salaries than the highest starting salary of the liberal-arts graduates, and the lowest starting salary of the accounting graduates exceeds more than half of the starting salaries of the liberal-arts graduates. Also, the variation in starting salaries of the accounting graduates is somewhat smaller than that of the liberal-arts graduates.

3.95 Bell-shaped

Uniform

Right skewed

3.97 (a) To begin, we store the exam score data in a column named SCORES. To obtain a modified boxplot, we

■   Now choose **Graphs ▶ Boxplot...**

■   Select SCORES in the **Y** text box for Graph 1. Leave the **X** text box blank.

■   Click **OK**

 (b) To determine the smallest and largest data values and the quartiles of the data, we

■   Choose **Stat ▶ Basic Statistics ▶ Display Descriptive Statistics...**

■   Select SCORES in the **Variables** text box

■   Click **OK**

The output for part (a) as shown in the Graph Window and that for part (b) shown in the Sessions Window are, respectively:

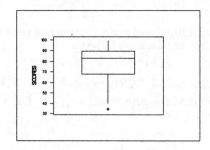

	N	MEAN	MEDIAN	TRMEAN	STDEV	SEMEAN
SCORES	20	77.75	83.00	78.94	17.61	3.94

	MIN	MAX	Q1	Q3
SCORES	34.00	100.00	67.75	89.75

Thus, the smallest and largest data values are 34 and 100, respectively. The first, second, and third quartiles are 67.75, 83.00, and 89.75, respectively.

3.99 (a) Using Minitab, we begin by storing the price data in a column named PRICES. To obtain a modified boxplot, we

■   Now choose **Graphs ▶ Boxplot...**

■   Select PRICES in the **Y** text box for **Graph 1**. Leave the **X** text box blank.

■   Click **OK**

(b) To determine the smallest and largest data values and the quartiles of the data, we

■   Choose **Stat ▶ Basic Statistics ▶ Display Descriptive Statistics...**

■   Select PRICES in the **Variables** text box, and click **OK**.

The output for part (a) as shown in the Graph Window and that for part (b) shown in the Sessions Window are, respectively:

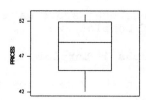

Variable	N	Mean	Median	TrMean	StDev	SE Mean
PRICES	15	48.400	49.000	48.538	3.501	0.904

Variable	Minimum	Maximum	Q1	Q3
PRICES	42.000	53.000	45.000	52.000

The five number summary consists, in order, of Min, Q1, Median, Q3, Max.

3.101 (a)   Using Minitab, we begin by storing the price data in a column named HEIGHTS. To obtain a modified boxplot, we

■   Now choose **Graphs ▶ Boxplot...**

■   Select HEIGHTS in the **Y** text box for **Graph 1**. Leave the **X** text box blank.

■   Click **OK**

(b)   To determine the smallest and largest data values and the quartiles of the data, we

■   Choose **Stat ▶ Basic Statistics ▶ Display Descriptive Statistics...**

■   Select HEIGHTS in the **Variables** text box

■   Click **OK**

The output for part (a) as shown in the Graph Window and that for part (b) shown in the Sessions Window are, respectively:

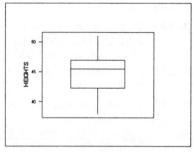

Variable	N	Mean	Median	TrMean	StDev	SEMean
HEIGHTS	20	44.850	45.500	44.889	3.392	0.758

Variable	Min	Max	Q1	Q3
HEIGHTS	38.000	51.000	42.250	47.000

The five number summary consists, in order, of Min, Q1, Median, Q3, Max.

3.103 (a)   The modified boxplot for the beef-consumption data in Printout 3.6 shows that there are 10 units between tick marks on the vertical axis.

The lower and upper ends of the box represent the first and third quartiles. The line across the middle of the box represents the median. Also, the three smallest values are marked here by asterisks (*), indicating that they are potential outliers.

Thus, we see that the smallest data value is about 0, the first quartile is about 56, the median is about 62, the third quartile is about 72, and the largest data value is about 90.

Regarding the variation in the data set, observe that the two boxes in the modified boxplot indicate the spread of the second and third quarters of the data. Also observe that the two whiskers, i.e., the solid vertical lines below and above the box, as well as the denoted potential outliers, indicate the spread of the first and fourth quarters. Thus, we see the least variation occurs in the second quarter of the data , the next least in the third quarter, the next least in the fourth quarter and the greatest in the first quarter.

(b)   Yes. 0, 8, and 20 are potential outliers. Assuming that there are no recording errors, the potential outliers may be data from vegetarians, people who do not eat beef for some other reason, or people on low-beef diets.

(c)   In Printout 3.7, the smallest and largest data values in the data set are found in the last row. The smallest data value is the entry in this row under Min, and the largest data value is the entry in this row under Max. Thus, the smallest and largest data values in the data set are 0 and 89, respectively. (These answers are close to the approximate values obtained in part (a)).

The first and third quartiles are also found in the last row under Q1 and Q3, namely 54.50 and 72.75 respectively. (The answers in part (a) compare well with the exact values.)

The median is found in the second row of the table under Median and has the value 62.00.(The approximate value of the median in part (a), i.e., 62, is exactly the same.)

Exercises 3.5

3.105 The ultimate objective of dealing with sample data in inferential studies is to describe the entire population.

3.107 (a) $\overline{x}$ (b) s

3.109 The z-score corresponding to an observed value of a variable tells us <u>how many standard deviations the observation is from the mean and, by its sign, what direction it is from the mean.</u>

3.111 The number 242.6 is a parameter since it is the mean of the entire population.

3.113 If we consider the heights as a sample:

(a) $\overline{x} = \dfrac{\Sigma x}{n} = \dfrac{375}{5} = 75$ inches

(b) $s = \sqrt{\dfrac{\Sigma (x-\overline{x})^2}{n-1}} = \sqrt{\dfrac{156}{4}} = 6.2$ inches.

If we consider the heights as a population:

(c) $\mu = \dfrac{\Sigma x}{N} = \dfrac{375}{5} = 75$ inches

(d) $\sigma = \sqrt{\dfrac{\Sigma (x-\mu)^2}{N}} = \sqrt{\dfrac{156}{5}} = 5.6$ inches.

(e)   The two means are equal because they are computed in the same way. We sum the data and then divide by the total number of pieces of data.

(f)   The two standard deviations are different because they are computed differently. In the defining formula for the sample standard deviation *s*, we divide by one less than the total number of pieces of data. In the defining formula for σ, we divide by the total number of pieces of data.

3.115 (a) $\mu = \dfrac{\Sigma x}{N} = \dfrac{1165}{12} = \$97.08$

(b) $\quad \sigma = \sqrt{\dfrac{\Sigma(x-\mu)^2}{N}} = \sqrt{\dfrac{3234.92}{12}} = \$16.4.$

3.117 (a) $\quad \mu = \dfrac{\Sigma x}{N} = \dfrac{603}{12} = \$50.2$ million

(b) $\quad \sigma = \sqrt{\dfrac{\Sigma x^2}{N} - \mu^2} = \sqrt{\dfrac{57935}{12} - (50.25)^2} = \$48.0$ million

3.119 (a) The standardized version of y is z = (y - 356)/1.63.

(b) The mean and standard deviation of z are 0 and 1 respectively.

(c) For y = 352, the z score is (352 - 356)/1.63 = -2.45

For x = 361, the z score is (361 - 356)/1.63 = 3.07

(d) The value 352 is 2.45 standard deviations below the mean 356.

The value 361 is 3.07 standard deviations above the mean 356.

(e)

$\overline{x}-3s$	$\overline{x}-2s$	$\overline{x}-s$	$\overline{x}$	$\overline{x}+s$	$\overline{x}+2s$	$\overline{x}+3s$	
•						|•	
351.11	352.74	354.37	356.00	357.63	359.26	360.89	x
-3	-2	-1	0	1	2	3	Z

3.121 (a) The mean weight of 1997-98 Dallas Cowboys is 242.6 lbs and the standard deviation is 48.4 lbs. Therefore the z-score for Stepfret Williams's weight of 170 lbs is z = (170 - 242.6)/48.4 = -1.50.

(b) At least $1 - 1/(1.5)^2 = 5/9 = 55.6\%$ of the weights lie within 1.5 standard deviations of the mean. Therefore, at least 55.6% of the weights are greater than 170 lbs.

3.123 We are given μ = 25 mpg, σ = 1.15 mpg, and x = 21.4 mpg. The z-score and relative standing calculations are:

x	z = (x-μ)/σ	let k=|z|; $1-1/k^2$ =
21.4	(21.4-25)/1.15 = -3.13	$1-1/3.13^2 = 0.898$

(a) A gas mileage of 21.4 mpg is 3.13 standard deviations below the mean.

(b) A gas mileage of 21.4 mpg is less than at least 89.8% of the gas mileages for all cars of this model.

(c) It does appear that this car is getting unusually low gas mileage since we are guaranteed that at least 89.8% of the cars get greater mileage.

3.125 The formulas used here to compute s and σ are:

$$s = \sqrt{\frac{\sum (x-\overline{x})^2}{n-1}} \qquad \text{and} \qquad \sigma = \sqrt{\frac{\sum (x-\mu)^2}{N}} \,.$$

For each data set, the relevant calculations are:

Data Set	Number of observations	$\sum(x-\text{mean})^2$	s	σ
1	4	14.00	2.16	1.87
2	7	24.86	2.04	1.88
3	10	36.40	2.01	1.91

(a)   The sample standard deviations are found in the fourth column of the previous table.

(b)   The population standard deviations are found in the fifth column of the previous table.

(c)   Comparing s and σ for a given data set, the two measures will tend to be closer together if the data set is large.

3.127 (a)   Using Minitab, we enter the data in a column named HEIGHTS.   The data values in inches are

70	70	76	74	75	68	69	69	71	71	71	73
70	71	73	70	73	72	71	74	74	76	70	74
74	74	74	75	75	75	75	79	77	77	75	75
76	78	72	74	77	71	76	73	74	79	79	78
78	78	78	65	65							

To obtain the mean,

■   Choose **Stat ▶ Basic Statistics ▶ Display Descriptive Statistics...**

■   Select HEIGHTS in the **Variables** text box

■   Click **OK**

Variable	N	Mean	Median	TrMean	StDev	SEMean
HEIGHTS	53	73.604	74.000	73.745	3.370	0.463

Variable	Min	Max	Q1	Q3
HEIGHTS	65.000	79.000	71.000	76.000

(b)   The population standard deviation requires adjustment from the sample standard deviation given in the output above. It is easy to do this within Minitab.

■   Choose **Calc ▶ Mathematical Expressions...**

■   Type K1 in the **Variable (New or Modified)** text box

■   In the **Mathematical Expressions** text box, type
   sqrt(52/53)*stdev('HEIGHTS')

■   Click **OK**

■   Choose **File ▶ Display Data...**

■   Select K1 in the **Columns, Constants, and Matrices to Display** text box. The result in the Sessions Window is

Data Display

K1   3.33840

K1 is the value of the population standard deviation.

Alternatively, you could enter the data in cells A1-A53 of the Excel spreadsheet. In A54, enter the formula =AVERAGE(A1:A53), and in A55, enter the formula =STDEVP(A1:A53). The result shown in the cells will be 73.60 and 3.338397, the same as given by Minitab.

3.129 Using Minitab, we enter the data in a column named AID. To obtain the mean,

■   Choose **Stat ▶ Basic Statistics ▶ Display Descriptive Statistics...**

■   Select AID in the **Variables** text box

■   Click **OK**

The population mean is the same as the sample mean in the output below:

Variable	N	Mean	Median	TrMean	StDev	SEMean
AID	12	50.3	37.0	42.5	50.1	14.5

Variable	Min	Max	Q1	Q3
AID	6.0	172.0	14.8	55.3

The population standard deviation requires adjustment from the sample standard deviation given in the output above. It is easy to do this within Minitab.

■   Choose **Calc ▶ Mathematical Expressions...**

■   Type K1 in the **Variable (New or Modified)** text box

■   In the **Mathematical Expressions** text box, type

        sqrt(11/12)*stdev('AID')

■   Click **OK**

■   Choose **File ▶ Display Data...**

■   Select K1 in the **Columns, Constants, and Matrices to Display** text box. The result in the Sessions Window is

K1   47.9881

K1 is the value of the population standard deviation.

Alternatively, you could enter the data in cells A1-A12 of the Excel spreadsheet. In A13, enter the formula =AVERAGE(A1:A12), and in A14, enter the formula =STDEVP(A1:A12). The result shown in the cells will be 50.25 and 47.98806, the same as given by Minitab.

REVIEW TEST FOR CHAPTER 3

1. (a)   Descriptive measures are numbers used to describe data sets.

    (b)   Measures of center indicate where the center or most typical value of a data set lies.

    (c)   Measures of variation indicate how much variation or spread the data set has.

2. The two most commonly used measures of center for quantitative data are the mean and the median.   The mean uses all of the data, but can be influenced by the presence of a few outliers.   The median is computed from the one or two center values in an ordered list of the data.   It does not make use of all of the data, but it has the advantage that it is not influenced by the presence of a few outliers.

3. The only measure of center, among those we discussed, that is appropriate for qualitative data is the mode.

4. (a)   standard deviation

    (b)   interquartile range

5. (a) $\bar{x}$ (b)   s   (c) μ (d) σ

6. (a)   Not necessarily true.

    (b)   This is necessarily true.

7. Almost all of the observations in any data set lie within <u>3</u> standard deviations of the mean.

8. (a)   The components of the five-number summary are the minimum, Q1, median, Q3, and the maximum.

    (b)   The median is a measure of the center.   The interquartile range which is found as Q3 - Q1, and the range, which is the difference between the maximum and the minimum are measures of variation.

    (c)   The boxplot is based on the five-number summary.

9. (a)   An outlier is an observation that falls well outside the overall pattern of the data.

    (b)   First compute the interquartile range IQR = Q3 - Q1.   Then compute two 'fences' as Q1 - 1.5IQR and Q3 + 1.5IQR.   Observations which are lower than the first fence or higher than the second fence are potential outliers and require further study.

10. (a)   A z-score for an observation x is obtained by subtracting the mean of the data set from x and then dividing the result by the standard deviation; i.e., $z = (x - \mu)/\sigma$ for population data or $z = (x - \bar{x})/s$ for sample data.

    (b)   A z-score indicates how many standard deviations an observation is below or above the mean of the data set.

    (c)   An observation with a z-score of 2.9 is likely to be greater than all or almost all of the other data values.

11. (a)   The mean is calculated as $\bar{x} = \Sigma x/n$.   For the sample of 10 Germans, $\bar{x} = 114/10 = 11.4$kg; for the sample of 15 Russians, $\bar{x} = 241/15 = 16.1$ kg.

    (b)   The median is found by ordering the observations from lowest to highest and selecting the observation in the middle.   For Germans, the median is 11.5 kg.   This is the average of the observations appearing in the fifth and sixth ordered positions; i.e., it is the average of the observations 11 and 12.   For Russians, the median is 17 kg.   This is the observation appearing in the eighth ordered position.

(c)   The mode is the most frequently occurring data value or values. For Germans, the modes are 8 kg and 12 kg. For Russians, the modes are 12 kg, 16kg, 19kg, and 23 kg.

12.   Many more marriages are characterized as being of short duration than other durations. Since the median is ordinarily preferred for data sets that have exceptional (very large or small) values, the median is more appropriate than the mean as a measure of central tendency for data on the duration of marriages.

13.   Regarding death certificates, we are interested in the most frequent cause of death. Causes of death are qualitative. There is no way to compute a mean or median for such data. The mode is the only measure of central tendency that can be used for qualitative data.

14.   (a) $\bar{x} = \dfrac{\Sigma x}{n} = \dfrac{48}{12} = 4.0$ minutes   (b)   range = $15 - 1 = 14$ minutes

(c) $s = \sqrt{\dfrac{\Sigma x^2 - (\Sigma x)^2/n}{(n-1)}} = \sqrt{\dfrac{374 - (48)^2/12}{(12-1)}} = 4.1$ minutes

15.   (a)

$\bar{x}-3s$	$\bar{x}-2s$	$\bar{x}-s$	$\bar{x}$	$\bar{x}+s$	$\bar{x}+2s$	$\bar{x}+3s$
18.3	31.7	45.1	58.5	71.9	85.3	98.7

(b)   At least 27 of the 36 millionaires are between 31.7 and 85.3 years old.

(c)   At least 89% of the 36 millionaires are between 18.3 and 98.7 years old.

16.   (a)   The position of the first quartile is $(n + 1)/4 = (36 + 1)/4 = 9.25$. Thus Q_1 is .25 of the distance from the ninth data value (48) to the tenth value (also 48) and therefore $Q_1 = 48$.

The second quartile is the median of the entire data set. The number of pieces of data is 36, and so the position of the median is $(36 + 1)/2 = 18.5$, halfway between the eighteenth and nineteenth data value. Thus the median of the entire data set is $(59 + 60)/2 = 59.5$. That is, $Q_2 = 59.5$.

The position of the third quartile is $3(n + 1)/4 = 3(36 + 1)/4 = 27.75$. Thus Q_3 is .75 of the distance from the 27th data value (68) to the 28th value (69), and therefore $Q_3 = 68 + .75(69 - 68) = 68.75$.

Interpreting our results, we conclude that 25% of the ages are less than 48 years; 25% of the ages are between 48 and 59.5 years; 25% of the ages are between 59.5 and 68.75 years; and 25% of the ages are greater than 68.75 years.

(b)   The IQR = $Q_3 - Q_1 = 68.75 - 48 = 20.75$. Thus, the middle 50% of the ages has a range of 20.75 years.

(c)   Min = 31, $Q_1 = 48$, $Q_2 = 59.5$, $Q_3 = 68.75$, Max = 79

(d)   The limits are given by:

Lower limit = $Q_1 - 1.5(IQR) = 48.0 - 1.5(20.75) = 16.875$

Upper limit = $Q_3 + 1.5(IQR) = 68.75 + 1.5(20.75) = 99.875$

(e)   No potential outliers

(f)   A boxplot is constructed easily using the information in part (a).

   (i)   low data value = 31 years

   (ii)   high data value = 79 years

   (iii) Q_1 = 48.0 years

   (iv) Q_3 = 68.75 years

   (v)   median = Q_2 = 59.5 years.

These values are used to construct the boxplot as follows:

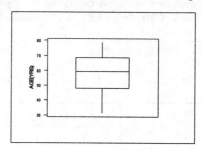

17.   (a) $\mu = \dfrac{\Sigma x}{n} = \dfrac{165.3}{9} = 18.37$ thousand

   (b) $\sigma = \sqrt{\dfrac{\Sigma x^2}{N} - \mu^2} = \sqrt{\dfrac{3835.07}{9} - (18.37)^2} = 9.42$ thousand

   (c) $z = (x - \mu)/\sigma = (x - 18.37)/9.42$

   (d)   The mean of z is 0, and the standard deviation of z is 1.

   (e)

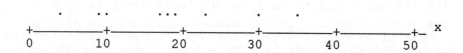

Converting each x value to its corresponding z-score results in the dotplot below. The relative positions of the points remain unchanged, but the scale is different.

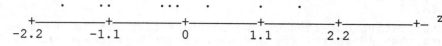

   (f)

x	$z = (x - \mu)/\sigma$
34.9	$(34.9 - 18.37)/9.42 = 1.75$
18.1	$(18.1 - 18.37)/9.42 = -0.03$

The enrollment at UCLA is 1.75 standard deviations above the mean; the enrollment at UCSD is 0.03 standard deviation below the mean.

18.   (a)   The mean price given here is a sample mean. This is because it is the mean price per gallon for the sample of 10,000 gasoline service stations.

(b)   The letter used to designate the mean of $1.35 is $\bar{x}$.

(c)   The mean price given here is a statistic. It is a descriptive measure for a sample.

19.   To begin, we store the age data in a column named AGES. To employ Minitab's Descriptive statistics procedure, we

■   Choose **Stat ▶ Basic Statistics ▶ Display Descriptive Statistics...**

■   Select <u>AGES</u> in the **Variables** text box

■   Click **OK**

The results are shown in the Sessions window as

Descriptive Statistics

Variable	N	Mean	Median	TrMean	StDev	SEMean
AGES	36	58.53	59.50	58.75	13.36	2.23

Variable	Min	Max	Q1	Q3
AGES	31.00	79.00	48.00	68.75

20.   (a)   With the age data already stored in a column named AGES, we

■   Choose **Graph ▶ Boxplot...**

■   Select <u>AGES</u> in the **Y** text box for **Graph 1**

■   Click **OK**

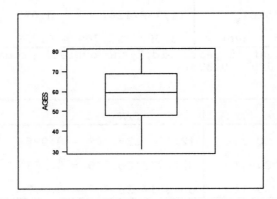

(b)   The 5 number summary may be obtained from the descriptive statistics printout in Exercise 19.

Minimum = 31.0, Maximum = 79.0, Median = 59.5, Q_1 = 48.0, Q_3 = 68.75

21.   (a)   The least variation occurs in the third quarter of the age data, the next least in the fourth quarter, the next least in the second quarter, and the greatest in the first quarter. Also, minimum ≈ 31, maximum ≈ 79, Q_1 ≈ 48, Q_2 ≈ 60, Q_3 ≈ 69.

(b)   There are no potential outliers in the data.

(c)   Exact five-number summary is 31, 48, 59.5, 68.75, and 79.

CHAPTER 4 ANSWERS

Exercises 4.1

4.1 An experiment is an action the result of which cannot be predicted with certainty. An event is a specified result that may or may not occur when the experiment is performed.

4.3 There is no difference.

4.5 The frequentist interpretation of probability is that the probability of an event is the proportion of times the event occurs in a large number of repetitions of the experiment.

4.7 The following could not possibly be probabilities:

(b)   -0.201: A probability cannot be negative.

(e)   3.5: A probability cannot exceed 1.

4.9 The total number of units (in thousands) is N = 109,456.

(a) The probability that the unit has 4 rooms is f/N = 20,789/109,456 = 0.190 (to three decimal places)

(b) The probability that the unit has more than 4 rooms is f/N = (24328 + 22151 + 14283 + 15555)/109456 = 76217/109456 = 0.696.

(c) The probability that the unit has 1 or 2 rooms is f/N = (862 + 1422)/109456 = 2284/109456 = 0.021.

(d) The probability that the unit has fewer than one room is f/N = 0/109456 = 0.000.

(e) The probability that the unit has one or more rooms is f/N = 109456/109456 = 1.000.

4.11 (a) For ages between 25-34 years, f/N = 32,077/126,709 = 0.253.

(b)

Age (yrs)	f/N
45-54	25,514/126,709 = 0.201
55-64	11,739/126,709 = 0.093
65 & over	3,690/126,709 = 0.029

For "at least 45 years old," the overall probability is 0.201 + 0.093 + 0.029 = 0.323.

(c)

Age (yrs)	f/N
20-24	12,138/126,709 = 0.096
25-34	32,077/126,709 = 0.253
35-44	35,051/126,709 = 0.277

For "between 20 and 44 years old," the overall probability is 0.096 + 0.253 + 0.277 = 0.626.

(d)

Age (yrs)	f/N
16-19	6,500/126,709 = 0.051
55-64	11,739/126,709 = 0.093
65 & over	3,690/126,709 = 0.029

For "under 20 or over 54," the overall probability is 0.051 + 0.093 + 0.029 = 0.173.

4.13 (a) There are five ways in which the 36 possibilities sum to 6. Thus, f/N = 5/36 = 0.139.

(b) There are 18 ways in which the 36 possibilities provide a sum that is even. Thus, f/N = 18/36 = 0.500.

(c) There are six ways in which the 36 possibilities sum to 7 and two ways in which the 36 possibilities sum to 11. Thus, f/N = (6 + 2)/36 = 0.222.

(d) There is one way in which the 36 possibilities sum to 2, two ways in which they sum to 3, and one way in which they sum to 12. Thus, f/N = (1 + 2 + 1)/36 = 0.111.

4.15 N = 16,154 + 1,493 + 4,342 = 21,989.

(a) proprietorship: f/N = 16,154/21,989 = 0.735.

(b) not a partnership: 1 - partnership = 1 - 1,493/21,989

= 1 - 0.068 = 0.932.

(c) corporation: f/N = 4,342/21,989 = 0.197.

4.17 The event in part (d), that the housing unit has fewer than one room, is impossible. The event in part (e), that the housing unit has one or more rooms, is a certainty.

4.19 We will use RF, RR, LF, LR to denote right front, right rear, left front, and left rear, respectively. The list of possibilities includes all of the possible pairs of tires that the two students could name. These are

RF,RF	RF,RR,	RF,LF	RF,LR
RR,RF	RR,RR	RR,LF	RR,LR
LF,RF	LF,RR	LF,LF	LF,LR
LR,RF	LR,RR	LR,LF	LR,LR

There are 16 possible pairs, and there are 4 pairs in which the students name the same tire. Thus the probability that they say the same tire is f/N = 4/16 = 1/4 or 0.25.

4.21 When tossing a balanced coin twice and observing the total number of heads, there are four possible equally-likely outcomes, not three. The probability of getting two heads is 1/4, not 1/3.

4.23 The frequentist interpretation construes the probability of an event to be the proportion of times the event occurs in a large number of repetitions of the experiment. Some 'experiments' can not be run more than once under identical conditions - a horse race, a World Series, a super Bowl football game - so it makes no sense to think about a large number of repetitions of the experiment. For these experiments, the frequentist interpretation cannot be used, and therefore it cannot be used as an overall definition of probability.

Exercises 4.2

4.25 Venn diagrams are useful for portraying events and relationships between events.

4.27 Two events are mutual exclusive if they cannot occur at the same time, i.e., they have no outcomes in common. Three events are mutually exclusive if no two of them can occur at the same time, i.e., no pair of the events has any outcomes in common.

4.29   A = {2, 4, 6}; B = {4, 5, 6}; C = {1, 2}; D = {3}

4.31   A = {HHTT, HTHT, HTTH, THHT, THTH, TTHH}

B = {TTHH, TTHT, TTTH, TTTT}

C = {HHHH, HHHT, HHTH, HHTT, HTHH, HTHT, HTTH, HTTT}

D = {HHHH, TTTT}

4.33   (a)   (not A) = {1, 3, 5}

(not A) is the event the die comes up odd.

(b)   (A & B) = {4, 6}

(A & B) is the event the die comes up four or six.

(c)   (B or C) = {1, 2, 4, 5, 6}

(B or C) is the event the die does *not* come up three.

4.35   (a)   (not B) = {HHHH, HHHT, HHTH, HHTT, HTHH, HTHT, HTTH,

HTTT, THHH, THHT, THTH, THTT}

(not B) is the event that at least one of the first two tosses is heads.

(b)   (A & B) = {TTHH}

(A & B) is the event that the first two tosses are tails and the last two are heads.

(c)   (C or D) = {HHHH, HHHT, HHTH, HHTT, HTHH, HTHT, HTTH, HTTT, TTTT}

(C or D) is the event that the first toss is a head or all four tosses are tails.

4.37   (a)   (not A) is the event that the unit has more than 4 rooms. There are 24,328 + 22,151 + 14,183 + 15,555 = 76,217 (thousand) such units.

(b)   (A&B) is the event that the unit has at most 4 rooms and at least 2 rooms, i.e., has 2 or 3 or 4 rooms. There are 1,422 + 10,166 20,789 = 32,377 (thousand) such units.

(c)   The event (C or D) is the event that the unit has between 5 and 7 rooms inclusive or more than seven rooms, i.e., has 5 or more rooms. This is the same as the event (not A) and therefore there are 76,217 (thousand) such units.

4.39   (a)   (not C) is the event that the person selected is 45 years old or older. There are 40,943 thousand such people.

(b)   (not B) is the event that the person selected is either under 20 or over 54. There are 21,929 thousand such people.

(c)   (B & C) is the event that the person selected is between 20 and 44, inclusive. There are 79,266 thousand such people.

(d)   (A or D) is the event that the person selected is either under 20 or over 54. There are 21,929 thousand such people. [*Note:* From part (b), we see that (not B) = (A or D).]

4.41   (a)   Events A and B are not mutually exclusive. They have a four and a six in common.

(b)   Events B and C are mutually exclusive. They have no outcomes in common.

(c)   Events A, C, and D are not mutually exclusive. The outcome two is common to A and C.

(d)   Among A, B, C, and D, there are three mutually exclusive events. These are B, C, and D. Among B, C, and D, there are no outcomes in common. There are not, however, four mutually exclusive events. The outcome two is common to A and C, four is common to A and B, and six is common to A and B.

4.43 (a)   Events C and D are mutually exclusive. They have no outcomes in common.

(b)   Events B and C are not mutually exclusive. They have the ages 20 through 44 in common.

(c)   Events A, B, and D are mutually exclusive. They have no outcomes in common.

(d)   Events A, B, and C are not mutually exclusive. The ages 16 through 19 are common to A and C; the ages 20 through 44 are common to B and C.

(e)   Events A, B, C, and D are not mutually exclusive for the same reasons as those presented in part (d).

4.45 In the following Venn diagram, events A, B, and C are mutually exclusive; events A, B, and D are mutually exclusive; but no other three of the four events are mutually exclusive.

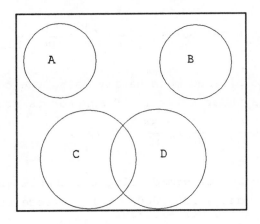

Exercises 4.3

4.47 Two of the 10 marbles are white. Let E be the event that the marble selected is white. Thus, the probability of selecting a white marble is $P(E) = f/N = 2/10 = 0.2$.

4.49 (a) $P(S) = f/N = (1 + 14 + 41)/100 = 0.56$.

(b)   S = (A or B or C)

(c) $P(A) = 1/100 = 0.01$; $P(B) = 14/100 = 0.14$;

$P(C) = 41/100 = 0.41$.

(d) $P(S) = P(A \text{ or } B \text{ or } C) = P(A) + P(B) + P(C) = 0.01 + 0.14 + 0.41 = 0.56$. This result is identical to the one found in part (a).

4.51 The parts to this problem are an application of the special addition rule of probability.

(a)   P(receipts under $100,000) = P(A) + P(B) + P(C)

= 0.601 + 0.081 + 0.088 = 0.770

(b)   P(receipts at least $500,000) = P(E) + P(F)

= 0.036 + 0.041 = 0.077

(c)   P(receipts between $25,000 and $499,999)

= P(B) + P(C) + P(D) = 0.081 + 0.088 + 0.153 = 0.322

(d)   The interpretation of each of the results above in terms of percentages is as follows:

(i)   77.7% of U.S. businesses had receipts under $100,000.

(ii)   7.7% of U.S. businesses had receipts of at least $500,000.

(iii) 32.2% of U.S. businesses had receipts between $25,000 and $499,999.

4.53 (a) $P(\text{senator is at least } 40) = \dfrac{14}{100} + \dfrac{41}{100} + \dfrac{27}{100} + \dfrac{17}{100} = \dfrac{99}{100} = 0.99$

This is accomplished more easily using the complementation rule:
P(senator is at least 40) = 1 - P(senator is under 40)

$= 1 - \dfrac{1}{100} = 0.99$.

(b) $P(\text{senator is under } 60) = \dfrac{1}{100} + \dfrac{14}{100} + \dfrac{41}{100} = \dfrac{56}{100} = 0.56$

This is accomplished more easily using the complementation rule:
P(senator is under 60) = 1 - P(senator is at least 60)

$= 1 - (\dfrac{27}{100} + \dfrac{17}{100}) = \dfrac{56}{100} = 0.56$

4.55 This problem is to be performed using the complementation rule.

(a)   P(receipts under $1,000,000) = 1 - P(receipts $1,000,000 or more)
= 1 - P(F)= 1 - 0.041 = 0.959.

(b)   P(receipts at least $50,000) = 1 - P(receipts under $50,000)
= 1 - P(A) - P(B)= 1 - 0.601 - 0.081 = 0.318.

4.57 (a) $P(A) = \dfrac{6}{36} = 0.167$; $P(B) = \dfrac{2}{36} = 0.056$; $P(C) = \dfrac{1}{36} = 0.028$;

$P(D) = \dfrac{2}{36} = 0.056$; $P(E) = \dfrac{1}{36} = 0.028$; $P(F) = \dfrac{5}{36} = 0.139$;

$P(G) = \dfrac{6}{36} = 0.167$

(b)   P(7 or 11)= P(A) + P(B) = 0.167 + 0.056 = 0.223.

(c)   P(2 or 3 or 12) = P(C) or P(D) or P(E).
= 0.028 + 0.056 + 0.028 = 0.112

(d)   P(8 or doubles):   Using Figure 4.1: $\frac{10}{36}$ = 0.278

(e)   P(8) + P(doubles) - P(8 & doubles) = 0.139 + 0.167 - 0.028= 0.278.

4.59   (a)   P(F) = 0.520;   P(D) = 0.095;   P(F & D) = 0.054.

(b)   P(F or D) = 0.520 + 0.095 - 0.054 = 0.561.

56.1% of U.S. adults are either female or divorced.

(c)   P(M) = 1 - 0.520 = 0.480.

4.61   Recalling the general addition rule and making the appropriate substitutions, we have:

P(A or B) = P(A) + P(B) - P(A & B)

or

1/2   =   1/3 + P(B) - 1/10 .

Rearranging terms and solving for P(B), we find:

P(B) = 1/2 - 1/3 + 1/10 = 15/30 - 10/30 + 3/30 = 8/30 = 0.267

4.63   Let:

J = person enjoys job

L = person enjoys personal life

not J = person doesn't enjoy job

not L = person doesn't enjoy personal life

We are given: P((not J) & (not L)) = 0.15,

P(J & (not L)) = 0.80, and

P(J & L) = 0.04.

(a)   We must find P(J or L), which can be expressed in terms of the general addition rule as:

P(J or L) = P(J) + P(L) - P(J & L)   .

Notice that we are given P(J & L) = 0.04.

We find P(J) and P(L) as follows:

(i)   P(J) = P(J & (not L)) + P(J & L) = 0.80 + 0.04 = 0.84.

(ii)   P((not L)) = P((not J) & (not L)) + P(J & (not L))

= 0.15 + 0.80 = 0.95.

Since P(not L) = 0.95, P(L) = 1 - P(not L) = 1 - 0.95

= 0.05.

Thus, P(J or L) = 0.84 + 0.05 - 0.04 = 0.85.

(b)   We must find P(L & (not J)) = P((not J) & L). We use the relationship:

P(L) = P(J & L) + P((not J) & L)

and make the appropriate substitutions for items already solved:

0.05 = 0.04 + P((not J) & L).

Rearranging terms, we have

P((not J) & L) = 0.05 - 0.04 = 0.01.

4.65 $P(A \text{ or } B \text{ or } C \text{ or } D) = P(A) + P(B) + P(C) + P(D) - P(A\&B) - P(A\&C) - P(A\&D) - P(B\&C) - P(B\&D) - P(C\&D) + P(A\&B\&C) + P(A\&B\&D) + P(A\&C\&D) + P(B\&C\&D) - P(A\&B\&C\&D)$

Exercises 4.4

4.67   The total number of observations of bivariate data can be obtained from the frequencies in a contingency table by summing the counts in the cells, summing the row totals, or summing the column totals.

4.69   (a)   Data obtained by observing values of one variable of a population are called <u>univariate</u> data.

   (b)   Data obtained by observing values of two variables of a population are called <u>bivariate</u> data.

4.71   (a)   This contingency table has eight cells.

   (b)   The total number of institutions of higher education in the United States is 3274.

   (c)   There are 863 institutions of higher education in the Midwest.

   (d)   There are 1471 public institutions.

   (e)   There are 502 private schools in the South.

4.73   (a)   The number of office-based surgeons is 73,970.

   (b)   The number of ophthalmologists is 15,540.

   (c)   The number of office-based ophthalmologists is 12,328.

   (d)   The number of surgeons who are either office based or ophthalmologists is $73,970 + 15,540 - 12,328 = 77,182$.

   (e)   The number of hospital-based general surgeons is 12,225.

   (f)   The number of hospital-based OB/GYNs is 6,734.

   (g)   The number of surgeons who are not hospital-based is $103,601 - 25,901 = 77,700$.

4.75   (a)   The missing entries in the second, third, and fourth rows are, respectively, 130, 183, and 27.

   (b)   This contingency table has 15 cells.

   (c)   There are 554 thousand farms having under 50 acres.

   (d)   There are 217 thousand farms that are tenant operated.

   (e)   There are 111 thousand farms that are operated by part owners and have between 500 and 1000 acres.

   (f)   There are $596 + 217 = 813$ thousand farms that are not full-owner operated.

   (g)   There are $55 + 27 + 24 = 106$ thousand farms that are tenant-operated and have at least 180 acres.

4.77   (a) T_2: the institution selected is private;

   R_3: the institution selected is in the South;

   $(T_1 \& R_4)$: the institution selected is a public school in the West.

   (b) $P(T_2) = 1803/3274 = 0.551$; $P(R_3) = 1035/3274 = 0.316$;

   $P(T_1 \& R_4) = 313/3274 = 0.096$.

   The interpretation of each item above is: 55.1% of institutions of higher education are private; 31.6% are in the South; 9.6% are public schools in the West.

(c)

| | Type | | |
Region	Public T_1	Private T_2	$P(R_i)$
Northeast R_1	0.081	0.170	0.251
Midwest R_2	0.110	0.154	0.264
South R_3	0.163	0.153	0.316
West R_4	0.096	0.074	0.170
$P(T_j)$	0.449	0.551	1.000

(d)   Begin with the table in part (c). Summing the cell entries in each row results in the marginal probability in the respective row. Summing the cell entries in each column results in the marginal probability in the respective column.

4.79 (a) B_1: the base of practice for the surgeon is an office;

S_3: the surgeon's specialty is orthopedics;

$(B_1 \& S_3)$: the surgeon has a specialty in orthopedics and is based in an office.

(b) $P(B_1) = 73,970/103,601 = 0.714$; $P(S_3) = 18,026/103,601 = 0.174$;

$P(B_1 \& S_3) = 13,364/103,601 = 0.129.$

The interpretation of each item above is: 71.4% of surgeons base their practice in an office; 17.4% of surgeons have a specialty in orthopedics; 12.9% of surgeons specialize in orthopedics and are based in an office.

(c) $P(B_1 \text{ or } S_3) = (24128 + 24150 + 13364 + 12328 + 4248 + 414)/103601$
$= 78632/103601 = 0.759$

(d) $P(B_1 \text{ or } S_3) = P(B_1) + P(S_3) - P(B_1 \& S_3) = 0.714 + 0.174 - 0.129$
$= 0.759$

(e)   Base of Practice

	Office B_1	Hospital B_2	Other B_3	$P(S_i)$
General Surgery S_1	0.233	0.118	0.016	0.367
Obstetrics/ gynecology S_2	0.233	0.065	0.011	0.309
Orthopedics S_3	0.129	0.041	0.004	0.174
Ophthalmology S_4	0.119	0.026	0.005	0.150
$P(B_j)$	0.714	0.250	0.036	1.000

4.81 (a)   (i) A_3: the farm selected has between 180 and 500 acres, inclusive.

(ii) T_2: the farm selected is part-owner operated;

(c) $(T_1 \& A_5)$: the farm selected is full-owner operated with at least 1,000 acres.

(b)   (i) $P(A_3) = 428/1925 = 0.222$;

(ii) $P(T_2) = 596/1925 = 0.310$;

(iii) $P(T_1 \& A_5) = 35/1925 = 0.018$.

(c)   Tenure of operator

Acreage	Full Owner T_1	Part Owner T_2	Tenant T_3	Total
Under 50 A_1	23.1	3.0	2.7	28.8
50-179 A_2	20.5	6.8	3.1	30.3
180-499 A_3	9.9	9.5	2.9	22.2
500-999 A_4	2.5	5.8	1.4	9.7
1000+ A_5	1.8	5.9	1.2	9.0
Total	57.8	31.0	11.3	100.0

4.83 The categories for each of the characteristics in the cross classification are selected so that no member of the population (or sample) can belong to more than one of the categories.

Exercises 4.5

4.85 (a)   Conditional probability is the probability that an event will occur given that another event has occurred.

(b)   In the notation $P(B|A)$, the given event is A.

4.87 (a) $P(B) = 4/52 = 0.077$; the probability of selecting a king from an ordinary deck of 52 playing cards is 0.077.

(b) $P(B|A) = 4/12 = 0.333$; given that the selection is made from among (the 12) face cards, the probability of selecting a king is 0.333.

(c) $P(B|C) = 1/13 = 0.077$; given that the selection is made from among (the 13) hearts, the probability of selecting a king is 0.077.

(d) $P(B|(\text{not } A)) = 0/40 = 0$; given that the selection is made from among (the 40) non-face cards, the probability of selecting a king is 0.

(e) $P(A) = 12/52 = 0.231$; the probability of selecting a face card from an ordinary deck of 52 playing cards is 0.231.

(f) $P(A|B) = 4/4 = 1$; given that the selection is made from among (the four) kings, the probability of selecting a face card is 1.

(g) $P(A|C) = 3/13 = 0.231$; given that the selection is made from among (the 13) hearts, the probability of selecting a face card is 0.231.

(h) $P(A|(\text{not } B)) = 8/48 = 0.167$; given that the selection is made from among (the 48) non-kings, the probability of selecting a face card is 0.167.

4.89 (a) $P(\text{exactly 4 rooms}) = 20{,}789/109{,}456 = 0.190$

(b) $P(\text{exactly 4 rooms} \mid \text{at least 2 rooms}) = f/N$

$= 20{,}789/(109{,}456 - 862) = 20{,}789/108{,}594 = 0.191$

(c) $P(\text{at most 4 rooms} \mid \text{at least 2 rooms}) = f/N$

$= (1422 + 10166 + 20789)/108{,}594 = 32377/108594 = 0.298$

(d)   (i)   19.0% of the units have exactly 4 rooms;

(ii)   Of those with at least 2 rooms, 19.1% have exactly 4 rooms;

(iii) Of those with at least 2 rooms, 29.8% have at most 4 rooms.

4.91 The probability that the institution selected is:

(a)   in the Northeast is $821/3274 = 0.251$.

(b)   in the Northeast, given that it is a private school, is $555/1803 = 0.308$.

(c)   a private school, given that it is in the Northeast, is $555/821 = 0.676$.

(d)   Each of the above is interpreted, respectively, as the following percentage:

(i)   25.1% of all institutions of higher education are in the Northeast;

(ii)   30.8% of all private institutions of higher education are in the Northeast;

(iii) 67.6% of all institutions of higher education in the Northeast are private schools.

4.93 (a) $P(T_3) = 217/1925 = 0.1127$;

(b) $P(T_3 \ \& \ A_3) = 55/1925 = 0.0286$;

(c) $P(A_3|T_3) = 55/217 = 0.2535$;

(d) $P(A_3|T_3) = P(T_3 \ \& \ A_3)/P(T_3) = 0.0286/0.1127 = 0.2538$.

(e)   Parts (a) - (c) are interpreted, respectively, as follows:

(i)   the probability is 0.1127 that an operator is a tenant;

(ii)   the probability is 0.0286 that an operator is a tenant and runs a farm between 180 and 499 acres.

(iii) of those who are tenant operators, the probability is 0.2535 that this type of operator runs a farm between 180 and 499 acres.

4.95 The probability that the member selected is a:

(a)   senator is 0.187.

(b)   Republican senator is 0.103.

(c)   Republican, given that he or she is a senator is 0.103/0.187 = 0.551.

(d)   senator, given that he or she is a Republican is 0.103/0.527 = 0.195.

(e)   Each of the above is interpreted, respectively, as the following percentage:

(i)   18.7% of the members of the 105th Congress are senators;

(ii)   10.3% of the members of the 105th Congress are Republican senators;

(iii) 55.1% of the senators in the 105th Congress are Republicans;

(iv)   19.5% of the Republicans in the 105th Congress are senators.

4.97 Let: A = African-American;

W = Woman.

We are given: $P(A) = 0.126$ and $P(A \& W) = 0.066$. The percentage of African-Americans who are women is:

$P(W|A) = P(W \& A)/P(A) = 0.066/0.126 = 0.524$ or 52.4% .

4.99 (a)   Answers will vary, but any experiment in which all possible outcomes are equally likely and can be counted will be included. For example, if two cards are dealt from a standard deck, P(two kings|at least one king) can be computed both directly and using the conditional probability rule.

(b)   Answers will vary, but for any experiment in which outcomes are not equally likely or cannot be counted, only the conditional probability rule can be used. For example, P(Fleetfeet will win a horse race|Fleetfeet will finish third or better) cannot be computed directly since the possible outcomes are not equally likely.

Exercises 4.6

4.101 (a)   General multiplication rule: $P(A\&B) = P(A) \cdot P(B|A)$

Conditional-probability rule: $P(B|A) = P(A\&B)/P(A)$

(b)   Multiplying both sides of the conditional probability rule by $P(A)$ results in the general multiplication rule.

(c)   When the marginal and conditional probabilities are known, we can use the general multiplication rule to obtain a joint probability. When the joint and marginal probabilities are known, we can use the conditional probability rule to obtain conditional probabilities.

4.103 Let W = event that chosen person is a woman H = event that the chosen person suffers from holiday depression. We are given that $P(W) = .52$ and $P(H|W) = .44$. The desired probability is $P(H\&W) = P(H|W) \cdot P(W) = (.44)(.52) = 0.2288$. Thus 22.88% of U.S. adults are women who suffer from holiday depression.

4.105 (a)   Let: SIX1 = the event that the first card selected is numbered 6.
There is only one card numbered 6; also, N = 10.   Thus:

$$P(SIX1) = 1/10 = 0.1.$$

(b)   With the first card selected numbered 6, let:

NINE2 = the event that the second card selected is
numbered 9.

There is only one card numbered 9; also N = 9.   Thus:

$$P(NINE2|SIX1) = 1/9 = 0.111.$$

(c)   To find the probability of selecting a 6 and then a 9, we apply the general multiplication rule:

$$P(SIX1 \& NINE2) = P(SIX1) \cdot P(NINE2|SIX1)$$

and use the results presented in parts (a) and (b) above:

$$P(SIX1 \& NINE2) = (0.1) \cdot (0.111) = 0.011.$$

(d)   Of the ten original cards, 5 of them are over 5.   These are the cards numbered 6, 7, 8, 9, 10.   Let:

OFIVE1 = the event that the first card selected is over 5.

Thus:

$$P(OFIVE1) = 5/10 = 0.5 .$$

Of the nine remaining cards, 4 of them are over 5.   Let:

OFIVE2 = the event that the second card selected is over 5.

Thus:

$$P(OFIVE2|OFIVE1) = 4/9 = 0.444.$$

To find the probability that both cards selected are over 5, we apply the general multiplication rule:

$$P(OFIVE1 \& OFIVE2) = P(OFIVE1) \cdot P(OFIVE2|OFIVE1)$$

and use the two results presented before this previous result:

$$P(OFIVE1 \& OFIVE2) = (0.5) \cdot (0.444) = 0.222 .$$

4.107 Let: D_i = Democrat on selection i, where i = 1, 2;

R_i = Republican on selection i, where i = 1,2;

I_i = Independent on selection i, where i = 1,2.

(a) $P(R_1 \& D_2) = (32/50)(17/49) = 0.222.$

(b) $P(R_1 \& R_2) = (32/50)(31/49) = 0.405.$

(c)

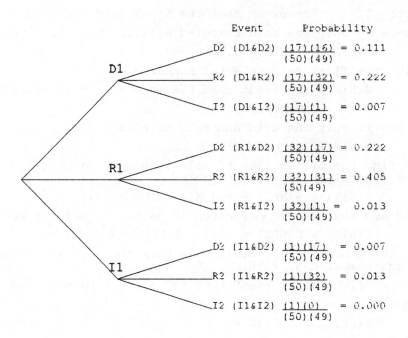

	Event	Probability
D2 (D1&D2)	$\frac{(17)(16)}{(50)(49)}$ = 0.111
R2 (D1&R2)	$\frac{(17)(32)}{(50)(49)}$ = 0.222
I2 (D1&I2)	$\frac{(17)(1)}{(50)(49)}$ = 0.007
D2 (R1&D2)	$\frac{(32)(17)}{(50)(49)}$ = 0.222
R2 (R1&R2)	$\frac{(32)(31)}{(50(49)}$ = 0.405
I2 (R1&I2)	$\frac{(32)(1)}{(50)(49)}$ = 0.013
D2 (I1&D2)	$\frac{(1)(17)}{(50)(49)}$ = 0.007
R2 (I1&R2)	$\frac{(1)(32)}{(50)(49)}$ = 0.013
I2 (I1&I2)	$\frac{(1)(0)}{(50)(49)}$ = 0.000

(d)   The probability that both governors selected belong to the same party or are both independents is:

$$P(R_1 \,\&\, R_2) + P(D_1 \,\&\, D_2) + P(I_1 \,\&\, I_2)$$
$$= 0.405 + 0.111 + .000 = 0.516.$$

(e)   P(one governor is Republican and one is Democrat) =

$$P(R_1 \,\&\, D_2) + P(D_1 \,\&\, R_2) = 0.222 + 0.222 = 0.444$$

4.109 (a) $P(C_1) = 9.3/61.4 = 0.151$

(b) $P(C_1 | S_2) = 1.3/25.8 = 0.050$

(c)   Events C_1 and S_2 are independent if

$$P(C_1 | S_2) = P(C_1) \;.$$

Using the results presented in parts (a) and (b) previously, we see that $P(C_1 | S_2) \neq P(C_1)$; thus, events C_1 and S_2 are not independent.

(d)   Let: S_1 = an injured person is male;

C_2 = an injured person was hurt at home.

To conclude that S_1 is independent of C_2, the following rule of independence must hold:

$$P(S_1 | C_2) = P(S_1) \;.$$

From the table, $P(S_1 | C_2) = 9.8/21.4 = 0.458$. Also, $P(S_1) = 35.6/61.4 = 0.58$. Since $P(S_1 | C_2) \neq P(S_1)$, S_1 and C_2 are not independent.

4.111 (a) $P(A) = 4/8 = 0.5;$ $P(B) = 4/8 = 0.5;$ $P(C) = 3/8 = 0.375.$

(b) $P(B | A) = 2/4 = 0.5$

(c)   A and B are independent events because $P(B|A) = P(B)$.

(d) $P(C|A) = 1/4 = 0.25$

(e)   A and C are not independent events because $P(C|A) \neq P(C)$.

4.113 (a) $P(P_1) = 0.469$; $P(C_2) = 0.187$; $P(P_1 \& C_2) = 0.084$.

(b)   The special multiplication rule states that two events P_1 and C_2 are independent if:

$$P(P_1 \& C_2) = P(P_1) \cdot P(C_2) .$$

Using the information presented in part (a),

$$P(P_1) \cdot P(C_2) = (0.469) \cdot (0.187) = 0.088.$$

Since this product does not equal $P(P_1 \& C_2) = 0.084$, the events P_1 and C_2 are not independent.

4.115 Let A_i = draw an ace on selection i, where i = 1,2.

(a)   If the first card is replaced before the second card is drawn:

$$P(A_1 \& A_2) = (4/52)(4/52) = 0.006 .$$

(b)   If the first card is not replaced before the second card is drawn:

$$P(A_1 \& A_2) = (4/52)(3/51) = 0.005 .$$

4.117 The exercise states that E and F are independent. This means that:

$$P(E \& F) = P(E) \cdot P(F) .$$

Also given is $P(E) = 1/3$ and $P(F) = 1/4$.

(a)   To find $P(E \& F)$, implement the independence rule above:

$$P(E \& F) = (1/3) \cdot (1/4) = 1/12 = 0.083.$$

(b)   To find $P(E$ or $F)$, implement the general addition rule:

$$P(E \text{ or } F) = P(E) + P(F) - P(E \& F)$$

and substitute the calculations above into the formula:

$$P(E \text{ or } F) = (1/3) + (1/4) - (1/12)$$
$$= 0.333 + 0.250 - 0.083 = 0.5 .$$

4.119 Let:

not F_i = non-failure of "criticality 1" item i, for i = 1, 2, ..., 748;

F_i = failure of "criticality 1" item i, for i = 1, 2, ..., 748.

Also, $P(\text{not } F_i) = 0.9999$ and $P(F_i) = 0.0001$.

(a)   The probability that none of the "criticality 1" items would fail is:

$$P(\text{not } F_1) \cdot (P(\text{not } F_2) \cdot \ ... \ \cdot P(\text{not } F_{748}) = (0.9999)^{748} = 0.928 .$$

(b)   The probability that at least one "criticality 1" item would fail is:

$$1 - P(\text{none of the "criticality 1" items would fail})$$
$$= 1 - 0.928 = 0.072.$$

(c)   There was a 7.2% chance that at least one "criticality 1" item would fail. In other words, on the average, at least one "criticality 1" item will fail in 7.2 out of every 100 such missions.

4.121 The probability of a home-buyer purchasing a resale home is 0.773. By the complementation rule, the probability is $1 - 0.773 = 0.227$ that a home-buyer will purchase a new home.

(a) Letting R1, R2, and R3 be the events that the first, second, and third purchases, respectively, will be resale homes and letting N4 be the event that the fourth purchase will be a new home:

$$P(R1 \ \& \ R2 \ \& \ R3 \ \& \ N4) = P(R1) \cdot P(R2) \cdot P(R3) \cdot P(N4)$$

$$= (0.773) \cdot (0.773) \cdot (0.773) \cdot (0.227) = 0.105 \ .$$

(b) Letting R1, R3, and R4 be the events that the first, third, and fourth purchases, respectively, will be resale homes and letting N2 be the event that the second purchase will be a new home:

$$P(R1 \ \& \ N2 \ \& \ R3 \ \& \ R4) = P(R1) \cdot P(N2) \cdot P(R3) \cdot P(R4)$$

$$= (0.773) \cdot (0.227) \cdot (0.773) \cdot (0.773) = 0.105 \ .$$

Compare this answer with the answer in part (a). Reordering the items comprising each outcome results in the same probability. These two examples illustrate two ways of ordering three R's and one N.

(c) Letting R1 and R4 be the events that the first and fourth purchases, respectively, will be resale homes and letting N2 and N3 be the events that the second and third purchases, respectively, will be new homes:

$$P(R1 \ \& \ N2 \ \& \ N3 \ \& \ R4) = P(R1) \cdot P(N2) \cdot P(N3) \cdot P(R4)$$

$$= (0.773) \cdot (0.227) \cdot (0.227) \cdot (0.773) = 0.031 \ .$$

(d) There are four possible ways in which exactly three of the four home purchases will be resales (the new home purchased could be first, second, third, or fourth), with two of the four ways presented in parts (a) and (b). The four ways are

R1	&	R2	&	R3	&	N4
R1	&	R2	&	N3	&	R4
R1	&	N2	&	R3	&	R4
N1	&	R2	&	R3	&	R4 .

The joint probability of any one of the four combinations above is 0.105, as illustrated in either part (a) or part (b). With all of this as background, the probability that exactly three of the four will be resales is equal to the sum of the joint probabilities of the four combinations above, or 0.105 + 0.105 + 0.105 + 0.105 = 0.420.

4.123 Let: M = male;   F = female;   L = activity limitation.

Also, $P(L|M) = 0.136$ and $P(L|F) = 0.144$. Note that these probabilities are different from each other. This means that activity limitation is influenced by a person's gender.

If gender and activity limitation were statistically independent, the following would have to hold:

$$P(L|M) = P(L)$$

and

$$P(L|F) = P(L) \ .$$

But $P(L)$ is different depending upon a person's gender. Thus, gender and activity limitation are not statistically independent.

4.125 (a) If two events A and B are mutually exclusive, they cannot both happen, so $P(A\&B) = 0$.

(b)   If neither A nor B is impossible, then P(A) > 0 and P(B) > 0. Since A and B are independent, P(A&B) = P(A)· P(B), but this cannot be zero since neither A nor B has a probability of 0.

(c)   Answers will vary. If A and B are two events such that P(A) = 0.4, P(B) = 0.3, and P(A&B) = 0.2, then A and B are not mutually exclusive since they can happen simultaneously, and they are not independent since P(A&B) ≠ P(A)· P(B).

4.127 (a)   We have

$$P(A) = \frac{18}{36} = \frac{1}{2}$$

$$P(B) = \frac{18}{36} = \frac{1}{2}$$

$$P(C) = \frac{18}{36} = \frac{1}{2}$$

$$P(A \& B) = \frac{9}{36} = \frac{1}{4}$$

$$P(A \& C) = \frac{9}{36} = \frac{1}{4}$$

$$P(B \& C) = \frac{9}{36} = \frac{1}{4}$$

$$P(A \& B \& C) = \frac{9}{36} = \frac{1}{4}$$

Consequently,

$$P(A \& B) = P(A)P(B)$$
$$P(A \& C) = P(A)P(C)$$
$$P(B \& C) = P(B)P(C)$$

However,

$$P(A \& B \& C) = \frac{1}{4} \neq \frac{1}{2} \cdot \frac{1}{2} \cdot \frac{1}{2} = P(A)P(B)P(C)$$

Thus, the events A, B, and C are *not* independent.

(b)   We have

$$P(D) = \frac{18}{36} = \frac{1}{2}$$

$$P(E) = \frac{18}{36} = \frac{1}{2}$$

$$P(F) = \frac{4}{36} = \frac{1}{9}$$

$$P(D \& E) = \frac{6}{36} = \frac{1}{6}$$

$$P(D \& F) = \frac{3}{36} = \frac{1}{12}$$

$$P(E \& F) = \frac{2}{36} = \frac{1}{18}$$

$$P(D \& E \& F) = \frac{1}{36}$$

Consequently,

$$P(D \& E \& F) = \frac{1}{36} = \frac{1}{2} \cdot \frac{1}{2} \cdot \frac{1}{9} = P(D)P(E)P(F)$$

However, since

$$P(D \& E) = \frac{1}{6} \neq \frac{1}{2} \cdot \frac{1}{2} = P(D)P(E)$$

the events D, E, and F are *not* independent.

Exercises 4.7

4.129 Four events are exhaustive if one or more of them must occur.

4.131 (a) $P(R_3)$

   (b) $P(S|R_3)$

   (c) $P(R_3|S)$

4.133 (a)   P(believe in aliens) = 0.54 · 0.48 + 0.33 · 0.52 = 0.431, so 43.1% of U.S. adults believe in aliens.

   (b)   P(believe in aliens|woman) = 0.33; so 33% of women believe in aliens.

   (c)   P(woman|believe in aliens) = (0.33· 0.52)/.431 = 0.398; so 39.8% of adults who believe in aliens are women.

4.135 Let:

A = 18-24	P(A) = 0.127	P(M|A) = 0.83
B = 25-34	P(B) = 0.207	P(M|B) = 0.54
C = 35-44	P(P) = 0.220	P(M|C) = 0.43
D = 45-54	P(D) = 0.165	P(M|D) = 0.37
E = 55-64	P(E) = 0.109	P(M|E) = 0.27
F = 65 & over	P(F) = 0.172	P(M|F) = 0.20
M = movie goer		

(a) P(movie goer) = P(A)P(M|A) + P(B)P(M|B) + P(C)P(M|C) + P(D)P(M|D) + P(E)P(M|E) + P(F)P(M|F)

= (0.127 · 0.83) + (0.207 · 0.54) + (0.22 · 0.43) + (0.165 · 0.37) + (0.109 · 0.27) + (0.172 · 0.20) = 0.437

(b) P(25-34 years old|movie goer) = (0.207 · 0.54)/0.437 = 0.256

(c) 43.7% of adults are movie goers, and of those who are movie goers, 25.6% are between 25 and 34 years old.

4.137 Let: M = sell more than projected P(M) = 0.10 P(R|M) = 0.70

C = sell close to projected   P(C) = 0.30 P(R|C) = 0.50

L = sell less than projected P(L) = 0.60 P(R|L) = 0.20

R = revised for a second edition

(a) P(R) = P(M)·P(R|M) + P(C)·P(R|C) + P(L)·P(R|L)

= (0.10 · 0.70) + (0.30 · 0.50) + (0.60 · 0.20) = 0.34 or 34%

(b) $P(L|R) = \frac{P(L \& R)}{P(R)} = \frac{P(L) \cdot P(R|L)}{P(R)} = \frac{0.60 \cdot 0.20}{0.34} = 0.353$ or 35.3%

4.139 Let: P = poisoning   P(P|M) = 0.133   P(P|F) = 0.364

H = hanging/strangling P(H|M) = 0.154   P(H|F) = 0.126

G = firearms   P(G|M) = 0.651   P(G|F) = 0.408

O = other   P(O|M) = 0.062   P(O|F) = 0.102

M = male commits suicide   P(M) = 0.811

F = female commits suicide   P(F) = 0.189

(a) P(G) = P(M and G) + P(F and G)

= P(M) · P(G|M) + P(F) · P(G|F)

= (0.811 · 0.651) + (0.189 · 0.408) = 0.605

(b) P(F) = 0.189

(c) $P(F|G) = \frac{P(F \& G)}{P(G)} = \frac{P(F) \cdot P(G|F)}{P(G)} = \frac{0.189 \cdot 0.408}{0.605} = 0.127$

(d) The interpretation of each item above is as follows:

(i) Firearms accounted for 60.5% of all suicides;

(ii) Females accounted for 18.9% of all suicides;

(iii) Females accounted for 12.7% of those suicides that resulted from use of a firearm.

4.141 Let: 0B = zero broken eggs in carton   P(0B) = 0.785

1B = one broken egg in carton   P(1B) = 0.192

2B = two broken eggs in carton   P(2B) = 0.022

3B = three broken eggs in carton   P(3B) = 0.001

B = selected egg is broken

4.141 Let: 0B = zero broken eggs in carton   P(0B) = 0.785

1B = one broken egg in carton   P(1B) = 0.192

2B = two broken eggs in carton   P(2B) = 0.022

3B = three broken eggs in carton   P(3B) = 0.001

 B = selected egg is broken

$$P(1B|B) = \frac{P(1B) \cdot P(B|1B)}{P(0B) \cdot P(B|0B) + P(1B) \cdot P(B|1B) + P(2B) \cdot P(B|2B) + P(3B) \cdot P(B|3B)}$$

$$= \frac{0.192 \cdot (1/12)}{0.785 \cdot (0/12) + 0.192 \cdot (1/12) + 0.022 \cdot (2/12) + 0.001 \cdot (3/12)} = 0.803$$

4.143 Let: N = enjoy neither their job nor their personal life   P(N) = 0.15

O = enjoy their job but not their personal life   P(O) = 0.80

B = enjoy both their job and their personal life   P(B) = 0.04

J = people interviewed that enjoy their job

I = people interviewed that enjoy their job who also enjoy their personal life

(a)   P(J) = P(O) + P(B) = 0.80 + 0.04 = 0.84

84% of the people interviewed enjoy their jobs.

(b)   P(I) = P(B)/P(J) = 0.04/0.84 = 0.048

4.8% of the people interviewed who enjoy their jobs also enjoy their personal lives.

4.145 (a)

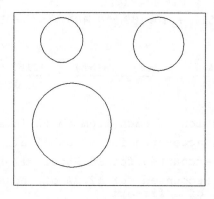

(b)   In baseball, in one time at bat, a batter might strike out, get a base on balls, or get a hit. These are mutually exclusive, but not exhaustive since there are other things that could happen such as being hit by a pitch, getting on base on an error or a fielder's choice, or executing a sacrifice.

(c)   The two rectangles, A and B, and the circle C are exhaustive, but not mutually exclusive.

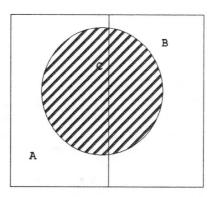

(d)   A person selected at random could be male, female, or statistician. These are exhaustive, but since a statistician is either male or female, they are not mutually exclusive.

Exercises 4.8

4.147 The BCR is often referred to as the multiplication rule because the total number of possibilities in a multi-step process is the product of the numbers of ways each action can be carried out (as long as the number of ways for each action is constant and not dependent on how other actions were performed).

4.149 (a)

Model	Elevation	Outcome
P	A	PA
	B	PB
	C	PC
	D	PD
V	A	VA
	B	VB
	C	VC
	D	VD
M	A	MA
	B	MB
	C	MC
	D	MD

(b)   There are 12 choices for the selection of a home.

(c)   There are 3 ways to choose the model and 4 ways to choose the elevation, so there are 3·4 = 12 ways to choose the model and elevation.

4.151 (a)   There are 9·10·10 = 900 different area codes possible.

(b)   There are 9·10·10·10·10·10·10 = 9,000,000 local telephone numbers possible for each area code.

(c)   There are 900 x 9,000,000 = 8,100,000,000 possible telephone numbers.

4.153 There are $6 \cdot 8 \cdot 5 \cdot 19 \cdot 16 \cdot 14 = 1,021,440$ possibilities for answering all six questions.

4.155 (a) $(4)_3 = 4 \cdot 3 \cdot 2 = 24$ (b) $(15)_4 = 15 \cdot 14 \cdot 13 \cdot 12 = 32,760$

(c) $(6)_2 = 6 \cdot 5 = 30$ (d) $(10)_0 = 1$

(e) $(8)_8 = 8 \cdot 7 \cdot 6 \cdot 5 \cdot 4 \cdot 3 \cdot 2 \cdot 1 = 40,320$

4.157 There are $30 \cdot 29 \cdot 28 \cdot 27 = 657,720$ ways to make the four investments.

4.159 (a)   There are $10 \cdot 9 \cdot 8 \cdot 7 \cdot 6 \cdot 5 \cdot 4 \cdot 3 \cdot 2 \cdot 1 = 3,628,800$ possibilities for the order in which the subject writes down the numbers.

(b)   With 3,628,800 possibilities in total, the subject has a one in 3,628,800 chance of guessing the numbers in the correct order.

4.161 (a) $\binom{4}{3} = \dfrac{4!}{3!\,(4-3)!} = \dfrac{4!}{3!\,1!} = \dfrac{4 \cdot 3!}{3!\,1!} = \dfrac{4}{1} = 4$

(b) $\binom{15}{4} = \dfrac{15!}{4!\,(15-4)!} = \dfrac{15!}{4!\,11!} = \dfrac{15 \cdot 14 \cdot 13 \cdot 12 \cdot 11!}{4!\,11!} = \dfrac{15 \cdot 14 \cdot 13 \cdot 12}{24} = 1,365$

(c) $\binom{6}{2} = \dfrac{6!}{2!\,(6-2)!} = \dfrac{6!}{2!\,4!} = \dfrac{6 \cdot 5 \cdot 4!}{2!\,4!} = \dfrac{6 \cdot 5}{2} = 15$

(d) $\binom{10}{0} = \dfrac{10!}{0!\,(10-0)!} = \dfrac{10!}{1 \cdot 10!} = 1$

(e) $\binom{8}{8} = \dfrac{8}{8!\,(8-8)!} = \dfrac{8!}{8!\,0!} = \dfrac{8!}{8!\,1} = 1$

4.163 (a)

$$\binom{100}{5} = \dfrac{100!}{5!\,(100-5)!} = \dfrac{100!}{5!\,95!}$$

$$= \dfrac{100 \cdot 99 \cdot 98 \cdot 97 \cdot 96 \cdot 95!}{5!\,95!} = \dfrac{100 \cdot 99 \cdot 98 \cdot 97 \cdot 96}{5 \cdot 4 \cdot 3 \cdot 2 \cdot 1} = 75,287,520$$

(b) $\binom{50}{5} \cdot 2^5 = \dfrac{50!}{5!\,(50-5)!} \cdot 32 = 67,800,320$

(c)   The probability that no state will have both of its senators on the committee is the answer in part (b) divided by the answer in part (a), which is 67,800,320/75,287,520 = 0.901.

4.165 $\begin{pmatrix} 45 \\ 6 \end{pmatrix} = \dfrac{45!}{6!(45-6)!} = \dfrac{45!}{6!\,39!} = \dfrac{45\cdot44\cdot43\cdot42\cdot41\cdot40\cdot39!}{6!\,39!} = \dfrac{45\cdot44\cdot43\cdot42\cdot41\cdot40}{6\cdot5\cdot4\cdot3\cdot2\cdot1} = 8,145,060$

4.167 (a) The probability that exactly one of the TVs selected is defective is:

$$\frac{\begin{pmatrix}6\\1\end{pmatrix}\begin{pmatrix}94\\4\end{pmatrix}}{\begin{pmatrix}100\\5\end{pmatrix}} = \frac{\dfrac{6!}{1!(6-1)!}\cdot\dfrac{94!}{4!(94-4)!}}{75,287,520} = \frac{6\cdot3,049,501}{75,287,520} = 0.243$$

(b) The probability that at most one of the TVs selected is defective is the sum of the probabilities for zero and one defectives. The probability of zero defectives is:

$$\frac{\begin{pmatrix}6\\0\end{pmatrix}\begin{pmatrix}94\\5\end{pmatrix}}{\begin{pmatrix}100\\5\end{pmatrix}} = \frac{\dfrac{6!}{0!(6-6)!}\cdot\dfrac{94!}{5!(94-5)!}}{\begin{pmatrix}100\\5\end{pmatrix}} = \frac{54,891,018}{75,287,520} = 0.729$$

The probability of one defective is 0.243, as calculated in part (a). Thus, the probability that at most one of the TVs selected is defective is 0.243 + 0.729 = 0.972.

(c) The probability that at least one of the TVs selected is defective is:

1 - probability of zero defectives = 1 - 0.729 = 0.271.

4.169 (a) The probability that you win the jackpot is:

$$\frac{\begin{pmatrix}6\\6\end{pmatrix}\begin{pmatrix}36\\0\end{pmatrix}}{\begin{pmatrix}42\\6\end{pmatrix}} = \frac{\dfrac{6!}{6!(6-6)!}\cdot\dfrac{36!}{0!(36-0)!}}{\dfrac{42!}{6!(42-6)!}} = \frac{1\cdot1}{5,245,786} = 0.000000191.$$

(b) The probability that your ticket contains exactly four winning numbers is:

$$\frac{\begin{pmatrix}6\\4\end{pmatrix}\begin{pmatrix}36\\2\end{pmatrix}}{\begin{pmatrix}42\\6\end{pmatrix}} = \frac{\dfrac{6!}{4!(6-4)!}\cdot\dfrac{36!}{2!(36-2)!}}{5,247,786} = \frac{15\cdot630}{5,245,786} = 0.00180145.$$

(c)   The probability that you don't win a prize is:

$$\frac{\binom{6}{0}\binom{36}{6} + \binom{6}{1}\binom{36}{5} + \binom{6}{2}\binom{36}{4}}{\binom{42}{6}} = \frac{1,947,792 + 2,261,952 + 883,575}{5,245,786} = 0.971.$$

4.171 Approximately 50% of U.S. adults are Democrats. Ten U.S. adults are selected at random.

(a)   The probability that exactly five are Democrats is:

$$\binom{10}{5} / 2^{10} = 252/1024 = 0.246.$$

(b)   The required calculations for determining the approximate probability that at least eight are Democrats are:

$$\binom{10}{8} / 1024 = 0.0439$$

$$\binom{10}{9} / 1024 = 0.0097$$

$$\binom{10}{10} / 1024 = 0.0010 .$$

The approximate probability that at least eight are Democrats is the sum of the three quantities above, or 0.0546.

Note: The answers are exact for sampling with replacement. Since the population size is large relative to the sample size, the answers are excellent approximations when the sampling is without replacement.

4.173 (a)   The number of ways that n items can be selected from a collection of N items is:

$$\binom{N}{n} = \frac{N!}{n! \, (N-n)!} .$$

To determine f, defined as the number of ways that any particular sample of size n is the one selected, we begin by partitioning the observations comprising the sample of size n into 2 groups; namely, a subset containing the desired n observations and a subset containing none of the remaining N-n observations.

There are n items in the subset containing the desired n observations. Of these n items, n are to be selected. This can be done in

$$\binom{n}{n} = \frac{n!}{n!\,(n-n)!} = \frac{n!}{n!\,0!} = 1 \text{ way.}$$

There are N-n items in the subset containing none of the desired n observations. Of these N-n items, n-n (or zero) are to be selected. This can be done in

$$\binom{N-n}{n-n} = \frac{(N-n)!}{(n-n)!\,[(N-n) - (n-n)]!} = \frac{(N-n)!}{0!\,(N-n)!} = 1 \text{ way.}$$

Using the two results above in the context of the fundamental counting rule, there is a total of

$$1 \cdot 1 = 1$$

outcome in which exactly n of the n items are selected. Thus, f = 1. Now, by the f/N-rule, the probability that any particular sample of size n is selected from a population of size N equals

$$\frac{f}{N} = \frac{1}{\dfrac{N!}{n!\,(N-n)!}} = \frac{1}{\binom{N}{n}}\;.$$

(b)   The probability that any specified member of the population is included in the sample is a simple extension of part (a) above. First, the number of ways that a sample of n items can be selected from a collection of N items is again

$$\binom{N}{n} = \frac{N!}{n!\,(N-n)!}\;.$$

To determine f, defined as the number of ways that any (one) specified member of the population is included in the sample, we begin by partitioning the observations comprising the sample of size n into 2 groups; namely, a subset containing the desired single observation and a subset containing the remaining N-1 observations.

There is 1 item in the subset containing the desired 1 observation. Of this 1 item, 1 is to be selected. This can be done in

$$\binom{1}{1} = \frac{1!}{1!\,(1-1)!} = 1 \text{ way.}$$

There are N-1 items in the subset containing the remaining observations. Of these N-1 items, n-1 are to be selected. This can be done in

$$\binom{N-1}{n-1} = \frac{(N-1)!}{(n-1)! \, [(N-1)-(n-1)]!} = \frac{(N-1)!}{(n-1)! \, (N-n)!} \text{ ways.}$$

Using the two results above in the context of the fundamental counting rule, there is a total of

$$1 \cdot \frac{(N-1)!}{(n-1)! \, (N-n)!}$$

outcomes in which exactly 1 of the n items is selected. This is also the value of f. Now, by the f/N-rule, the probability that any (one) specified member of the population is included in the sample equals

$$\frac{f}{N} = \frac{(N-1)!}{(n-1)! \, (N-n)!} \div \frac{(N!)}{n! \, (N-n)!}$$

$$= \frac{(N-1)!}{(N-1)! \, (N-n)!} \cdot \frac{n! \, (N-n)!}{N!} = \frac{n!}{(n-1)!} \cdot \frac{(N-1)!}{N!} \cdot \frac{(N-n)!}{(N-n)!} = \frac{n}{N}$$

Notice that the expression before the last equality sign simplifies to n/N because n!/(n-1)! = n and (N-1)!/N! = 1/N.

(c)   To determine f for the probability that any k specified members of the population are included in the sample is

$$f = \binom{k}{k} \cdot \binom{N-k}{n-k} = 1 \cdot \frac{(N-k)!}{(n-k)! \, [(N-k)-(n-k)]!} = \frac{(N-k)!}{(n-k)! \, (N-n)!}$$

So, the probability that any k specified members of the population are included in the sample is

REVIEW TEST FOR CHAPTER 4

1.   Probability theory enables us to control and evaluate the likelihood that a statistical inference is correct, and it provides the mathematical basis for statistical inference.

2.   (a)   The equal-likelihood model is used for computing probabilities when an experiment has N possible outcomes, all equally likely.

      (b)   If an experiment has N equally likely outcomes and an event can occur in f ways, then the probability of the event is f/N.

3.   In the frequentist interpretation of probability, the probability of an event is the relative frequency of the event in a large number of experiments.

4.   The numbers (b) -0.047 cannot be a probability because it is negative and (c) 3.5 cannot be a probability because it is greater than 1.

5.   The Venn diagram is a common graphical technique for portraying events and relationships between events.

6.   Two or more events are mutually exclusive if no two of the events have any outcomes in common, i.e., no two of the events can occur at the same time.

7.   (a)   P(E)

      (b)   P(E) = 0.436

8.   (a)   False. For any two events, the probability that one or the other occurs equals the sum of the two individual probabilities minus the probability that both events occur simultaneously.

      (b)   True. Either the event occurs or it doesn't. Since the probability that something in the sample space occurs is 1, the probability that the event occurs plus the probability that it doesn't occur equals 1. Thus the probability that it occurs is 1 minus the probability that it doesn't occur.

9.   Frequently, it is quicker to compute the probability of the complement of an event than it is to compute the probability of the event itself. The complement rule allows one to use the probability of the complement to obtain the probability of the event itself.

10.   (a)   Data obtained by observing values of one variable of a population are called <u>univariate</u> data.

      (b)   Data obtained by observing values of two variables of a population are call <u>bivariate</u> data.

      (c)   A frequency distribution for bivariate data is call a <u>contingency table</u>.

11.   The sum of the joint probabilities in a row or column of a joint probability distribution equals the <u>marginal</u> probability in that row or column.

12.   (a)   P(B|A)

      (b)   A

13.   Conditional probabilities can sometimes be computed directly as f/N or by using the formula P(B|A) = P(A&B)/P(A).

14.   The joint probability of two independent events is the product of their marginal probabilities.

15.   If two or more events have the property that at least one of them must occur when the experiment is performed, then the events are said to be <u>exhaustive</u>.

16.   If r actions are to be performed in a definite order and there are m_1 possibilities for the first action, m_2 possibilities for the second action, ..., and m_r possibilities for the rth action, then there are $m_1 m_2 m_3 ... m_r$ possibilities altogether for the r actions.

17.   (a)   ABC   ACB   BAC   BCA   CAB   CBA

             ABD   ADB   BAD   BDA   DAB   DBA

             ACD   ADC   CAD   CDA   DAC   DCA

             BCD   BDC   CBD   CDB   DBC   DCB

      (b)   ABC   ABD   ACD   BCD

      (c) $(4)_3$ = 24 $\binom{4}{3}$ = 4 ways.

18.   (a) $P(A) = \dfrac{10,547}{87,620} = 0.120$

   (b) $P(D) + P(E) + P(F) = \dfrac{11,931 + 8,992 + 17,878}{87,620} = 0.443$

   (c)

Event	Probability
A	0.120
B	0.191
C	0.195
D	0.136
E	0.103
F	0.204
G	0.051

19.   (a)   (not J) is the event that the return selected shows an adjusted gross income of at least $100,000. There are 4,508 thousand such returns.

   (b)   (H & I) is the event that the return selected shows an adjusted gross income of between $20,000 and $50,000. There are 37,988 thousand such returns.

   (c)   (H or K) is the event that the return selected shows an adjusted gross income of at least $20,000. There are 60,374 thousand such returns.

   (d)   (H & K) is the event that the return selected shows an adjusted gross income of between $50,000 and $100,000. There are 17,878 thousand such returns.

20.   (a)   Events H and I are not mutually exclusive. Both have events C, D, and E in common.

   (b)   Events I and K are mutually exclusive. Both have nothing in common.

   (c)   Events H and (not J) are mutually exclusive. Both have nothing in common.

   (d)   Events H, (not J), and K are not mutually exclusive. While H and (not J) have nothing in common, a part of K (i.e., event F) is common to event H, and the other part of K (i.e., event G) is common to event (not J).

21.   (a)

$$P(H) = \frac{17,065 + 11,931 + 8,992 + 17,878}{87,620} = 0.638$$

$$P(I) = 1 - \frac{17,878 + 4,508}{87,620} = 1 - 0.255 = 0.745$$

$$P(J) = 1 - \frac{4,508}{87,620} = 1 - 0.051 = 0.949$$

$$P(K) = \frac{17,878 + 4,508}{87,620} = 0.255$$

(b)   H =   (C or D or E or F)
   I =   (A or B or C or D or E)
   J =   (A or B or C or D or E or F)
   K =   (F or G)

(c)   P(H) = P(C) + P(D) + P(E) + P(F) = 0.195 + 0.136 + 0.103 + 0.204
      = 0.638
   P(I) = P(A) + P(B) + P(C) + P(D) + P(E)
      = 0.120 + 0.191 + 0.195 + 0.136 + 0.103 = 0.745
   P(J) = P(A) + P(B) + P(C) + P(D) + P(E) + P(F)
      = 0.120 + 0.191 + 0.195 + 0.136 + 0.103 + 0.204 = 0.949
   P(K) = P(F) + P(G) = 0.204 + 0.051 = 0.255

22.   (a)

$$P(\text{not } J) = \frac{4,508}{87,620} = 0.051$$

$$P(H \& I) = \frac{17,065 + 11,931 + 8,992}{87,620} = 0.434$$

$$P(H \text{ or } K) = \frac{17,065 + 11,931 + 8,992 + 17,878 + 4,508}{87,620} = 0.689$$

$$P(H \& K) = \frac{17,878}{87,620} = 0.204$$

(b)   P(J) = 1 - P(not J) = 1 - 0.050 = 0.949
(c)   P(H or K) = P(H) + P(K) - P(H & K) = 0.638 + 0.255 - 0.204
         = 0.689.
(d)   The answer in (c) agrees with that in (a).

23.   (a)   The contingency table has six cells.

(b)   There are 13,615 thousand students in high school.

(c)   48,778 thousand students attend public schools.

(d)   2,562 thousand students attend private colleges.

24.   (a)   (i) L_3 = the student selected is in college;

(ii) T_1 = the student selected attends a public school;

(iii) $P(T_1 \& L_3)$ = the student selected attends a public college.

(b) $P(L_3) = \dfrac{12,174}{56,340} = 0.216$

$P(T_1) = \dfrac{48,778}{56,340} = 0.866$

$P(T_1 \& L_3) = \dfrac{9,612}{56,340} = 0.171$

The interpretation of each result above as a percentage is as follows: 21.6% of students attend college, 86.6% attend public schools, and 17.1% attend public colleges.

(c)

	Type		
Level	Public T_1	Private T_2	$P(L_i)$
Elementary L_1	0.478	0.064	0.542
High School L_2	0.217	0.025	0.242
College L_3	0.171	0.045	0.216
$P(T_j)$	0.866	0.134	1.000

(d) $P(T_1 \text{ or } L_3) = P(T_1) + P(L_3) - P(T_1 \& L_3) =$

(i) $\dfrac{48,778 + 12,174 - 9,612}{56,340} = 0.911$.

(ii) $0.866 + 0.216 - 0.171 = 0.911$.

25. (a) $P(L_3|T_1) = \dfrac{9,612}{48,778} = 0.197$.

19.7% of students attending public schools are in college.

(b) $P(L_3|T_1) = \dfrac{P(L_3 \& T_1)}{P(T_1)} = \dfrac{0.171}{0.866} = 0.197$

26. (a) $P(T_2) = \dfrac{7,562}{56,340} = 0.134$; $P(T_2|L_2) = \dfrac{1,400}{13,615} = 0.103$

(b) For L_2 and T_2 to be independent, $P(T_2|L_2)$ must equal $P(T_2)$. From part (a), however, we see that this is not the case. In terms of percentages, each of these probabilities, respectively, means that 10.3% of high school students attend private schools, whereas 13.4% of all students attend private schools.

(c) Events L_2 and T_2 are not mutually exclusive. From Table 4.170 we see that both events can occur simultaneously. There are 1,400 thousand students who attend a private high school.

(d) Let: L_1 = a student is in elementary school;

T_1 = a student attends public school.

For L_1 and T_1 to be independent, $P(L_1|T_1)$ must equal $P(L_1)$. But, $P(L_1) = 0.542$ and $P(L_1|T_1) = 0.478/0.866 = 0.552$. Since $P(L_1|T_1) \neq P(L_1)$, the event that a student is in elementary school is *not* independent of the event that a student attends public school.

27. Let MA1, MP1, and MS1 denote, respectively, the events that the first student selected received a master of arts, a master of public administration, and a master of science; and let MA2, MP2, and MS2 denote, respectively, the events that the second student selected received a master of arts, a master of public administration, and a master of science.

(a) $P(\text{MA1 \& MS2}) = \dfrac{3}{50} \cdot \dfrac{19}{49} = 0.023$

(b) $P(\text{MP1 \& MP2}) = \dfrac{28}{50} \cdot \dfrac{27}{49} = 0.309$

(c)

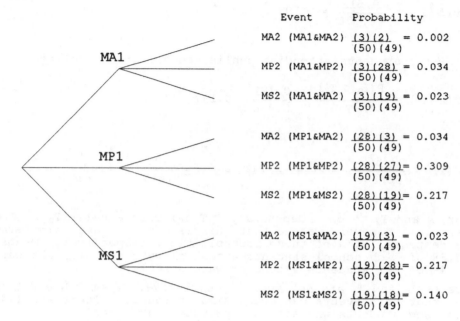

	Event	Probability	
	MA2 (MA1&MA2)	$\frac{(3)(2)}{(50)(49)}$	= 0.002
	MP2 (MA1&MP2)	$\frac{(3)(28)}{(50)(49)}$	= 0.034
	MS2 (MA1&MA2)	$\frac{(3)(19)}{(50)(49)}$	= 0.023
	MA2 (MP1&MA2)	$\frac{(28)(3)}{(50)(49)}$	= 0.034
	MP2 (MP1&MP2)	$\frac{(28)(27)}{(50)(49)}$	= 0.309
	MS2 (MP1&MS2)	$\frac{(28)(19)}{(50)(49)}$	= 0.217
	MA2 (MS1&MA2)	$\frac{(19)(3)}{(50)(49)}$	= 0.023
	MP2 (MS1&MP2)	$\frac{(19)(28)}{(50)(49)}$	= 0.217
	MS2 (MS1&MS2)	$\frac{(19)(18)}{(50)(49)}$	= 0.140

(d)   P(MA1 & MA2) + P(MP1 & MP2) + P(MS1 & MS2)

= 0.002 + 0.309 + 0.140 = 0.451 .

28.   (a)   Let C_i be the event that the ith man selected is color blind.

P(none color blind) = P{(not C_1)& (not C_2)& (not C_3)& (not C_4)} =

P(not C_1)P(not C_2)P(not C_3)P(not C_4) = 0.91^4 = 0.686

(b)   P{(not C_1)& (not C_2)& (not C_3)& (C_4)} =

P(not C_1)P(not C_2)P(not C_3)P(C_4) = 0.91· 0.91· 0.91· 0.09 = 0.068

(c)   There are four like sequences like the one in part (b) in which the fourth man is color blind. The others have the first man color blind and the others not; the second color blind and the others not; and the third color blind and the others not. Each sequence has the same probability as the one in part (b). Thus P(exactly one man is color blind) = 4 · 0.068 = 0.272

29.   (a)   No. A&B has a probability of 0.2. Thus A and B can occur together.

(b)   Yes.   P(A&B) = P(A)P(B).

30.   Let: N = person answered no

W = woman   P(W) = $\frac{1,025}{1,497}$ = 0.685

M = man $P(M) = \frac{472}{1,497}$ = 0.315

(a) $P(N|W) = 0.45$

(b) $P(N) = P(N \& W) + P(N \& M) = P(W) \cdot P(N|W) + P(M) \cdot P(N|M)$

$= 0.685 \cdot 0.45 + 0.315 \cdot 0.63 = 0.507$.

(c) $P(W) = 0.685$

(d) $P(W|N) = \dfrac{P(W \& N)}{P(N)} = \dfrac{0.30825}{0.507} = 0.608$

(e)   Each result above is interpreted as follows: 45% of the women surveyed answered 'no' to the question; 50.7% of the people surveyed answered 'no' to the question; 68.5% of the people surveyed were women; 60.8% of the people who answered 'no' to the question were women.

(f)   The probabilities in parts (b) and (c) are prior; those in parts (a) and (d) are posterior.

31.   (a) $\dbinom{12}{2} = \dfrac{12!}{2! \ (12-2)} = \dfrac{12 \cdot 11}{2 \cdot 1} = 66$

(b) $(12)_3 = \dfrac{12!}{(12-3)!} = 12 \cdot 11 \cdot 10 = 1320$

(c)   (i) $\dbinom{8}{2} = \dfrac{8!}{2! \ (8-2)!} = \dfrac{8 \cdot 7}{2} = 28$

(ii) $(8)_3 = \dfrac{8!}{(8-3)!} = 8 \cdot 7 \cdot 6 = 336$

32.   (a) $\dbinom{52}{13} = \dfrac{52!}{13! \ (52-13)!} = 635,013,559,600$

(b) $\dfrac{\dbinom{4}{2}\dbinom{48}{11}}{\dbinom{52}{13}} = \dfrac{\dfrac{4!}{2! \ (4-2)!} \cdot \dfrac{48!}{11! \ (48-11)!}}{\dfrac{52!}{13! \ (52-13)!}} = 0.213$

(c) With four choices for the eight-card suit, three choices for the four-card suit, and two choices for the one-card suit, the probability of being dealt an 8-4-1 distribution is:

$$4 \cdot 3 \cdot 2 \cdot \frac{\binom{13}{8}\binom{13}{4}\binom{13}{1}\binom{13}{0}}{\binom{52}{13}} = 24 \cdot \frac{\dfrac{13!}{8!(13-8)!} \cdot \dfrac{13!}{4!(13-4)!} \cdot \dfrac{13!}{1!(13-1)!} \cdot \dfrac{13}{0!(13-0)!}}{635,013,559,600}$$

$$= 24 \cdot \frac{1287 \cdot 715 \cdot 13 \cdot 1}{635,013,559,600} = 24 \cdot \frac{11,962,665}{635,013,559,600}$$

$$= 24 \cdot 0.000018838 = 0.00045$$

(d) Initially, two suits from among the four are to be selected, from which five cards from each suit are drawn. From the remaining two suits, one is to be selected from which two cards are drawn. Finally, only one suit remains from which to draw the final card. The probability of being dealt a 5-5-2-1 distribution is:

$$\binom{4}{2}\binom{2}{1}\binom{1}{1} \cdot \frac{\binom{13}{5}\binom{13}{5}\binom{13}{2}\binom{13}{1}}{\binom{52}{13}} = 12 \cdot \frac{\dfrac{13!}{5!(13-5)!} \cdot \dfrac{13!}{5!(13-5)!} \cdot \dfrac{13!}{2!(13-2)!} \cdot \dfrac{13}{1!(13-1)!}}{635,013,559,600}$$

$$= 12 \cdot \frac{1287 \cdot 1287 \cdot 78 \cdot 13}{635,013,559,600} = 12 \cdot \frac{1,679,558,166}{635,013,559,600}$$

$$= 12 \cdot 0.00264492 = 0.032$$

(e) The probability of being dealt a hand void in a specified suit is:

$$\frac{\binom{39}{13}\binom{13}{0}}{\binom{52}{13}} = \frac{\dfrac{39!}{13!(39-13)!} \dfrac{13!}{0!(13-0)!}}{\dfrac{52!}{13!(52-13)!}} = 0.013$$

CHAPTER 5 ANSWERS

Exercises 5.1

5.1 (a) probability

(b) probability

5.3 The notation {X=3} denotes an event that occurs when X=3, whereas P(X=3) denotes the probability that the event {X=3} will occur. Another way of thinking about the difference is that the first is a set and the second is a number between zero and one.

5.5 This table will resemble the probability distribution of the random variable.

5.7 (a) X =1, 2, 3, 4, 5, 6, 7

(b) {X=5}

(c) P(X=5) = 6.1/83.6 = 0.073

7.3% of U.S. households consist of exactly five people.

(d)   (e)

Number of persons x	Probability P(X=x)
1	0.232
2	0.317
3	0.175
4	0.154
5	0.073
6	0.030
7	0.019

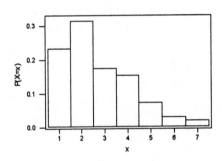

5.9 (a) Y = 2, 3, 4, 5, 6, 7, 8, 9, 10, 11, 12

(b) {Y = 7}

(c) P(Y = 7) = 6/36 = 0.167

(d)   (e)

Sum of dice y	Probability P(Y=y)
2	1/36
3	1/18
4	1/12
5	1/9
6	5/36
7	1/6
8	5/36
9	1/9
10	1/12
11	1/18
12	1/36

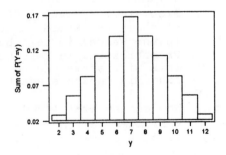

5.11 (a) $\{X = 4\}$ (b) $\{X \geq 2\}$ (c) $\{X < 5\}$ (d) $\{2 \leq X < 5\}$

(e) $P(X = 4) = 0.212$

(f) $P(X \geq 2) = P(X = 2) + P(X = 3) + P(X = 4) + P(X = 5) + P(X = 6)$
$$= 0.078 + 0.155 + 0.212 + 0.262 + 0.215 = 0.922$$

(g) $P(X < 5) = P(X = 0) + P(X = 1) + P(X = 2) + P(X = 3) + P(X = 4)$
$$= 0.029 + 0.049 + 0.078 + 0.155 + 0.212 = 0.523$$

(h) $P(2 \leq X < 5) = P(X = 2) + P(X = 3) + P(X = 4)$
$$= 0.078 + 0.155 + 0.212 = 0.445$$

5.13 $P(Z \leq 1.96) + P(Z > 1.96) = 1$.

Since $P(Z > 1.96) = 0.025$,
$$P(Z \leq 1.96) + 0.025 = 1, \text{ or}$$
$$P(Z \leq 1.96) = 1 - 0.025 = 0.975.$$

5.15 (a) $P(X \leq c) + P(X > c) = 1.$

Since $P(X > c) = \alpha$,
$$P(X \leq c) + \alpha = 1, \text{ or}$$
$$P(X \leq c) = 1 - \alpha.$$

(b) $P(Y < -c) + P(-c \leq Y \leq c) + P(Y > c) = 1.$

Since $P(Y < -c) = P(Y > c) = \alpha/2$:
$$\alpha/2 + P(-c \leq Y \leq c) + \alpha/2 = 1, \text{ or}$$
$$P(-c \leq Y \leq c) = 1 - \alpha/2 - \alpha/2 = 1 - \alpha.$$

(c) $P(T < -c) + P(-c \leq T \leq c) + P(T > c) = 1.$

Since $P(-c \leq T \leq c) = 1 - \alpha$:
$$P(T < -c) + (1 - \alpha) + P(T > c) = 1, \text{ or}$$
$$P(T < -c) + P(T > c) = 1 - 1 + \alpha = \alpha.$$

Since $P(T < -c) = P(T > c)$:
$$2 \cdot [P(T > c)] = \alpha \quad \text{or} \quad P(T > c) = \alpha/2.$$

5.17 (a) After storing the probability distribution from Table 5.2 in two columns named x and P(X=x), choose **Calc ▶ Random data ▶ Discrete...** , type <u>2000</u> in the **Generate rows of data** text box, click in the **Store in column(s)** text box and type <u>NUMSIBS</u>, click in the **Values in** text box and specify x, click in the **Probabilities in** text box and specify "P(X=x)", and click **OK**. This will store the numbers of siblings for 2000 observations of a randomly chosen student.

(b) Now choose **Stat ▶ Tables ▶ Tally...**, specify NUMSIBS in the **Variables** text box, select **Counts** and **Percents** from the list of display check boxes, and click **OK**. Our results are

NUMSIBS	Count	Percent
0	409	20.45
1	822	41.10
2	577	28.85
3	154	7.70
4	38	1.90
N=	2000	

The percentages are very close to those of the probability distribution.

(c)   To create a histogram of the proportions, name a new column PROP, and enter the proportions from your output for 0, 1, 2, 3, and 4

siblings. Then choose **Graph ▶ Chart...**, click in the **Y** text box for **Graph 1** and specify PROP, click in the **X** text box for **Graph 1** and specify x, click the **Edit attributes...** button, click in the **Bar Width** text box for **Graph 1** and type <u>1</u>, and click **OK**. Click the **Frame** stand-alone pop-up-menu button, select **Min and Max...**, click in the **Minimum for Y** text box and type <u>0</u>, and click **OK**. Click the **Frame** stand-alone pop-up-menu button, select **Axis...**, click in the **Label** text box for **1** and type <u>Number of siblings</u>, click in the **Label** text box for **2** and type <u>Probability</u>, click **OK**, and click **OK**. The result is

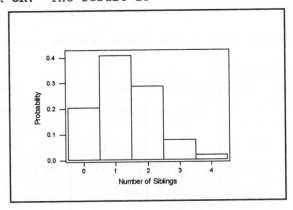

(d)   Parts (b) and (c) illustrate that the proportions of times each possible value occurs will approximate the probability distribution itself, and the histogram of proportions will approximate the probability histogram for X.

Exercises 5.2

5.19   The required calculations are:

x	P(X=x)	xP(X=x)	x^2	x^2P(X=x)
1	0.232	0.232	1.000	0.232
2	0.317	0.634	4.000	1.268
3	0.175	0.525	9.000	1.575
4	0.154	0.616	16.000	2.464
5	0.073	0.365	25.000	1.825
6	0.030	0.180	36.000	1.080
7	0.019	0.133	49.000	0.931
		2.685		9.375

(a) $\mu_x = \Sigma x P(X=x) = 2.685$. The average number of persons in a U.S. household is about 2.685.

(b) $\sigma_x = \sqrt{\Sigma x^2 P(X=x) - \mu_x^2} = \sqrt{9.375 - (2.685)^2} = 1.47$

(c)

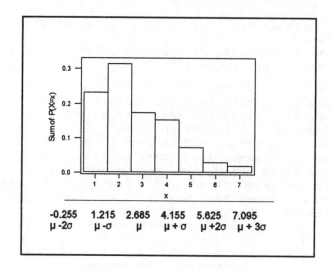

5.21 The required calculations are:

y	P(Y=y)	yP(Y=y)	$y-\mu_y$	$(y-\mu_y)^2$	$(y-\mu_y)^2 P(Y=y)$	y^2	$y^2 P(Y=y)$
2	1/36	2/36	-5	25	25/36	4	4/36
3	2/36	6/36	-4	16	32/36	9	18/36
4	3/36	12/36	-3	9	27/36	16	48/36
5	4/36	20/36	-2	4	16/36	25	100/36
6	5/36	30/36	-1	1	5/36	36	180/36
7	6/36	42/36	0	0	0/36	49	294/36
8	5/36	40/36	1	1	5/36	64	320/36
9	4/36	36/36	2	4	16/36	81	324/36
10	3/36	30/36	3	9	27/36	100	300/36
11	2/36	22/36	4	16	32/36	121	242/36
12	1/36	12/36	5	25	25/36	144	144/36
		252/36=7			210/36		1,974/36

(a) $\mu_y = \Sigma yP(Y=y) = 252/36 = 7$. When rolling two balanced dice, the average of their sum is 7.

(b) $\sigma_y = \sqrt{\Sigma(y-\mu_y)^2 P(Y=y)} = \sqrt{210/36} = 2.415$

or

$\sigma_y = \sqrt{\Sigma y^2 P(Y=y) - \mu_y^2} = \sqrt{1974/36 - (7)^2} = 2.415$

(c)

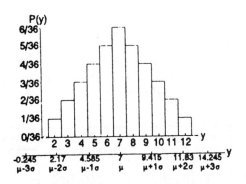

5.23   (a)

$$P(X=1) = \frac{18}{38} = 0.474;$$

$$P(X=-1) = \frac{20}{38} = 0.526; \quad 0.474 + 0.526 = 1.000.$$

(b) $\mu = \Sigma x \cdot P(X=x) = 1(0.474) - 1(0.526) = -0.052$

(c)   On the average, you will lose 5.2¢ per play.

(d)   If you bet $1 on red 100 times, you can expect to lose $100 \cdot 0.052$ = $5.20. If you bet $1 on red 1000 times, you can expect to lose $52.00.

(e)   Roulette is not a profitable game for a person to play. Parts (c) and (d) demonstrate that, no matter how much you play, you can expect to lose. Also, part (a) shows that a higher probability is associated with losing rather than winning.

5.25

w	P(W=w)	w·P(w)	w^2	w^2·P(W=w)
0	0.80	0.00	0	0.00
1	0.15	0.15	1	0.15
2	0.05	0.10	4	0.20
		0.25		0.35

(a)

$$\mu_w = \sum w \cdot P(W=w) = 0.25$$

$$\sigma_w = \sqrt{\sum w^2 \cdot P(W=w) - \mu_w^2} = \sqrt{0.35 - (0.25)^2} = \sqrt{0.2875} = 0.536$$

(b)   On the average, there are 0.25 breakdowns per day.

(c)   Assuming 250 work days per year, the number of breakdowns expected per year is $(0.25)(250) = 62.5$.

5.27

500w	P(500w)	500w·P(500w)	(500w)^2	(500w)^2·P(500w)
0	0.80	0.00	0	0
500	0.15	75.00	250,000	37000
1000	0.05	50.00	1,000,000	50000
		125.00		87500

(a)   Columns 1 and 2 comprise the probability distribution of the random variable 500W. We have used P(500y) for P(500Y = 500y) in the table headings.

(b)   Column 3 of the previous table provides the calculations for the mean: $\mu_{500w} = \$125.00$.

(c) $\mu_{500w} = 500\mu_w$

(d)   Columns 4 and 5 of the previous table provide the calculations for the standard deviation.

$$\sigma_{500w} = \sqrt{87,500 - (125)^2} = 268.1 = 500(0.536).$$

(e) $\sigma_{500w} = 500\sigma_w$.

(f)   The mean of a constant times a random variable equals the constant times the mean of the random variable. The standard deviation of a constant times a random variable equals the absolute value of the constant times the standard deviation of the random variable.

Exercises 5.3

5.29 Bernoulli trials are repeated identical trials for which each trial has two possible outcomes (commonly called success and failure), the trials are independent, and the probability of a success remains the same from trial to trial.

5.31 "bi" signifies "two". ("Bi-nomial" signifies "two names", success and failure.)

5.33 $7! = 7 \cdot 6 \cdot 5 \cdot 4 \cdot 3 \cdot 2 \cdot 1 = 5,040$

$8! = 8 \cdot 7 \cdot 6 \cdot 5 \cdot 4 \cdot 3 \cdot 2 \cdot 1 = 40,320$

$9! = 9 \cdot 8 \cdot 7 \cdot 6 \cdot 5 \cdot 4 \cdot 3 \cdot 2 \cdot 1 = 362,880$

5.35 (a) $\dbinom{5}{2} = \dfrac{5!}{2!(5-2)!} = \dfrac{5\cdot4\cdot3\cdot2\cdot1}{2\cdot1\cdot3\cdot2\cdot1} = 1$

(b) $\dbinom{7}{4} = \dfrac{7!}{4!(7-4)!} = \dfrac{7\cdot6\cdot5\cdot4\cdot3\cdot2\cdot1}{4\cdot3\cdot2\cdot1\cdot3\cdot2\cdot1} = 3$

(c) $\dbinom{10}{3} = \dfrac{10!}{3!(10-3)!} = \dfrac{10\cdot9\cdot8\cdot7\cdot6\cdot5\cdot4\cdot3\cdot2\cdot1}{3\cdot2\cdot1\cdot7\cdot6\cdot5\cdot4\cdot3\cdot2\cdot1} = 12$

(d) $\dbinom{12}{5} = \dfrac{12!}{5!(12-5)!} = \dfrac{12\cdot11\cdot10\cdot9\cdot8\cdot7\cdot6\cdot5\cdot4\cdot3\cdot2\cdot1}{5\cdot4\cdot3\cdot2\cdot1\cdot7\cdot6\cdot5\cdot4\cdot3\cdot2\cdot1} = 79$

5.37 (a) $\dbinom{5}{3} = \dfrac{5!}{3!(5-3)!} = \dfrac{5\cdot4\cdot3\cdot2\cdot1}{3\cdot2\cdot1\cdot2\cdot1} = 1$

(b) $\dbinom{10}{0} = \dfrac{10!}{0!(10-0)!} = \dfrac{10\cdot9\cdot8\cdot7\cdot6\cdot5\cdot4\cdot3\cdot2\cdot1}{1\cdot10\cdot9\cdot8\cdot7\cdot6\cdot5\cdot4\cdot3\cdot2\cdot1} =$

(c) $\dbinom{10}{10} = \dfrac{10!}{10!(10-10)!} = \dfrac{10\cdot9\cdot8\cdot7\cdot6\cdot5\cdot4\cdot3\cdot2\cdot1}{10\cdot9\cdot8\cdot7\cdot6\cdot5\cdot4\cdot3\cdot2\cdot1\cdot1} =$

(d) $\dbinom{9}{5} = \dfrac{9!}{5!(9-5)!} = \dfrac{9\cdot8\cdot7\cdot6\cdot5\cdot4\cdot3\cdot2\cdot1}{5\cdot4\cdot3\cdot2\cdot1\cdot4\cdot3\cdot2\cdot1} = 12$

5.39 (a) Each trial consists of observing whether the child is cured of the pinworm infestation by pyrantel pamoate. Each is either cured or not cured. The results are independent assuming that the children are not in the same families. Since a success *s* is that the child is cured, the success probability p is 0.90.

(b)

Outcome	Probability
sss	(0.90)(0.90)(0.90)=0.729
ssf	(0.90)(0.90)(0.10)=0.081
sfs	(0.90)(0.10)(0.90)=0.081
sff	(0.90)(0.10)(0.10)=0.009
fss	(0.10)(0.90)(0.90)=0.081
fsf	(0.10)(0.90)(0.10)=0.009
ffs	(0.10)(0.10)(0.90)=0.009
fff	(0.10)(0.10)(0.10)=0.001

(c)

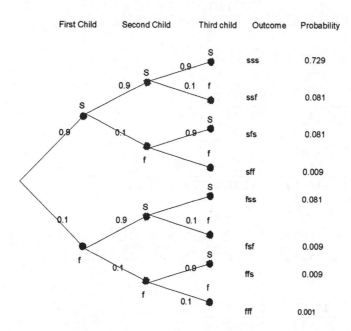

(d)   The outcomes in which exactly two of the three children are cured
      are ssf, sfs, and fss.
(e)   Each outcome in part (d) has the probability 0.081. This
      probability is the same for each outcome because each probability
      is obtained by multiplying two success probabilities of 0.90 and
      one failure probability of 0.10.
(f)   P(exactly two children are cured) = P(ssf) + P(sfs) + P(fss)
                                        = 0.081 + 0.081 + 0.081 = 0.243
(g)   P(exactly one child is cured) = P(sff) + P(fsf) + P(ffs)
                                    = 0.009 + 0.009 + 0.009 = 0.027
      P(exactly three children are cured) = P(sss) = 0.729, and
      P(exactly zero children are cured) = P(fff) = 0.001.
      Thus the complete probability distribution of X, the number of
      children out of three that are cured is

x	0	1	2	3
P(X=x)	0.001	0.027	0.243	0.729

5.41   (a)   p = 0.2

(b)

Outcome	Probability
ssss	(0.2)(0.2)(0.2)(0.2)=0.0016
sssf	(0.2)(0.2)(0.2)(0.8)=0.0064
ssfs	(0.2)(0.2)(0.8)(0.2)=0.0064
ssff	(0.2)(0.2)(0.8)(0.8)=0.0256
sfss	(0.2)(0.8)(0.2)(0.2)=0.0064
sfsf	(0.2)(0.8)(0.2)(0.8)=0.0256
sffs	(0.2)(0.8)(0.8)(0.2)=0.0256
sfff	(0.2)(0.8)(0.8)(0.8)=0.1024
fsss	(0.8)(0.2)(0.2)(0.2)=0.0064
fssf	(0.8)(0.2)(0.2)(0.8)=0.0256
fsfs	(0.8)(0.2)(0.8)(0.2)=0.0256
fsff	(0.8)(0.2)(0.8)(0.8)=0.1024
ffss	(0.8)(0.8)(0.2)(0.2)=0.0256
ffsf	(0.8)(0.8)(0.2)(0.8)=0.1024
fffs	(0.8)(0.8)(0.8)(0.2)=0.1024
ffff	(0.8)(0.8)(0.8)(0.8)=0.4096

(c)

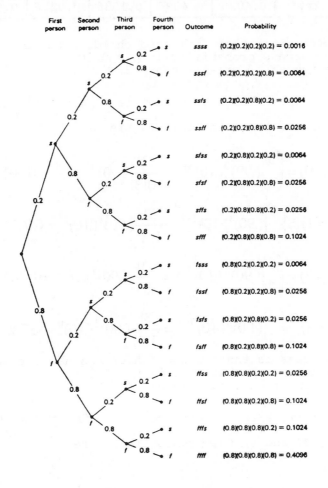

(d)   The outcomes in which exactly three of the four people have a psychiatric disorder are sssf, ssfs, sfss, and fsss.

(e)   Each outcome in part (d) has probability 0.0064. This probability is the same for each outcome because each probability is obtained by multiplying three success probabilities of 0.2 and one failure probability of 0.8.

(f)   P(exactly three have a psychiatric disorder)

= P(sssf) + P(ssfs) + P(sfss) + P(fsss)

= 0.0064 + 0.0064 + 0.0064 + 0.0064 = 0.0256

(g)   P(exactly zero have a psychiatric disorder) = P(ffff) = 0.4096

P(exactly one has a psychiatric disorder)

= P(fffs) + P(ffsf) + P(fsff) + P(sfff)

= 0.1024 + 0.1024 + 0.1024 + 0.1024 = 0.4096

P(exactly two have a psychiatric disorder)

= P(ssff) + P(sfsf) + P(sffs) + P(fssf) + P(fsfs) + P(ffss)

= 0.0256 + 0.0256 + 0.0256 + 0.0256 + 0.0256 + 0.0256 = 0.1536

P(exactly four have a psychiatric disorder) = P(ssss) = 0.0016

Thus

y	0	1	2	3	4
P(Y=y)	0.4096	0.4096	0.1536	0.0256	0.0016

5.43   Step 1: A success is that a treated child is cured.
Step 2: The success probability is p = 0.90.
Step 3: The number of trials is n = 3.

Step 4: The formula for x successes is $P(X = x) = \binom{3}{x}(.90)^x (.10)^{3-x}$. For x = 0, 1, 2, and 3, the probabilities are

$$P(X = 0) = \binom{3}{0}(.90)^0 (.10)^{3-0} = \frac{3!}{0!3!}(.90)^0 (.10)^3 = 0.001$$

$$P(X = 1) = \binom{3}{1}(.90)^1 (.10)^{3-1} = \frac{3!}{1!2!}(.90)^1 (.10)^2 = 0.027$$

$$P(X = 2) = \binom{3}{2}(.90)^2 (.10)^{3-2} = \frac{3!}{2!1!}(.90)^2 (.10)^1 = 0.243$$

$$P(X = 3) = \binom{3}{3}(.90)^3 (.10)^{3-3} = \frac{3!}{3!0!}(.90)^3 (.10)^0 = 0.729$$

5.45   Step 1: A success is that an adult American has a psychiatric disorder.
Step 2: The success probability is p = 0.20.
Step 3: The number of trials is n = 4.

Step 4: The formula for y successes is $P(Y = y) = \binom{4}{y}(.20)^y (.80)^{4-y}$. For y = 0, 1, 2, 3, and 4, the probabilities are

$$P(Y = 0) = \binom{4}{0}(.20)^0(.80)^{4-0} = \frac{4!}{0!4!}(.20)^0(.80)^4 = 0.4096$$

$$P(Y = 1) = \binom{4}{1}(.20)^1(.80)^{4-1} = \frac{4!}{1!3!}(.20)^1(.80)^3 = 0.4096$$

$$P(Y = 2) = \binom{4}{2}(.20)^2(.80)^{4-2} = \frac{4!}{2!2!}(.20)^2(.80)^2 = 0.1536$$

$$P(Y = 3) = \binom{4}{3}(.20)^3(.80)^{4-3} = \frac{4!}{3!1!}(.20)^3(.80)^1 = 0.0256$$

$$P(Y = 4) = \binom{4}{4}(.20)^4(.80)^{4-4} = \frac{4!}{4!0!}(.20)^4(.80)^0 = 0.0016$$

5.47 (a)   Since the graph is symmetric, p = 0.5.

(b)   Since the graph is right skewed, p is less than 0.5.

5.49 The calculations required to answer all parts of this exercise are:

$$P(0) = \binom{5}{0}(0.67)^0(0.33)^5 = 0.004 \qquad P(3) = \binom{5}{3}(0.67)^3(0.33)^2 = 0.328$$

$$P(1) = \binom{5}{1}(0.67)^1(0.33)^4 = 0.040 \qquad P(4) = \binom{5}{4}(0.67)^4(0.33)^1 = 0.332$$

$$P(2) = \binom{5}{2}(0.67)^2(0.33)^3 = 0.161 \qquad P(5) = \binom{5}{5}(0.67)^5(0.33)^0 = 0.135$$

(a)   P(2) = 0.161   (b)   P(4) = 0.332

(c)   P(X $\geq$ 4) = P(4) + P(5) = 0.332 + 0.135 = 0.467.

(d)   P(2 $\leq$ X $\leq$ 4) = P(2) + P(3) + P(4) = 0.161 + 0.328 + 0.332 = 0.821

(e)   (g)

x	P(X=x)
0	0.004
1	0.040
2	0.161
3	0.328
4	0.332
5	0.135

(f)   Left skewed

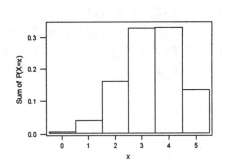

(h)

x	P(X=x)	xP(X=x)	x^2	x^2P(X=x)
0	0.004	0.000	0	0.000
1	0.040	0.040	1	0.040
2	0.161	0.322	4	0.644
3	0.328	0.984	9	2.952
4	0.332	1.328	16	5.312
5	0.135	0.675	25	3.375
		3.349		12.323

$\mu = 3.349$

$\sigma^2 = 12.323 - 3.349^2 = 1.107; \quad \sigma = \sqrt{1.107} = 1.052$

(i) $\mu = np = 5(0.67) = 3.35$

$\sigma^2 = np(1-p) = 5(0.67)(0.33) = 1.1055$

$\sigma = \sqrt{1.1055} = 1.051$

(j)   Out of any five races, the average number of favorites that finish in the money is 3.35.

5.51   The calculations required to answer all parts of this exercise are

$$P(0) = \binom{4}{0} (0.25)^0 (0.75)^4 = 0.316$$

$$P(1) = \binom{4}{1} (0.25)^1 (0.75)^3 = 0.422$$

$$P(2) = \binom{4}{2} (0.25)^2 (0.75)^2 = 0.211$$

$$P(3) = \binom{4}{3} (0.25)^3 (0.75)^1 = 0.047$$

$$P(4) = \binom{4}{4} (0.25)^4 (0.75)^0 = 0.004$$

(a) $P(2) = 0.211$

(b) $P(Y \leq 2) = P(0) + P(1) + P(2) = 0.316 + 0.422 + 0.211 = 0.949$

(c) $P(1 \leq Y \leq 3) = P(1) + P(2) + P(3) = 0.422 + 0.211 + 0.047 = 0.680$

(d) $P(Y < 1 \text{ or } Y > 2) = P(0) + P(3) + P(4) = 0.316 + 0.047 + 0.004$
$= 0.367$

(e)

y	P(Y=y)
0	0.316
1	0.422
2	0.211
3	0.047
4	0.004

(f) μ = np = 4(0.25) = 1.00

On the average, the number of children out of four that are not living with both parents is 1.00.

(g) σ^2 = np(1-p) = 4(0.25)(0.75) = 0.75

$$\sigma = \sqrt{0.75} = 0.866$$

(h)   The probability distribution is only approximately correct because the success probability p = 0.25 is based on a sample. Since sampling is without replacement, the exact distribution is a hypergeometric distribution.

5.53   n = 12;   p = 0.10

(a)   P(2) = 0.230

(b)   P(X ≤ 2) = P(0) + P(1) + P(2) = 0.282 + 0.377 + 0.230 = 0.889

(c)   P(X ≥ 2) = 1 - P(X < 2) = 1 - P(0) - P(1) = 1 - 0.282 - 0.377
                 = 0.341

(d)   P(1 ≤ X ≤ 3) = P(1) + P(2) + P(3) = 0.377 + 0.230 + 0.085 = 0.692

5.55   Summing the probabilities for 3, 4, 5, and 6 winning numbers, we determine that the probability of winning a prize (based upon purchasing one ticket) is 0.0290647. If an individual buys one *Lotto* ticket per week for a year, the probability of winning a prize remains the same from week to week. Notice also that the probability of not winning a prize is 1 - 0.0290647 = 0.9709353, and that the weekly trials of purchasing a ticket are identical and independent. All of these characteristics permit the use of the binomial probability formula to determine the probability that an individual wins a prize at least once in the 52 tries.

From the above, we have p = 0.0290647, n = 52, and the possible successes are X = 0, 1, 2, 3, ..., 52. Thus,

$$P(X \geq 1) = 1 - P(X=0) = 1 - \binom{52}{0} (0.0290647)^0 (0.9709353)^{52}$$

$$= 1 - (0.215722) = 0.784278 \ .$$

5.57   The problem states that 4% of the parties making reservations at a restaurant never show up. This also means that 96% of the parties making reservations do show up. Alternatively, the probability is 0.96 that any one party making a reservation shows up. The probability is (0.96)·(0.96) = 0.9216 that two parties making reservations show up. The probability is (0.96)·(0.96)·(0.96) = 0.8847 that three parties all show up. It is (0.96)·(0.96)·(0.96)·(0.96) = 0.8493 that four all show up; it is 0.8154 that five all show up, and the probability is 0.7828 that six parties making reservations all show up. The owner wants to be at least 80% sure that all parties making a reservation will show up.

According to the calculations above, this means that the owner can take five reservations. Notice that taking six reservations results in only a 78% chance (which is less than 80%) that all parties making a reservation will show up.

5.59 (a) If sampling is done with replacement, the trials are independent because the results of one trial have no effect on the probabilities of the outcomes of any other trial. The success probability remains the same from trial to trial because it always represents the proportion of the population having the 'success' attribute.

(b) If sampling is without replacement, the trials are not independent because the outcome of one trial changes the probabilities of the possible outcomes in succeeding trials. Those probabilities vary from trial to trial because the proportion of successes left in the population changes whenever one of the outcomes in the population, whether it be a success or failure, is removed from the population.

5.61 (a) N = 250, n = 4, and p = 0.94. Note that Np = 235 and N(1-p) = 15.

(b)
$$P(X = x) = \frac{\binom{235}{x}\binom{15}{4-x}}{\binom{250}{4}} \quad \text{for } x = 0, 1, 2, 3, 4$$

(i)
$$P(X = x) = \binom{4}{x}(.94)^x(.06)^{4-x} \quad \text{for } x = 0, 1, 2, 3, 4$$

(j) The binomial distribution success probability will remain at 0.94 for each of the four fuses sampled, whereas in the hypergeometric distribution, either the number of defective or the number of non-defective fuses will be reduced by one (as will the population size) as each fuse is selected for purchase. For example, the probability of selecting two defective fuses, using the binomial distribution is

$$P(X = 2) = \binom{4}{2}(.94)^2(.06)^{4-2} = \frac{4!}{2!2!}(.94)(.94)(.06)(.06), \text{ whereas using the}$$

hypergeometric distribution, we get (after simplifying)

$$P(X = 2) = \frac{\binom{235}{2}\binom{15}{2}}{\binom{250}{4}} = \frac{4!}{2!2!}\frac{235}{250}\frac{234}{249}\frac{15}{248}\frac{14}{247} \quad .$$

Since 234/249 differs slightly from 0.94, and 15/248 and 14/247 each differ slightly from 0.06, the final results using the two distributions will be very close, but not identical.

5.63 (a) $P(X = 4) = e^{-6.9} \dfrac{6.9^4}{4!} = 0.0952$

(b) $P(X \leq 2) = e^{-6.9} \dfrac{6.9^0}{0!} + e^{-6.9} \dfrac{6.9^1}{1!} + e^{-6.9} \dfrac{6.9^2}{2!}$

$= 0.0010 + 0.0070 + 0.0952 = 0.1032$

$P(4 \leq X \leq 10) = e^{-6.9} \dfrac{6.9^4}{4!} + e^{-6.9} \dfrac{6.9^5}{5!} + e^{-6.9} \dfrac{6.9^6}{6!} + e^{-6.9} \dfrac{6.9^7}{7!} + e^{-6.9} \dfrac{6.9^8}{8!}$

(c) $+ e^{-6.9} \dfrac{6.9^9}{9!} + e^{-6.9} \dfrac{6.9^{10}}{10!}$

$= 0.0952 + 0.1314 + 0.1611 + 0.1489 + 0.1284 + 0.0985 + 0.0679 = 0.8314$

5.65 First store the possible values for X (0, 1, 2, 3, 4, 5) in a column named x. Then choose **Calc ▶ Probability distributions ▶ Binomial...**, select the **Probability** option button to get individual probabilities, type 5 in the **Number of trials** text box, type .67 in the **Probability of success** text box, select the **Input column** option button, click in the **Input column** text box and specify x, and click **OK**. The result is

Binomial with n = 5 and p = 0.670000

x	P(X = x)
0.00	0.0039
1.00	0.0397
2.00	0.1613
3.00	0.3275
4.00	0.3325
5.00	0.1350

(a) P(2) = 0.1613   (b)   P(4) = 0.3325

(c) P(X ≥ 4) = 0.3325 + 0.1350 = 0.4675

(d) P(2 ≤ X ≤ 4) = 0.1613 + 0.3275 + 0.3325 = 0.8213

(e) The distribution is presented previous to part (a).

5.67 (a) The number of trials is 6; i.e., n = 6.

(b) The success probability is 0.59; i.e., p = 0.59.

(c) P(X ≤ 3) = 0.4764

(d) P(X ≥ 3) = 1 - P(X ≤ 2) = 1 - 0.1933 = 0.8067

(e) P(3) = P(X ≤ 3) - P(X ≤ 2) = 0.4764 - 0.1933 = 0.2831

5.69 (a) P(X ≤ 3) = 0.8567

(b) P(X ≥ 3) = 1 - P(X ≤ 2) = 1 - 0.6282 = 0.3718

(c) P(3) = P(X ≤ 3) - P(X ≤ 2) = 0.8567 - 0.6282 = 0.2285

(d) P(2 ≤ X ≤ 4) = P(X ≤ 4) - P(X ≤ 1) = 0.9623 - 0.3193 = 0.6430

(e) P(X < 2 or X > 5) = P(X ≤ 1) + [1 - P(X ≤ 5)]

$= 0.3193 + 1 - 0.9936$

$= 0.3257$

Exercises 5.4

5.71   (a) $P(2) = \dfrac{e^{-3}(3)^2}{2!} = 0.224$

(b)
$$P(X \leq 3) = P(0) + P(1) + P(2) + P(3)$$
$$= \frac{e^{-3}3^0}{0!} + \frac{e^{-3}3^1}{1!} + \frac{e^{-3}3^2}{2!} + \frac{e^{-3}3^3}{3!}$$
$$= 0.647$$

(c) $P(X > 0) = 1 - P(X = 0)$ (d) $\mu_X = \lambda = 3$

$$= 1 - \frac{e^{-3}3^0}{0!} = 0.950$$

(e) $\sigma_X = \sqrt{\lambda} = \sqrt{3} = 1.732$

5.73   (a)
$$P(4) = \frac{e^{-3.87}(3.87)^4}{4!} = 0.195$$

(b) $P(X \leq 1) = P(0) + P(1) = e^{-3.87}[\dfrac{3.87^0}{0!} + \dfrac{3.87^1}{1!}] = 0.102$

(c) $P(2 \leq X \leq 5) = P(2) + P(3) + P(4) + P(5)$
$$= e^{-3.87}[\frac{(3.87)^2}{2!} + \frac{(3.87)^3}{3!} + \frac{(3.87)^4}{4!} + \frac{(3.87)^5}{5!}] = 0.704$$

5.75   (a) $P(1) = \dfrac{e^{-1.7}(1.7)^1}{1!} = 0.311$

(b)
$$P(X \leq 2) = P(0) + P(1) + P(2)$$
$$= e^{-1.7}[\frac{(1.7)^0}{0!} + \frac{(1.7)^1}{1!} + \frac{(1.7)^2}{2!}] = 0.757$$

(c) $P(X \geq 2) = 1 - P(X \leq 1) = 1 - [P(0) + P(1)]$

$$= 1 - [e^{-1.7}[\frac{(1.7)^0}{0!} + \frac{(1.7)^1}{1!}]] = 0.507$$

5.77 (a) $\mu_X = 3.87$; on the average 3.87 α particles reach the screen during the 8-minute interval.

(b) $\sigma_X = \sqrt{3.87} = 1.967$

5.79 (a)   The required calculations are

$$P(0) = \frac{e^{-1.7}(1.7)^0}{0!} = 0.183 \qquad P(1) = \frac{e^{-1.7}(1.7)^1}{1!} = 0.311$$

$$P(2) = \frac{e^{-1.7}(1.7)^2}{2!} = 0.264 \qquad P(3) = \frac{e^{-1.7}(1.7)^3}{3!} = 0.150$$

$$P(4) = \frac{e^{-1.7}(1.7)^4}{4!} = 0.064 \qquad P(5) = \frac{e^{-1.7}(1.7)^5}{5!} = 0.022$$

$$P(6) = \frac{e^{-1.7}(1.7)^6}{6!} = 0.006 \qquad P(7) = \frac{e^{-1.7}(1.7)^7}{7!} = 0.001$$

$$P(8) = \frac{e^{-1.7}(1.7)^8}{8!} = 0.000$$

(b)

x	P(X=x)
0	0.183
1	0.311
2	0.264
3	0.150
4	0.064
5	0.022
6	0.006
7	0.001
8	0.000

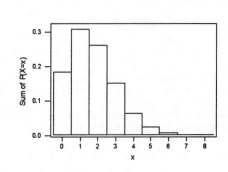

5.81 Step 1: Determine n and p; n=748, p=0.0001

Step 2: Check that n $\geq$ 100 and np $\leq$ 10.

From Step 1, we conclude that n=748 and np=(748)(0.0001)=0.0748.

Thus we see that n$\geq$100 and np$\leq$10.

Step 3: Use Poisson probability formula.

(a) $P(0) = \dfrac{e^{-0.0748}(0.0748)^0}{0!} = 0.928$

(b) $P(X \geq 1) = 1 - P(0) = 1 - 0.928 = 0.072$

5.83   Step 1:   Determine n and p; n=1000, p=0.002

Step 2:   Check that n $\geq$ 100 and np $\leq$ 10.

From Step 1, we conclude that n=1000 and np=(1000)(0.002)=2. Thus we see that n $\geq$ 100 and np $\leq$ 10.

Step 3:   Use Poisson probability formula.

(a) $P(2) = \dfrac{e^{-2}2^2}{2!} = 0.271$

(b) $P(X \leq 2) = P(0) + P(1) + P(2) = e^{-2}[\dfrac{2^0}{0!} + \dfrac{2^1}{1!} + \dfrac{2^2}{2!}] = 0.677$

(c) $P(X \geq 2) = 1 - [P(0) + P(1)] = 1 - e^{-2}[\dfrac{2^0}{0!} + \dfrac{2^1}{1!}] = 0.594$

5.85   If there is a Poisson distribution which provides a good approximation to the binomial distribution, the one with the same mean is a likely candidate since it will be 'centered' at the same place as the binomial distribution. Given that we only use this approximation when n is large and p is near zero, it will also be the case that 1-p is near one, and therefore np(1 - p) will be near np which we set equal to λ. Since λ is the variance of the Poisson distribution and np(1 - p) is the variance of the binomial distribution, the Poisson distribution with λ = np also has approximately the same variance (and hence standard deviation) as the binomial distribution it is approximating. With both the mean and standard deviation being approximately the same for both distributions, the approximating Poisson distribution should be very similar to the binomial distribution.

5.87   (a)   We choose **Calc ▶ Probability distributions ▶ Poisson...**   Then we select **Probability**, type 3.87 in the **Mean** text box, select **Input constant**, and type 4 to indicate that we want the probability for x = 4. Finally, click on **OK**.   The output shown in the Sessions window is

Probability Density Function
Poisson with μ = 3.87000

x	P(X = x)
4.00	0.1949

(b)   We choose **Calc ▶ Probability distributions ▶ Poisson...**Then we select **Cumulative Probability**, type <u>3.87</u> in the **Mean** text box, select **Input constant**, and type <u>1</u> to indicate that we want the probability for x ≤ 1. Finally, click on **OK**. The output shown in the Sessions window is

Cumulative Distribution Function
Poisson with μ = 3.87000

x	P(X <= x)
1.00	0.1016

(c) First name two columns x and P(X=x). Then enter 2, 3, 4, and 5 in

the X column. Choose **Calc ▶ Probability distributions ▶ Poisson...** Then we select **Probability**, type <u>3.87</u> in the **Mean** text box, click on **Input column** and select x in the text box, select P(x) for **Optional storage** and click on **OK**. The probabilities will now show in the P(X=x) column of the Data window. To get the sum

of these probabilities, we choose **Calc ▶ Column statistics...**, click on **Sum**, and select P(X=x) for the **Input variable**. The output shown in the Sessions window is
Column Sum
Sum of P(X=x)   =   0.70352

An alternative way of computing this probability is to use the cumulative probability procedure to find P(X ≤ 5) - P(X ≤ 1).

5.89 (a) $\lambda = 2.100$ (b) $P(X = 3) = 0.1890$

(c) $P(2 \leq X \leq 5) = P(2) + P(3) + P(4) + P(5)$
               $= 0.2700+0.1890+0.0992+0.0417=0.5999$

5.91 First name three columns x, BIN_P(x), and POI_P(x). Enter 1, 2, 3,...10

in the x column. Choose **Calc ▶ Probability distributions ▶ Binomial...** Then select **Probability**, type <u>1000</u> in the **Number of trials** text box, type <u>.002</u> in the **Probability of success** text box, click on **Input column** and select x in the text box, select BIN_P(x) for **Optional**

storage and click on **OK**. Next choose **Calc ▶ Probability distributions**

▶ Poisson... Then select **Probability**, type <u>2</u> in the **Mean** text box since np = 1000(.002) = 2, click on **Input column** and select x in the text box, select POI_P(x) for **Optional storage** and click on **OK**. The probabilities will now show in the POI_P(x) column of the Data window.

Use **Manip ▶ Display Data** and select x, BIN_P(x), AND POI_P(x) to display the results in the Sessions window shown as follows:

ROW	x	BIN_P(x)	POI_P(x)
1	0	0.135065	0.135335
2	1	0.270670	0.270671
3	2	0.270942	0.270671
4	3	0.180628	0.180447
5	4	0.090223	0.090224
6	5	0.036017	0.036089
7	6	0.011970	0.012030
8	7	0.003406	0.003437
9	8	0.000847	0.000859
10	9	0.000187	0.000191
11	10	0.000037	0.000038

5.93   (a)   The CDF command can be used to find the cumulative probabilities of 3 and 10 by specifying the Poisson distribution and cumulative probability with mean = 6.9. These cumulative probabilities for 3 and 10 can be stored in constants, K2 and K1, respectively. The $P(4 \leq X \leq 10)$ can then be found by letting K3 = K1-K2. The constant K3 will be the desired probability and can be displayed in the Sessions window.

(b)   Choose **Calc** ▶ **Probability distributions** ▶ **Poisson...** Then select **Cumulative Probability**, type <u>6.9</u> in the **Mean** text box, click on **Input constant** and type <u>10</u> in the text box, select K1 for

Optional storage and click on **OK**. Then choose **Calc** ▶ **Probability**

distributions ▶ **Poisson...** and select **Cumulative Probability**, type <u>6.9</u> in the **Mean** text box, click on **Input constant** and type <u>3</u> in the text box, select K2 for **Optional storage** and click on **OK**. To carry out the subtraction K1 - K2, store the result in K3, and

display K3, we choose **Calc** ▶ **Calculator...**, Type <u>K3</u> in the **Store result in variable** text box and type <u>K1 - K2</u> in the **Expression**

text box, and click on **OK**. Finally, choose **Manip** ▶ **Display Data...** and select K3 in the **Columns, constants, and matrices to display** text box and click on **OK**. The result shown in the Sessions window is

```
Data Display
K3      0.821296
```

REVIEW TEST FOR CHAPTER 5

1.   (a)   random variable

(b)   finite (or countably infinite)

2.   A probability distribution of a discrete distribution of a discrete random variable gives us a listing of the possible values of the random variable and their probabilities; or a formula for the probabilities.

3.   Probability histogram

4.   1

5.   (a) $P(X = 2) = 0.386$

(b)   38.6%

(c)   50(0.386) = 19.3 or about 19; 500(0.386) = 193

6.   3.6

7.   X is more likely to take a value close to its mean because it has less variation.

8.   The trials must have two possible outcomes (success and failure) per trial, must be independent, and have a probability of success p that remains constant for all trials.

9.   The binomial distribution is a probability distribution for the number of successes in a sequence of n Bernoulli trials.

10. $\dbinom{10}{3} = \dfrac{10!}{3!\,7!} = \dfrac{10(9)(8)}{3(2)(1)} = 120$

11.   Definition 5.4 for the mean and definition 5.5 for the standard deviation are applied to the binomial and Poisson distribution.   For example, in Definition 5.4, the formula for the binomial probability function is substituted for P(X = x) to get the mean.

12. (a) Binomial distribution

   (b) Hypergeometric distribution

   (c) The hypergeometric distribution may be approximated by the binomial distribution if the population size N is much greater than the sample size n. When this condition holds, the probability of a success does not change much from trial to trial.

13. (a) X = 1, 2, 3, 4   (b)   {X = 3}

   (c) P(X = 3) = 8,157/32,537 = 0.251.

      25.1% of the undergraduates at this university are juniors.

   (d)   (e)

| Class level | Probability |
x	P(X=x)
1	0.196
2	0.194
3	0.251
4	0.359

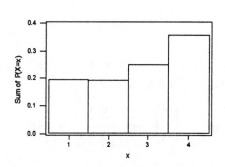

14. (a) {Y = 4}   (b)   {Y ≥ 4}
   (c) {2 ≤ Y ≤ 4} (d)   {Y ≥ 1}
   (e) P(Y = 4) = 0.174
   (f) P(Y ≥ 4) = P(4) + P(5) + P(6) = 0.174 + 0.105 + 0.043 = 0.322
   (g) P(2 ≤ Y ≤ 4) = P(2) + P(3) + P(4) = 0.232 + 0.240 + 0.174 = 0.646
   (h) P(Y ≥ 1) = 1 - P(Y ≤ 0) = 1 - 0.052 = 0.948

15. The required calculations are:

y	P(Y=y)	yP(Y=y)	y^2	y^2P(Y=y)
0	0.052	0.000	0	0.000
1	0.154	0.154	1	0.154
2	0.232	0.464	4	0.928
3	0.240	0.720	9	2.160
4	0.174	0.696	16	2.784
5	0.105	0.525	25	2.625
6	0.043	0.258	36	1.548
		2.817		10.199

   (a) $\mu_Y = \Sigma y P(Y=y) = 2.817$.

   (b) On the average, the number of busy lines is about 2.8.

   (c) $\sigma_Y = \sqrt{\Sigma (y - \mu_y)^2 P(Y=y)} = \sqrt{2.2635} = 1.50$, or

       $\sigma_y = \sqrt{\Sigma y^2 P(Y=y) - \mu_y^2} = \sqrt{10.199 - (2.817)^2} = 1.50$

(d)

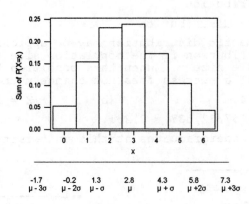

16. $0! = 1$ $3! = 3 \cdot 2 \cdot 1 = 6$ $4! = 4 \cdot 3 \cdot 2 \cdot 1 = 24$
      $7! = 7 \cdot 6 \cdot 5 \cdot 4 \cdot 3 \cdot 2 \cdot 1 = 5040$

17.   (a) $\binom{8}{3} = \dfrac{8!}{3! \, 5!} = 56$ (b) $\binom{8}{5} = \dfrac{8!}{5! \, 3!} = 56$

      (c) $\binom{6}{6} = \dfrac{6!}{6! \, 0!} = 1$ (d) $\binom{10}{2} = \dfrac{10!}{2! \, 8!} = 45$

      (e) $\binom{40}{4} = \dfrac{40!}{4! \, 36!} = 91{,}390$ (f) $\binom{100}{0} = \dfrac{100!}{0! \, 100!} = 1$

18.   (a) $p = 0.493$
      (b)

Outcome	Probability
sss	$(0.493)(0.493)(0.493) = 0.120$
ssf	$(0.493)(0.493)(0.507) = 0.123$
sfs	$(0.493)(0.507)(0.493) = 0.123$
sff	$(0.493)(0.507)(0.507) = 0.127$
fss	$(0.507)(0.493)(0.493) = 0.123$
fsf	$(0.507)(0.493)(0.507) = 0.127$
ffs	$(0.507)(0.507)(0.493) = 0.127$
fff	$(0.507)(0.507)(0.507) = 0.130$

(c)

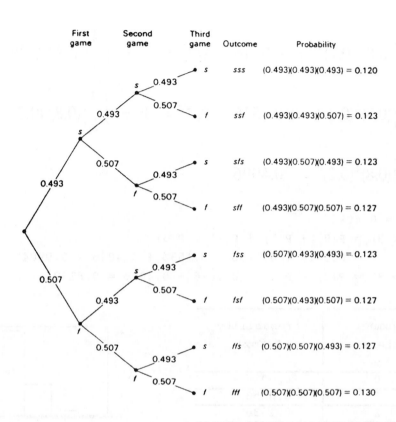

(d)   The outcomes in which the player wins exactly two out of the three times are ssf, sfs, and fss.

(e)   Each outcome in part (d) has probability 0.123. The probabilities are equal for each outcome because each probability is obtained by multiplying two success probabilities of 0.493 and one failure probability of 0.507.

(f)   P(wins 2 games) = P(ssf) + P(sfs) + P(fss) = 0.123 + 0.123 + 0.123

$$= 0.369$$

(g)   P(0 wins) = P(fff) = 0.130

P(1 win) = P(sff) + P(fsf) + P(ffs) = 0.127 + 0.127 + 0.127

$$= 0.381$$

P(3 wins) = P(sss) = 0.120

y	P(Y=y)
0	0.130
1	0.381
2	0.369
3	0.120

(h)   Binomial distribution

19.   The calculations required to answer all parts of this exercise are

$$P(X = 0) = \binom{4}{0}(0.8)^0(0.2)^4 = 0.0016 \quad P(X = 1) = \binom{4}{1}(0.8)^1(0.2)^3 = 0.0216$$

$$P(X = 2) = \binom{4}{2}(0.8)^2(0.2)^2 = 0.1536 \quad P(X = 3) = \binom{4}{3}(0.8)^3(0.2)^1 = 0.4096$$

$$P(X = 4) = \binom{4}{4}(0.8)^4(0.2)^0 = 0.4096$$

(a) $P(3) = 0.4096$

(b) $P(X \leq 3) = P(0) + P(1) + P(2) + P(3)$
         $= 0.0016 + 0.0256 + 0.1536 + 0.4096 = 0.5904$

(c) $P(X \geq 3) = P(3) + P(4) = 0.4096 + 0.4096 = 0.8192$

(d)

Number literate x	Probability P(X=x)
0	0.0016
1	0.0256
2	0.1536
3	0.4096
4	0.4096

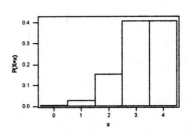

(e)   Left skewed, since p > 0.5.

(f)   See histogram at right.

(g)   The distribution is only approximate for several reasons: the actual distribution is hypergeometric, based on sampling without replacement; and the success probability p = 0.80 is probably based on a sample.

(h) $\mu = np = 4(.8) = 3.2$; On the average, 3.2 out of 4 Surinamese are literate.

(i) $\sigma^2 = np(1-p) = 4(.8)(.2) = 0.64$; $\sigma = \sqrt{0.64} = 0.8$

20.   (a)   p=0.5   (b)   p < 0.5

21.   (a) $P(X = 2) = e^{-1.75}\dfrac{1.75^2}{2!} = 0.266$

(b) $P(4 \leq X \leq 6) = e^{-1.75}\dfrac{1.75^4}{4!} + e^{-1.75}\dfrac{1.75^5}{5!} + e^{-1.75}\dfrac{1.75^6}{6!}$
         $= 0.068 + 0.024 + 0.007 = 0.099$

(c) $P(X \geq 1) = 1 - P(0) = 1 - e^{-1.75} \dfrac{1.75^0}{0!} = 1 - 0.174 = 0.826$

(d)   (e)

x	P(X=x)
0	0.174
1	0.304
2	0.266
3	0.155
4	0.068
5	0.024
6	0.007
7	0.002
8	0.000

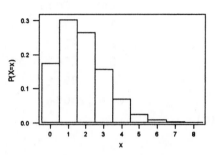

(f)   The distribution is right skewed. Yes, this is typical of Poisson distributions.

22.   (a) $\mu = \lambda = 1.75$; on the average, there are 1.75 calls to a wrong number per minute.

(b) $\sigma = \sqrt{\lambda} = \sqrt{1.75} = 1.323$

23.   Step 1:   Determine n and p; n = 10,000 and p = 0.00024.

Step 2:   Check if $n \geq 100$ and $np \leq 10$.

From part (a), we conclude that n = 10,000 and

np = 2.4.

Thus $n \geq 100$ and $np \leq 10$.

Step 3:   Use the Poisson probability formula.

(a) $\mu = np = 2.4$; you would expect to be dealt four of a kind roughly 2.4 times.

(b) $P(2) = \dfrac{e^{-2.4}(2.4)^2}{2!} = 0.2613$

(c)

$P(X \geq 2) = 1 - P(X \leq 1)$

$= 1 - [P(0) + P(1)] = 1 - [e^{-2.4}[\dfrac{(2.4)^0}{0!} + \dfrac{(2.4)^1}{1!}]] = 0.6916$

24.   (a)   Name three columns x, P(X=x), and CLASS. In the X column, enter the numbers 1, 2, 3, and 4. In the P(X=x) column, enter the probabilities .196, .194, .251, and .359. Now choose **Calc** ▶

Random Data ▶ **Discrete...**, type <u>2500</u> in the **Generate rows of data** text box, click in the **Store in column[s]** text box and type <u>CLASS</u>, click in the **Values in** text box and specify x, click in the **Probabilities in** text box and specify P(X=x), and click **OK**. The numbers of the classes obtained in 200 observations are now stored in CLASS.

(b)   Choose **Calc** ▶ **Tables** ▶ **Tally,** select CLASS in the **Variables** text box, check the **Counts** and **Percents Display** boxes, and click **OK.** Our result is

CLASS	Count	Percent
1	478	19.12
2	472	18.88
3	656	26.24
4	894	35.76
N=	2500	

(c)

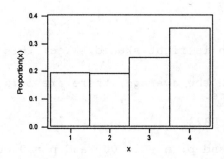

(d)   Parts (b) and (c) illustrate that samples tend to reflect the populations from which they are drawn.

25.   (a)   Name three columns Y, P(Y), and NUMLINE. In the Y column, enter the numbers 0, 1, 2, 3, 4, 5, and 6. In the P(Y) column, enter the probabilities .052, .154, .232, .240, .174, .105, and .043.

Now choose **Calc** ▶ **Random Data** ▶ **Discrete...,** type <u>200</u> in the **Generate rows of data** text box, click in the **Store in column[s]** text box and type <u>NUMLINE</u>, click in the **Values in** text box and specify Y, click in the **Probabilities in** text box and specify P(Y), and click **OK.** The numbers of busy lines obtained in 200 observations are now stored in NUMLINE.

(b)   To obtain the mean for the numbers of busy lines, we choose **Calc** ▶ **Column statistics...,** click on the **Mean** button and specify NUMLINE in the **Input Variable** text box, and click **OK.**

Our output was (yours may be different)

MEAN = 2.8600

The average value of the 200 observation was 2.86 which is quite close to the mean, $\mu_y = 2.817$.

(c)   Part (b) is illustrating that samples tend to reflect the properties of the populations from which they are drawn

26.   (a)   Choose **Calc** ▶ **Probability Distributions** ▶ **Binomial...** Then

click in the **Number of trials** text box and type $\underline{4}$, click in the **Probability** text box and type $\underline{.8}$, click on the **Input constant** button, click in its text box, and type $\underline{3}$. Finally, click   OK. The output shown in the Sessions window is

x	P(X = x)
3.00	0.4096

(b)   Choose **Calc ▶ Probability Distributions ▶ Binomial...** Then click in the **Number of trials** text box and type $\underline{4}$, click on the **Cumulative probability** button, click in the **Probability of success** text box and type $\underline{.8}$, click on the **Input constant** button, click in its text box, and type $\underline{3}$. Finally, click **OK**.   The output shown in the Sessions window is

x	P(X <= x)
3.00	0.5904

(c)   One way to find $P(X \geq 3)$ is by finding $1 - P(X \leq 2)$. $P(X \leq 2)$ is found using the cumulative probability for 2. We choose **Calc ▶ Probability Distributions ▶ Binomial...** Then click in the **Number of trials** text box and type $\underline{4}$, click on the **Cumulative probability** button, click in the **Probability of success** text box and type $\underline{.8}$, click on the **Input constant** button, click in its text box, and type $\underline{2}$. Finally, click **OK**.   The output shown in the Sessions window is

Cumulative Distribution Function

Binomial with n = 4 and p = 0.800000

x	P(X <= x)
2.00	0.1808

The desired probability is $1 - 0.1808 = 0.8192$

(d)   Name two columns x and P(X=x). In the x column, enter the numbers 0, 1, 2, 3, and 4. Then we choose **Calc ▶ Probability Distributions ▶ Binomial...** Click in the **Number of trials** text box and type $\underline{4}$, click on the **Probability** button, click in the **Probability of success** text box and type $\underline{.8}$, click on the **Input column** button, click in its text box and select X. Leave the **Option storage** text box empty and click **OK**.   The output shown in the Sessions window is

```
        Binomial with n = 4 and p = 0.800000
            x          P( X = x)
          0.00          0.0016
          1.00          0.0256
          2.00          0.1536
          3.00          0.4096
          4.00          0.4096
```

27.   (a)   The number of trials is 5; i.e., n = 5.

(b)   The success probability is 0.65; i.e., p = 0.65.

(c)   P(1) = 0.0488

(d)   P(1 $\leq$ X $\leq$ 3) = P(1) + P(2) + P(3)

= 0.0488 + 0.1811 + 0.3364 = 0.5663

(e)   P(X $\leq$ 1) = P(0) + P(1) = 0.0053 + 0.0488 = 0.0541

(f)   P(X $\geq$ 1) = 1 - P(0) = 1 - 0.0053 = 0.9947

28.   (a)   The number of trials is 8; i.e., n = 8.

(b)   The success probability is 0.57; i.e., p = 0.57.

(c)   P(X $\leq$ 4) = 0.4762

(d)   P(X $\geq$ 4) = 1 - P(X $\leq$ 3) = 1 - 0.2235 = 0.7765

(e)   P(4) = P(x $\leq$ 4) - P(X $\leq$ 3) = 0.4762 - 0.2235 = 0.2527

29.   (a)   P(2) = 0.0217

(b)   P(X $\geq$ 2) = 1 - P(X $\leq$ 1) = 1 - P(0) - P(1)

= 1 - 0.0002 - 0.0030 = 0.9968

(c)   P(X < 2) = P(0) + P(1) = 0.0002 + 0.0030 = 0.0032

(d)   P(2 $\leq$ X $\leq$ 5) = P(2) + P(3) + P(4) + P(5)

= 0.0217 + 0.0886 + 0.2169 + 0.3186 = 0.6458

30.   Name two columns x and P(X=x). In the x column, enter the numbers 0, 1, 2, 3, 4, 5, 6, 7, and 8. Then we choose **Calc** ▶ **Probability**

Distributions ▶ **Poisson**... Click in the **Mean** text box and type <u>1.75</u>, click on the **Probability** button, click on the **Input column** button, click in its text box and select x. Leave the **Option storage** text box empty and click **OK**. The output shown in the Sessions window is

```
        Poisson with μ = 1.75000
            x          P( X = x)
          0.00          0.1738
          1.00          0.3041
          2.00          0.2661
          3.00          0.1552
          4.00          0.0679
          5.00          0.0238
          6.00          0.0069
          7.00          0.0017
          8.00          0.0004
```

   (a)   P(2) = 0.2661
   (b)   P(4 ≤ X ≤ 6) = 0.0679 + 0.0238 + 0.0069 = 0.0986
   (c)   P(X ≥ 1) = 1 - P(0) = 1 - 0.1738 = 0.8262
   (d)   See table previous to part (a).

31.   (a) $\lambda = 2.400$
   (b)   P(X = 2) = P(X ≤ 2) - P(X ≤ 1) = 0.5697 - 0.3084
                  = 0.2613
   (c)   P(X ≥ 2) = 1 - P(X ≤ 1) = 1 - 0.3084 = 0.6916

CHAPTER 6 ANSWERS

Exercises 6.1

6.1 The histogram will be roughly bell-shaped.

6.3 Their distributions are identical. The mean and standard deviation completely determine the shape of a normal distribution. Thus if two normally distributed variables have the same mean and standard deviation, they also have the same distribution.

6.5 True. Both normal curves have the same shape. Spread (or shape) is represented by σ. For each normal curve, $\sigma = 3$.

6.7 True. The value of the parameter μ has no effect on the shape of a normal curve. The parameter μ affects only where the normal curve is centered. The shape of the normal curve is determined by the parameter σ.

6.9

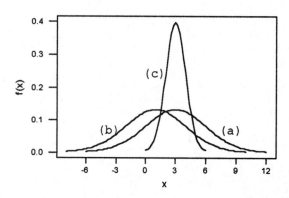

6.11 The percentage of all possible observations of a normally distributed variable that lie between 2 and 3 is the same as the area under the associated normal curve between 2 and 3. If the variable is only approximately normally distributed, then the percentage of all possible observations between 2 and 3 is only approximately the area under the associated normal curve between 2 and 3.

6.13 (a) The percentage of female students who are between 60 and 65 inches tall is 0.0450 + 0.0757 + 0.1170 + 0.1480 + 0.1713 = 0.5570 (55.70%).

     (b) The area under the normal curve with parameters $\mu = 64.4$ and $\sigma = 2.4$ between 60 and 65 is approximately 0.5570. This is only an estimate because the distribution of heights is only approximately normally distributed.

6.15 (a)

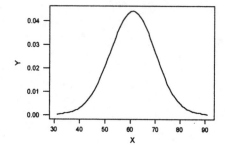

134

(b)   z = (x - 61)/9

(c)   z has a standard normal distribution (μ = 0 and σ = 1).

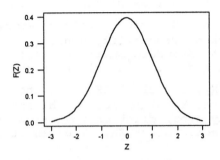

(d)   The percentage of finishers in the New York City 10-km run with times between 50 and 70 minutes is equal to the area under the standard normal curve between −1.22 and +1.00.

(e)   The percentage of finishers in the New York city 10-km run with times exceeding 75 minutes is equal to the area under the standard normal curve that lies to the right of 1.56.

6.17   (a) $y = \dfrac{1}{\sqrt{2\pi} \cdot 2} e^{-(x-5)^2/2\cdot 2^2}$ (b) $y = \dfrac{1}{\sqrt{2\pi}} e^{-z^2}$

6.19   Using Minitab we choose **Calc ▶ Make patterned data ▶ Simple set of numbers...**, enter X in the **Store patterned data** in text box, type 1 in the **From first value** text box, type 13 in the **To last value** text box, type .2 in the **In steps of** text box, and click **OK**. This will provide X values within 3 standard deviations on both sides of the mean. Now

choose **Calc ▶ Probability distribution ▶ Normal...**, click on **Probability density**, type 7 in the **Mean** text box and 2 in the **Standard deviation** text box, click in the **Input column** text box and select X, click in the **Optional storage** text box and type P(X), and click **OK**. Now

choose **Graph ▶ Plot...**, select P(X) in the **Y** column for **Graph1** and X for the **X** column. In the **Data display** area, click on the arrow to the right of **Display** and select **Connect**. Click **OK**. The result is

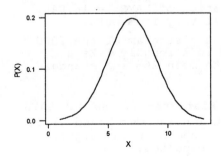

6.21 (a)   Using Minitab we choose **Calc ▶ Make patterned data ▶ Simple set of numbers...**, enter X in the **Store patterned data** in text box, type 218 in the **From first value** text box, type 314 in the **To last value** text box, type .5 in the **In steps of** text box, and click **OK**. This will provide X values within 3 standard deviations on both

sides of the mean.   Now choose **Calc ▶ Probability distribution ▶ Normal...**, click on **Probability density**, type 266 in the **Mean** text box and 16 in the **Standard deviation** text box, click in the **Input column** text box and select X, click in the **Optional storage** text

box and type P(X), and click **OK**.   Now choose **Graph ▶ Plot...**, select P(X) in the **Y** column for **Graph1** and X for the **X** column. In the **Data display** area, click on the arrow to the right of **Display** and select **Connect**.   Click **OK**.   The result is

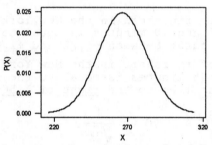

(b)   Following the procedure in Example 6.2, we choose **Calc ▶ Random**

Data ▶ Normal..., type 1000 in the **Generate row of data** text box, click in the **Store in column(s)** text box and type DAYS, click in the **Mean** text box and type 266, click in the **Standard deviation** text box and type 16, and click **OK**.

(c)   We would expect the sample mean and sample deviation to be approximately equal to the mean and standard deviation of the population, 266 and 16 respectively.   This is because we expect the sample to reflect approximately the characteristics of the population.

(d)   **Calc ▶ Column Statistics...**, click on the **Mean** button, click in the **Input variable** text box and select DAYS, and click **OK**.   The repeat the process, selecting the **Standard deviation** button.   The results are shown in the Session Window.   The results we obtained were

        Mean of DAYS = 266.88
    Standard deviation of DAYS = 16.048.
    Your results will vary from ours.

(f)   We would expect a histogram of the 1000 observations to look roughly like a normal curve with mean 266 and standard deviation 16.

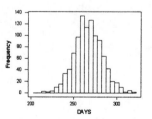

(g)   Choose **Graph ▶ Histogram...**, select DAYS in the box **Graph 1** box under **X** and click **OK**. The result is shown at the right. The result is as expected.

Exercises 6.2

6.23   Finding areas under the standard normal curve is important because for <u>any</u> normally distributed variable, we can obtain the percentage of all possible observations that lie within any specified range by first converting x values to z-scores and then finding the corresponding area under the standard normal curve.

6.25   The total area under the curve is 1, and the standard normal curve is symmetric about 0. Therefore, the area to the left of 0 is 0.5, and the area to the right of 0 is 0.5.

6.27   The area under the standard normal curve to the right of 0.43 is 1 - the area to the left of 0.43. The area to the left of 0.43 is 0.6664. Therefore, the area to the right of 0.43 is 1 - 0.6664 = 0.3336.

6.29   The area to the left of z = 3.00 is 0.9987 and the area to the left of - 3.00 is 0.0013. Therefore the area between -3.00 and 3.00 is 0.9987 - 0.0013 = 0.9974.

6.31   (a)   Locate the row (tenths digit) and column (hundredths digit) of the specified z-score. The corresponding table entry is the area under the standard normal curve that lies to the left of the z-score.

(b)   The area that lies under the standard normal curve to the right of a specified z-score is 1 - (area to the left of the z-score).

(c)   The area that lies under the standard normal score between two specified z-scores, say a and b, where a < b, is found by subtracting the area to the left of a from the area to the left of b.

6.33   (a)   (b)

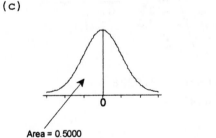

Area = 0.9875

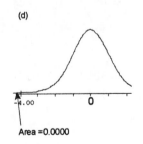

Area = 0.0594

(c)   (d)

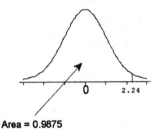

Area = 0.5000

(d)

Area = 0.0000

6.35   (a)

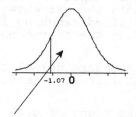

Area = 1.0 - 0.1423 = 0.8577

(b)

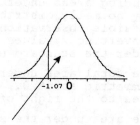

Area = 1.0 - 0.7257 = 0.2743

(c)

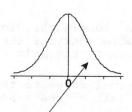

Area = 1.0 - 0.5 = 0.5

(d)

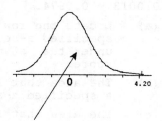

Area = 1.0 - 1.0 = 0.0

6.37   (a)

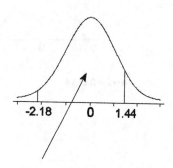

Area = 0.9251 - 0.0146
     = 0.9105

(b)

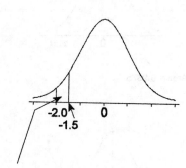

Area = 0.0668 - 0.0228
     = 0.0440

(c)   (d)

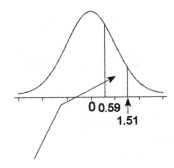

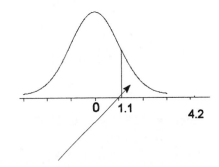

Area = 0.9345 - 0.7224   Area = 1.0 - 0.8643
     = 0.2121   = 0.1357

6.39 (a)   (b)

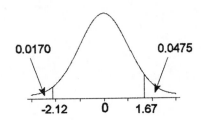

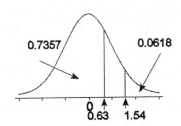

Area = 0.0170 + (1 - 0.9525) Area = 0.7357 + (1 - 0.9382)
     = 0.0645   = 0.7975

6.41 (a) The area to the left of z = 1.28 is 0.8997. The area to the left
               of z = -1.28 is 0.1003. The area between z = -1.28 and z = 1.28
               is 0.8997 - 0.1003 = 0.7994.

          (b) The area to the left of z = 1.64 is 0.9495. The area to the left
               of z = -1.64 is 0.0505. The area between z = -1.64 and z = 1.64
               is 0.9495 - 0.0505 = 0.8990.

          (c) The area to the left of z = -1.96 is 0.025. The area to the right
               of z = 1.96 is 1.0000 - 0.9750 = 0.025. The area either to the
               left of z = -1.96 or to the right of z = 1.96 is 0.025 + 0.025 =
               0.0500.

          (d) The area to the left of z = -2.33 is 0.0099. The area to the
               right of z = 2.33 is 1.0000 - 0.9901 = 0.0099. The area either to
               the left of z = -2.33 or to the right of z = 2.33 is 0.0099 +
               0.0099 = 0.0198.

6.43 (a)   (b)

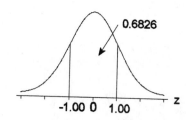

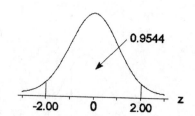

Area = 0.8413 − 0.1587   Area = 0.9772 − 0.0228
     = 0.6826   = 0.9544

(c)

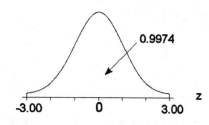

Area = 0.9987 − 0.0013 = 0.9974

6.45   **6.47**

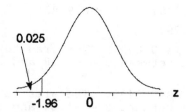

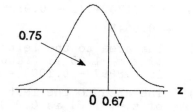

z = −1.96   z = 0.67

6.49

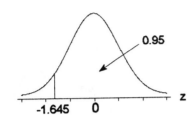

z = -1.645

6.51

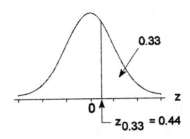

z = 0.44

6.53 (a)   (b)

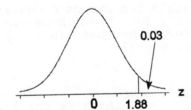   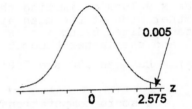

z = 1.88   z = 2.575

6.55

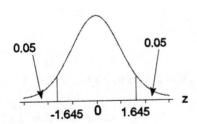

z = ± 1.645

6.57

$z_{0.10}$	$z_{0.05}$	$z_{0.025}$	$z_{0.01}$	$z_{0.005}$
1.28	1.645	1.96	2.33	2.575

6.59   (a) z_α (b) $-z_\alpha$ (c) $\pm\, z_{\alpha/2}$
(d)
          For part (a):   For part (b):   For part (c):

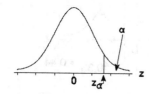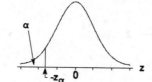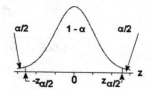

Exercises 6.3

6.61   The x values delimiting the interval within two standard deviations either side of the mean are $\mu - 2\sigma$ and $\mu + 2\sigma$. When these are standardized using $z = (x - \mu)/\sigma$, the former becomes $[(\mu - 2\sigma) - \mu]/\sigma = -2$ and the latter becomes $[(\mu + 2\sigma) - \mu]/\sigma = +2$.

6.63   (a)   Between 500 and 700:

          z-score computations: Area to the left of z:

$$x = 500 \Rightarrow z = \frac{500 - 586}{97} = -0.89 \qquad\qquad 0.1867$$

$$x = 700 \Rightarrow z = \frac{700 - 586}{97} = 1.18 \qquad\qquad 0.8810$$

          Total area = 0.8810 - 0.1867 = 0.6943 = 69.43%
(b)   At least 400:
          z-score computation: Area to the left of z:

$$x = 400 \Rightarrow z = \frac{400 - 586}{97} = -1.92 \qquad\qquad 0.0274$$

          Total area = 1.0000 - 0.0274 = 0.9726 = 97.26%

6.65   (a)   Between 190 and 210:

z-score computations: Area to the left of z:

$$x = 190 \Rightarrow z = \frac{190-175}{14} = 1.07$$ 0.8577

$$x = 210 \Rightarrow z = \frac{210-175}{14} = 2.50$$ 0.9938

Total area = 0.9938 − 0.8577 = 0.1361 = 13.61%

   (b)   Less than 150:

z-score computation: Area to the left of z:

$$x = 150 \Rightarrow z = \frac{150-175}{14} = -1.79$$ 0.0367

Total area = 0.0367 = 3.67%

6.67   (a)   P(x > 75):

z-score computation: Area to the left of z:

$$x = 75 \Rightarrow z = \frac{75-61}{9} = 1.56$$ 0.9406

Total area = 1.0000 − 0.9406 = 0.0594

   (b)   P(x < 50 or x > 70):

z-score computations: Area to the left of z:

$$x = 50 \Rightarrow z = \frac{50-61}{9} = -1.22$$ 0.1112

$$x = 70 \Rightarrow z = \frac{70-61}{9} = 1$$ 0.8413

Total area = 0.1112 + (1.0000 − 0.8413) = 0.2699

6.69

Part	Standard deviations to either side of the mean	Area under normal curve	Percent
(a)	1	0.3413 x 2 = 0.6826	68.26
(b)	2	0.4772 x 2 = 0.9544	95.44
(c)	3	0.4987 x 2 = 0.9974	99.74

6.71

Part	Percent	Lower bound	Upper bound
(a)	68.26	153.10 - 1(17.20) = 135.9	153.10 + 1(17.20) = 170.3
(b)	95.44	153.10 - 2(17.20) = 118.7	153.10 + 2(17.20) = 187.5
(c)	99.74	153.10 - 3(17.20) = 101.5	153.10 + 3(17.20) = 204.7

(d)

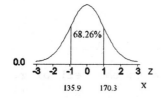
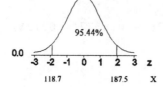
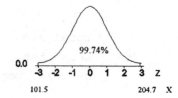

6.73 (a) $P_{15} = 175 + (-1.04)(14) = 160.44$ lb.

(b) $P_{98} = 175 + (2.05)(14) = 203.70$ lb.

(c) $Q_1 = 175 + (-0.67)(14) = 165.62$ lb.

$Q_2 = 175 + (0)(14) = 175.00$ lb.

$Q_3 = 175 + (0.67)(14) = 184.38$ lb.

6.75 (a) $Q_1 = 153.10 + (-0.67)(17.20) = \$141.58$

$Q_2 = 153.10 + (0)(17.20) = \$153.10$

$Q_3 = 153.10 + (0.67)(17.20) = \$164.62$

(b) $P_{30} = 153.10 + (-0.52)(17.20) = \$144.16$

(c) $P_{85} = 153.10 + (1.04)(17.20) = \$170.99$

6.77 (a)   95% of the population values lie within 1.96 standard deviations to either side of the mean.

(b)   89.9% of the population values lie within 1.64 standard deviations to either side of the mean.

6.79 (a)   99% of the population values lie within 2.575 standard deviations to either side of the mean.

(b)   80% of the population values lie within 1.28 standard deviations to either side of the mean.

6.81 (a)   Exact percentage between 62 and 63 = 0.1170 (11.70%)

z-score computations: Area to the left of z:

$$x = 62 \Rightarrow z = \frac{62 - 64.4}{2.4} = -1 \qquad\qquad 0.1587$$

$$x = 63 \Rightarrow z = \frac{63 - 64.4}{2.4} = -0.58 \qquad\qquad 0.2810$$

Area between x-values = 0.2810 - 0.1587 = 0.1223 (12.23%)

The two percentages are quite close.

(b)   Exact percentage between 65 and 70 = 0.1575 + 0.1100 + 0.0735

$$+ \; 0.0374 + 0.0199 = 0.3983 \; (39.83\%)$$

z-score computations: Area to the left of z:

$$x = 65 \Rightarrow z = \frac{65 - 64.4}{2.4} = 0.25 \qquad\qquad 0.5987$$

$$x = 70 \Rightarrow z = \frac{70 - 64.4}{2.4} = 2.33 \qquad\qquad 0.9901$$

Area between x-values = 0.9901 - 0.5987 = 0.3914 (39.14%)

The two percentages are quite close.

6.83 $P(\mu - z_{\alpha/2} \; \sigma < x < \mu + z_{\alpha/2}) = P(-z_{\alpha/2} \; \sigma < x - \mu < z_{\alpha/2} \; \sigma)$

$$= P(-z_{\alpha/2} < (x - \mu)/\sigma < z_{\alpha/2})$$

$$= P(-z_{\alpha/2} < z < z_{\alpha/2}) = (1 - \alpha/2) - \alpha/2$$

$$= 1 - \alpha$$

6.85 We will obtain the values needed for parts (a) and (b) simultaneously. Enter the three x values 190, 210, and 150 in Column C1. Then choose

Calc ▶ Probability distributions ▶ Normal..., select the **Cumulative probability** option button, click in the **Mean** text box and type 175, click in the **Standard deviation** text box and type 14, select the **Input constant** option button, click in the **Input column** text box and type C1, click in the **Optional storage** text box and type C2, and click **OK**. The areas to the left of each of the three x values will appear in C2 so that columns C1 and C2 will be

```
190    0.858012
210    0.993790
150    0.037073
```

(a)   The percentage of males between 18 and 24 years old who are between 190 and 210 pounds is 0.993790 - 0.858012 = 0.135778 (13.58%).

(b)   The percentage of males between 18 and 24 years old who are less than 150 pounds is 0.037073 (3.71%).

6.87 (a)   Choose **Calc ▶ Probability distributions ▶ Normal...**, select the **Inverse cumulative probability** option button, click in the **Mean** text box and type 175, click in the **Standard deviation** text box and type 14, select the **Input constant** option button, click in the **Input constant** text box and type 0.15, and click **OK**. The output in the Sessions window is

Normal with mean = 175.000 and standard deviation = 14.0000

```
P( X <= x)            x
  0.1500         160.4899
```

(b)   Choose **Calc ▶ Probability distributions ▶ Normal...**, click in the **Input constant** text box and type 0.98, and click **OK**. The output in the Sessions window is

Normal with mean = 175.000 and standard deviation = 14.0000

```
P( X <= x)            x
  0.9800         203.7525
```

(c)   Using Minitab, we choose **Calc ▶ Probability distributions ▶ Normal...**, click in the **Input constant** text box and type <u>0.25</u>, and click **OK**. The output in the Sessions window is

Normal with mean = 175.000 and standard deviation = 14.0000

P(X <= x)   x

    0.2500   165.5571

Choose **Calc ▶ Probability distributions ▶ Normal...**, click in the **Input constant** text box and type <u>0.75</u>, and click **OK**. The output in the Sessions window is

Normal with mean = 175.000 and standard deviation = 14.0000

P(X <= x)   x

    0.7500   184.4429

Exercises 6.4

6.89 Decisions about whether a variable is normally distributed often are important in subsequent analyses such as percentile calculations or in determining the type of statistical inference procedure to be used.

6.91 A normal probability plot is a plot of the actual observations versus the normal scores - the observations we would expect to get for a variable having a normal distribution. If the variable is normally distributed, then the normal probability plot should yield roughly a straight line. If the plot is roughly linear, then we accept as reasonable that the variable is approximately normally distributed. If it exhibits significant curvature, then we conclude that the variable is probably not normally distributed.

6.93 (a)

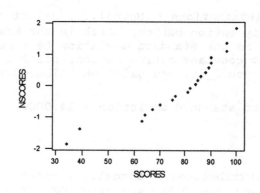

Exam Score x	Normal Score y
34	-1.87
39	-1.40
63	-1.13
64	-0.92
67	-0.74
70	-0.59
75	-0.45
76	-0.31
81	-0.19
82	-0.06
84	0.06
85	0.19
86	0.31
88	0.45
89	0.59
90	0.74
90	0.92
96	1.13
96	1.40
100	1.87

(b)   Based on the probability plot, there appear to be two outliers in the sample: 34 and 39.

(c)   Based on the probability plot, the sample does not appear to come from an approximately normally distributed population.

6.95   (a)

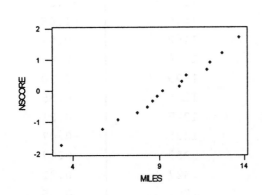

Miles driven	Normal Score
x	y
3.3	-1.74
5.7	-1.24
6.6	-0.94
7.7	-0.71
8.3	-0.51
8.6	-0.33
8.9	-0.16
9.2	0.0
10.2	0.16
10.3	0.33
10.6	0.51
11.8	0.71
12.0	0.94
12.7	1.24
13.7	1.74

(b)   Based on the probability plot, there do not appear to be any outliers in the sample.

(c)   Based on the probability plot, the sample appears to be from an approximately normally distributed population.

6.97   Assuming that the data are in a column named SCORES, choose **Calc ▶ Calculator...**, type <u>NSCORE</u> in the **Store result in variable** text box, scroll down to **Normal scores** in the **Functions** list box, click on the **Select** button, specify SCORES for **number** in the **Expression** text box by double-clicking on it, and click **OK**. Now choose **Graph ▶ Plot...**, select NSCORE for the **Y** variable for **Graph 1** and SCORES for the **X** variable for **Graph 1**, and click **OK**. The output in the Graph window follows.

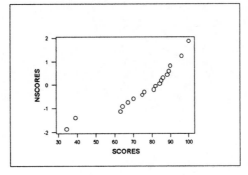

6.99 (a)

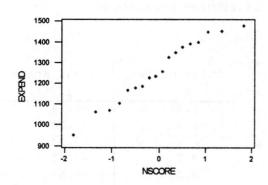

Expenditures	Normal Score
x	y
949	-1.82
1059	-1.35
1070	-1.06
1102	-0.84
1168	-0.66
1179	-0.50
1185	-0.35
1227	-0.21
1235	-0.07
1259	0.07
1327	0.21
1351	0.35
1376	0.50
1393	0.66
1400	0.84
1452	1.06
1456	1.35
1480	1.82

(b)   Based on the probability plot, there do not appear to be any outliers in the sample.

(c)   Based on the probability plot, the sample appears to be from an approximately normally distributed population.

6.101 (a)

Length of stay	Normal Score
x	y
1	-1.89
1	-1.43
3	-1.16
3	-0.95
4	-0.78
4	-0.63
5	-0.49
6	-0.36
6	-0.24
7	-0.12
7	0.0
9	0.12
9	0.24
10	0.36
12	0.49
12	0.63
13	0.78
15	0.95
18	1.16
23	1.43
55	1.89

(b)   Based on the probability plot, there appears to be one possible outlier in the sample: 55.

(c)   Based on the probability plot, the sample does not appear to be from an approximately normally distributed population.

6.103 (a)   We will generate four columns of fifty observations each by choosing **Calc ▶ Random data ▶ Normal...**, typing 50 in the **Generate rows of data** text box, clicking in the **Store in column(s)** text box and typing C1-C4, clicking in the **Mean** text box and typing 266, clicking in the **Standard deviation** text box and typing 16, and clicking **OK**. Now click in the worksheet column title row and name the four columns GEST1, GEST2, GEST3, and GEST4.

(b)   Next we create a column with the normal scores for the first sample (GEST1) and then create the normal probability plot. To do this, we choose **Calc ▶ Calculator...**, type NSCORE1 in the **Store result in variable** text box, scroll down to **Normal scores** in the **Functions** list box, click on the **Select** button, specify GEST1 for **number** in the **Expression** text box by double-clicking on it, and click **OK**. Now choose **Graph ▶ Plot...**, select NSCORE1 for the **Y** variable for **Graph 1** and GEST1 for the **X** variable for **Graph 1**, and click **OK**. Repeat this process for GEST2 and NSCORE2, GEST3 and NSCORE3, and GEST4 and NSCORE4. Our results follow. Yours will vary from ours.

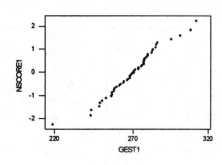

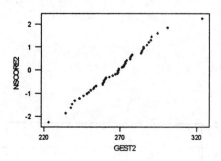

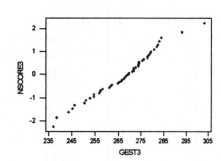

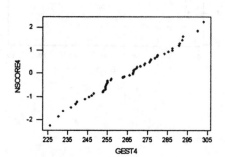

(c)   Since the data were generated from a normal distribution, we would expect the four plots to be roughly linear, as they are.

Exercises 6.5

6.105 It is not practical to use the binomial probability formula when the number of trials, *n*, is large.

6.107 (a)   (i) $P(x = 4 \text{ or } 5) = 0.2051 + 0.2461 = 0.4512$

(ii) $P(3 \leq x \leq 7) = 0.1172 + 0.2051 + 0.2461 + 0.2051 + 0.1172 = 0.8907$

(b)   (i) $P(x = 4 \text{ or } 5)$:

Step 1: $n = 10$; $p = 0.5$

Step 2: $np = 5$; $n(1 - p) = 5$.   Since both np and $n(1 - p)$ are at least 5, the normal approximation can be used.

Step 3: $\mu_x = np = 5$; $\sigma_x = \sqrt{np(1-p)} = 1.58$

Step 4:   Making the continuity correction, we find the area under the normal curve with parameters $\mu = 5$ and $\sigma = 1.58$ that lies between 3.5 and 5.5.

z-score computations: Area to the left of z:

$$x = 3.5 \Rightarrow z = \frac{3.5-5}{1.58} = -0.95 \qquad 0.1711$$

$$x = 5.5 \Rightarrow z = \frac{5.5-5}{1.58} = 0.32 \qquad 0.6255$$

Total area = 0.6255 - 0.1711 = 0.4544

(ii) $P(3 \le x \le 7)$:

Steps 1, 2 and 3 are the same as above.

Step 4: $x = 2.5$ and $x = 7.5$

z-score computations: Area to the left of z:

$$x = 2.5 \Rightarrow z = \frac{2.5-5}{1.58} = -1.58 \qquad 0.0571$$

$$x = 7.5 \Rightarrow z = \frac{7.5-5}{1.58} = 1.58 \qquad 0.9429$$

Total area = 0.9429 - 0.0571 = 0.8858

6.109 If n = 30 and p remains 0.5,

$$\mu = np = 15 \quad and \quad \sigma = \sqrt{np(1-p)} = 2.74.$$

Thus, the normal curve used to approximate probabilities for the number of correct guesses is that with parameters $\mu = 15$ and $\sigma = 2.74$.

6.111 For parts (a), (b), and (c), steps 1-3 are as follows:

Step 1:   n = 200; p = 0.67

Step 2:   np = 134; n(1 - p) = 66. Since both np and n(1 - p) are at least 5, the normal approximation can be used.

Step 3: $\mu_x = np = 134$; $\sigma_x = \sqrt{np(1-p)} = 6.65$

(a) $P(x = 140)$:

Step 4:   x = 139.5 and x = 140.5

z-score computations: Area to the left of z:

$$x = 139.5 \Rightarrow z = \frac{139.5-134}{6.65} = 0.83 \qquad 0.7967$$

$$x = 140.5 \Rightarrow z = \frac{140.5-134}{6.65} = 0.98 \qquad 0.8365$$

Total area = 0.8365 - 0.7967 = 0.0398

(b) $P(120 \le x \le 130)$:

Step 4:   x = 119.5 and x = 130.5

z-score computations: Area to the left of z:

$$x = 119.5 \Rightarrow z = \frac{119.5 - 134}{6.65} = -2.18 \qquad 0.0146$$

$$x = 130.5 \Rightarrow z = \frac{130.5 - 134}{6.65} = -0.53 \qquad 0.2981$$

Total area = 0.2981 - 0.0146 = 0.2835

(c) $P(x \geq 150)$:

Step 4: $x = 149.5$

z-score computation: Area to the left of z:

$$x = 149.5 \Rightarrow z = \frac{149.5 - 134}{6.65} = 2.33 \qquad 0.9901$$

Total area = 1.0000 - 0.9901 = 0.0099

6.113 For parts (a), (b), and (c), steps 1-3 are as follows:

Step 1: $n = 500$; $p = 0.383$

Step 2: $np = 191.5$; $n(1 - p) = 308.5$. Since both np and $n(1 - p)$ are at least 5, the normal approximation can be used.

Step 3: $\mu_x = np = 191.5$; $\sigma_x = \sqrt{np(1-p)} = 10.87$

(a) $P(x = 200)$:

Step 4: $x = 199.5$ and $x = 200.5$

z-score computations: Area to the left of z:

$$x = 199.5 \Rightarrow z = \frac{199.5 - 191.5}{10.87} = 0.74 \qquad 0.7704$$

$$x = 200.5 \Rightarrow z = \frac{200.5 - 191.5}{10.87} = 0.83 \qquad 0.7967$$

Total area = 0.7967 - 0.7704 = 0.0263

(b) $P(180 \leq x \leq 210)$:

Step 4: $x = 179.5$ and $x = 210.5$

z-score computations: Area to the left of z:

$$x = 179.5 \Rightarrow z = \frac{179.5 - 191.5}{10.87} = -1.10 \qquad 0.1357$$

$$x = 210.5 \Rightarrow z = \frac{210.5 - 191.5}{10.87} = 1.75 \qquad 0.9599$$

Total area = 0.9599 - 0.1357 = 0.8242

(c) $P(x \leq 225)$:

Step 4:   x = 225.5

z-score computation: Area to the left of z:

$$x = 225.5 \rightarrow z = \frac{225.5 - 191.5}{10.87} = 3.13 \qquad 0.9991$$

Total area = 0.9991

6.115 (a)   100 bets are made. The gambler is ahead if she has won more than 50 bets. So, we are concerned with finding $P(x > 50)$ or $P(x \geq 51)$.

Step 1:   n = 100; p = 18/38 = 0.47368

Step 2:   np = 47.368; n(1 - p) = 52.632. Since both np and n(1-p) are at least 5, the normal approximation can be used.

Step 3: $\mu_x = np = 47.368$; $\sigma_x = \sqrt{np(1-p)} = 4.993$

Step 4:   x = 50.5

z-score computation: Area to the left of z:

$$x = 50.5 \rightarrow z = \frac{50.5 - 47.368}{4.993} = 0.63 \qquad 0.7357$$

Total area = 1.0000 - 0.7357 = 0.2643

(b)   1000 bets are made. The gambler is ahead if she has won more than 500 bets. So, we are concerned with finding $P(x > 500)$ or $P(x \geq 501)$.

Step 1:   n = 1000; p = 18/38 = 0.47368

Step 2:   np = 473.68; n(1 - p) = 526.32. Since both np and n(1-p) are at least 5, the normal approximation can be used.

Step 3: $\mu_x = np = 473.68$; $\sigma_x = \sqrt{np(1-p)} = 15.789$

Step 4:   x = 500.5

z-score computation: Area to the left of z:

$$x = 500.5 \rightarrow z = \frac{500.5 - 473.68}{15.789} = 1.70 \qquad 0.9554$$

Total area = 1.0000 - 0.9554 = 0.0466

(c)   5000 bets are made. The gambler is ahead if she has won more than 2500 bets. So, we are concerned with finding $P(x > 2500)$ or $P(x \geq 2501)$.

Step 1:   n = 5000; p = 18/38 = 0.47368

Step 2:   np = 2368.4; n(1 - p) = 2631.6. Since both np and n(1-p) are at least 5, the normal approximation can be used.

154   CHAPTER 6

Step 3: $\mu_x = np = 2368.4;\quad \sigma_x = \sqrt{np(1-p)} = 35.306$

Step 4: $x = 2500.5$

z-score computation: Area to the left of z:

$$x = 2500.5 \Rightarrow z = \frac{2500.5 - 2368.4}{35.306} = 3.74 \qquad 0.9999$$

Total area = $1.0000 - 0.9999 = 0.0001$

REVIEW TEST FOR CHAPTER 6

1. Two primary reasons for studying the normal distribution are:
    (a)   it is often appropriate to use the normal distribution as the distribution of a population or random variable.
    (b)   the normal distribution is frequently employed in inferential statistics.

2. (a)   A variable is normally distributed if its distribution has the shape of a normal curve.
    (b)   A population is normally distributed if a variable of the population is normally distributed and it is the only variable under consideration.
    (c)   The parameters for a normal curve are the mean μ and the standard deviation σ.

3. (a)   False. There are many types of distributions that could have the same mean and standard deviation.
    (b)   True. The mean and standard deviation completely determine a normal distribution, so if two normal distributions have the same mean and standard deviation, then those two distributions are identical.

4. The percentages for a normally distributed variable and areas under the corresponding normal curve (expressed as a percentage) are identical.

5. The distribution of the standardized version of a normally distributed variable is the standard normal distribution, that is, a normal distribution with a mean of 0 and standard deviation of 1.

6. (a)   True.
    (b)   True.

7. (a)   The (second) curve with $\sigma = 6.2$ has the largest spread.
    (b)   The first and second curves are centered at $\mu = 1.5$.
    (c)   The first and third curves have the same shape because σ is the same for both.
    (d)   The third curve is centered farthest to the left because it has the smallest value of μ.
    (e)   The fourth curve is the standard normal curve because $\mu = 0$ and $\sigma = 1$.

8. Key fact 6.2.

9. (a)   The table entry corresponding to the specified z-score is the area to the left of that z-score.
    (b)   The area to the right of a specified z-score is found by subtracting the table entry from 1.

(c)   The area between two specified z-scores is found by subtracting the table entry for the smaller z-score from the table entry for the larger z-score.

10.   (a)   Find the table entry that is closest to the specified area. The z-score determined by locating the corresponding marginal values is the z-score that has the specified area to its left.

(b)   Subtract the specified area from 1. Find the entry in the table that is closest to the result of the subtraction. The z-score determined by locating the corresponding marginal values is the z-score that has the specified area to its right.

11. Z_α is the z-score that has area α to its right under the standard normal curve.

12.   The 68.26-95.44-99.74 rule states that for a normally distributed variable: 68.26% of all possible observations lie within one standard deviation to either side of the mean; 95.44% of all possible observations lie within two standard deviations to either side of the mean; 99.74% of all possible observations lie within three standard deviation to either side of the mean.

13.   The normal scores for a sample of observations are the observations we would expect to get for a sample of the same size for a variable having the standard normal distribution.

14.   If we observe the values of a normally distributed variable for a sample, then a normal probability plot should be roughly <u>linear</u>.

15.

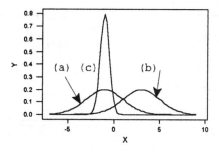

16.   (a)

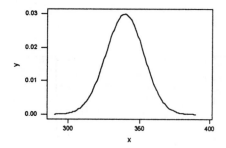

(b) $z = (x - 339.6)/13.3$

(c)

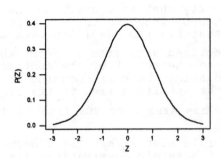

(d) $P(320 \le x \le 360) = 0.8672$

(e)   The percentage of Morgan mares that have gestation periods exceeding 369 days equals the area under the standard normal curve that lies to the <u>right</u> of <u>2.21</u>.

17.   (a)   The area to the right of 1.05 is $1 - 0.8531 = 0.1469$.

(b)   The area to the left of −1.05 is 0.1469.

(c)   The area between −1.05 and 1.05 is $0.8531 - 0.1469 = 0.7062$.

18.   (a)   Area = 0.0013   (b)   Area = $1 - 0.7291 = 0.2709$

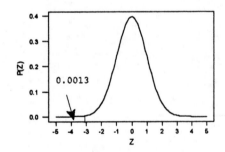

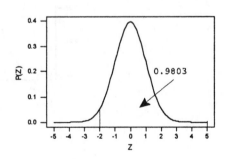

(c)   Area = $0.9970 - 0.8665 = 0.1305$ (d)   Area = $1.000 - 0.0197$
                                                                $= 0.9803$

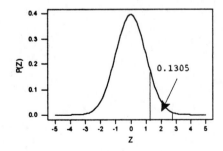

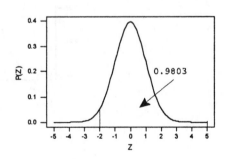

(e)   Area = 0.0668 - 0.0000 = 0.0668   (f) Area = 0.8413 + (1 - 0.9987) = 0.8426

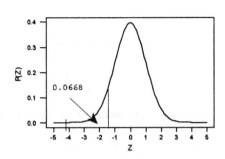

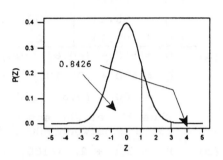

19.   (a) $z = -0.52$

    (b)   z will have 0.9 to its left: $z = 1.28$

    (c) $z_{0.025} = 1.96$; $z_{0.05} = 1.645$; $z_{0.01} = 2.33$; $z_{0.005} = 2.575$

    (d)   -2.575 and +2.575

20.   (a)   **Between 350 and 625:**

        z-score computations: **Area to the left of z:**

$$x = 350 \Rightarrow z = \frac{350 - 500}{100} = -1.50 \qquad 0.0668$$

$$x = 625 \Rightarrow z = \frac{625 - 500}{100} = 1.25 \qquad 0.8944$$

        Total area = 0.8944 - 0.0668 = 0.8276 = 82.76%

    (b)   **375 or greater:**

        z-score computation: **Area to the left of z:**

$$x = 375 \Rightarrow z = \frac{375 - 500}{100} = -1.25 \qquad 0.1056$$

        Total area = 1.0000 - 0.1056 = 0.8944 = 89.44%

    (c)   **Below 750:**

        z-score computation: **Area to the left of z:**

$$x = 750 \Rightarrow z = \frac{750 - 500}{100} = 2.5 \qquad 0.9938$$

        Total area = 0.9938 = 99.38%

21.

Part	Percent	Lower bound	Upper bound
(a)	68.26	500 - 1(100) = 400	500 + 1(100) = 600
(b)	95.44	500 - 2(100) = 300	500 + 2(100) = 700
(c)	99.74	500 - 3(100) = 200	500 + 3(100) = 800

22. (a) $Q_1 = 500 + (-0.67)(100) = 433$

$Q_2 = 500 + (0)(100) = 500$

$Q_3 = 500 + (0.67)(100) = 567$

Thus 25% of GRE scores are below 433, 25% are between 433 and 500, 25% are between 500 and 567, and 25% are above 567.

(b) $P_{99} = 500 + 2.33(100) = 733$

Thus, 99% of GRE scores are below 733 and 1% are above 733.

23. (a)

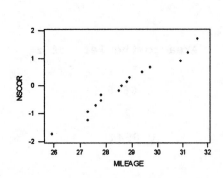

Mileage x	Normal Score y
25.9	-1.74
27.3	-1.24
27.3	-0.94
27.6	-0.71
27.8	-0.51
27.8	-0.33
28.5	-0.16
28.6	0.0
28.8	0.16
28.9	0.33
29.4	0.51
29.7	0.71
30.9	0.94
31.2	1.24
31.6	1.74

(b) Based on the probability plot, there do not appear to be any outliers in the sample.

(c) Based on the probability plot, the sample appears to be from an approximately normally distributed population.

24. For parts (a), (b), and (c), steps 1-3 are as follows:

Step 1: $n = 1500$; $p = 0.80$

Step 2: $np = 1200$; $n(1 - p) = 300$. Since both np and $n(1 - p)$ are at least 5, the normal approximation can be used.

Step 3:

$$\mu_x = np = 1200; \quad \sigma_x = \sqrt{np(1-p)} = 15.49$$

(a) $P(x = 1225)$:

Step 4:   x = 1224.5 and x = 1225.5

   z-score computations: Area to the left of z:

$$x = 1224.5 \Rightarrow z = \frac{1224.5 - 1200}{15.49} = 1.58$$ 0.9429

$$x = 1225.5 \Rightarrow z = \frac{1225.5 - 1200}{15.49} = 1.65$$ 0.9505

Total area = 0.9505 - 0.9429 = 0.0076

(b)   P(x ≥ 1175):

Step 4:   x = 1174.5

   z-score computation: Area to the left of z:

$$x = 1174.5 \Rightarrow z = \frac{1174.5 - 1200}{15.49} = -1.65$$ 0.0495

Total area = 1.0000 - 0.0495 = 0.9505

(c)   P(1150 ≤ x ≤ 1250):

Step 4:   x = 1149.5 and x = 1250.5

   z-score computations: Area to the left of z:

$$x = 1149.5 \Rightarrow z = \frac{1149.5 - 1200}{15.49} = -3.26$$ 0.0006

$$x = 1250.5 \Rightarrow z = \frac{1250.5 - 1200}{15.49} = 3.26$$ 0.9994

Total area = 0.9994 - 0.0006 = 0.9988

25.   (a)

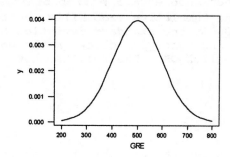

(b)   We will generate a column of 1000 observations by choosing **Calc ▶**
       Random data ▶ Normal..., typing <u>1000</u> in the **Generate rows of data**
       text box, clicking in the **Store in column(s)** text box and typing
       <u>GRE</u>, clicking in the **Mean** text box and typing <u>500</u>, clicking in the
       Standard deviation text box and typing <u>100</u>, and clicking **OK**.

(c)   We would expect the mean to be about 500 and the standard
       deviation to be about 100 since the sample is expected to reflect
       the characteristics of the population from which it is drawn.

(d)   Choose **Calc ▶ Column statistics...**, click on the **Mean** button, and
       select GRE in the **Input variable:** text box , and click **OK**. Then
       repeat the process, clicking on the **Standard deviation** button.
       The results are

       Mean of GRE = 500.69

       Standard deviation of GRE = 96.739

(e)   The histogram of the 1000 GRE scores should look roughly like a
       bell-shaped curve centered at 500 and most of the observations
       should be between 200 and 800.

(f)   Choose **Graph ▶ Histogram...**, select GRE for **Graph 1** for **X** and
       click **OK**. The result is

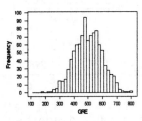

26.   Using Minitab, we first enter all of the values in parts a,b, and c of
       Exercise 20 (350, 625, 375, and 750) in a column (say C23) of the
       Minitab worksheet. Then choose **Calc ▶ Probability distributions ▶**
       Normal..., click on the **Cumulative Probability** button, type <u>500</u> in the
       Mean test box, type <u>100</u> in the **Standard deviation** text box, click on the
       Input column text box and type <u>C23</u>, click in the **Optional storage** text
       box and type <u>C24</u>, and click **OK**. The results shown in C23 and C24 are

       350   0.066807
       625   0.894350
       375   0.105650
       750   0.993790

The right hand column gives the probability that the GRE score is less
than the value in the left hand column.

(a)   Thus P(350 ≤ GRE < 625) = 0.894350 − 0.066807 = 0.827543 (82.75%)

(b)   P(GRE > 375) = 1 − 0.105650 = 0.895450 (89.545%)

(c)   P(GRE < 750) = 0.993790 (99.38%)

27.   Using Minitab, we first enter all of the percentile values from parts a and b of Exercise 22 (.25, .5, .75, and .99) in a column (say C25) of

the Minitab worksheet. Then choose **Calc ▶ Probability distributions ▶ Normal...**, click on the **Inverse Cumulative Probability** button, type <u>500</u> in the **Mean** test box, type <u>100</u> in the **Standard deviation** text box, click on the **Input column** text box and type <u>C25</u>, click in the **Optional storage** text box and type <u>C26</u>, and click **OK**. The results shown in C25 and C26 are

        0.25   432.551

        0.50   500.000

        0.75   567.449

        0.99   732.635

(a)   The first quartile is 432.551, the second quartile is 500, and the third quartile is 567.449. This means that 25% of the observations are below 432.551, 50% are below 500, and 75% are below 567.449.

(b)   The 99th percentile is 732.635. This means that 99% of the observations are less than 732.635.

28.   Using Minitab with the data in a column named MPG, we choose **Calc ▶ Calculator...**, click in the **Store result in variable:** text box and type <u>NSCORMPG</u>, click in the **Functions** box and select **Normal scores**, then

select MPG for **number**, and click **OK**. Then choose **Graph ▶ Plot...**, select NSCORMPG for the **Y** variable for **Graph 1** and MPG for the **X** variable, and click **OK**.

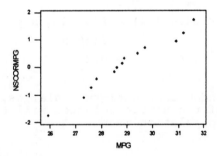

Minitab plots the identical points in the same place so that it appears that only 13 of the 15 data points are present. In the first graph (Exercise 23), the identical mileage values are plotted with different normal scores and aligned vertically.

Exercises 7.1

7.1 Sampling is often preferable to conducting a census because it is quicker, less costly, and sometimes it is the only practical way to get information.

7.3 (a) $\mu = \Sigma x/N = 322/5 = \$64.4$ thousand.

(b)

Sample	Salaries	$\overline{x}$
G,L	70, 63	66.5
G,S	70, 44	57.0
G,A	70, 75	72.5
G,T	70, 70	70.0
L,S	63, 44	53.5
L,A	63, 75	69.0
L,T	63, 70	66.5
S,A	44, 75	59.5
S,T	44, 70	57.0
A,T	75, 70	72.5

(c)

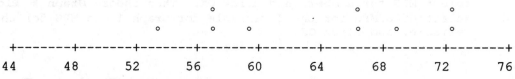

```
                                   o                   o            o
                       o        o     o             o     o      o
        +-------+-------+-------+-------+-------+-------+-------+-------+
        44      48      52      56      60      64      68      72      76
```

(d) $P(\overline{x} = \mu) = P(\overline{x} = 64.4) = 0.0$

(e) $P(64.4 - 4 \leq \overline{x} \leq 64.4 + 4) = P(60.4 \leq \overline{x} \leq 68.4)$
$$= P(66.5)$$
$$= 0.2$$

If we take a random sample of two salaries, there is a 20% chance that the mean of the sample selected will be within four (that is, $4,000) of the population mean.

7.5   (b)

Sample	Salaries	$\overline{x}$
G,L,S	70,63,44	59.00
G,L,A	70,63,75	69.33
G,L,T	70,63,70	67.67
G,S,A	70,44,75	63.00
G,S,T	70,44,70	61.33
G,A,T	70,75,70	71.67
L,S,A	63,44,75	60.67
L,S,T	63,44,70	59.00
L,A,T	63,75,70	69.33
S,A,T	44,75,70	63.00

(c)

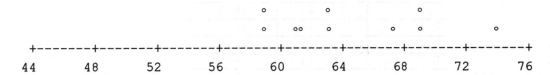

(d) $P(\overline{x} = \mu) = P(\overline{x} = 64.4) = 0.0$

(e) $P(64.4 - 4 \leq \overline{x} \leq 64.4 + 4) = P(60.4 \leq \overline{x} \leq 68.4)$

   $= P(60.67) + P(61.33) + P(63) + P(67.67)$

   $= 0.1 + 0.1 + 0.2 + 0.1 = 0.5$

   If we take a random sample of three salaries, there is a 50% chance that the mean of the sample selected will be within four (i.e., $4,000) of the population mean.

7.7   (b)

Sample	Salaries	$\overline{x}$
G,L,S,A,T	70,63,44,75,70	64.40

(c)

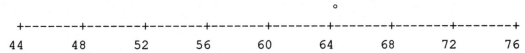

(d) $P(\overline{x} = \mu) = P(\overline{x} = 64.4) = 0$

(e) $P(64.4 - 4 \leq \overline{x} \leq 64.4 + 4) = P(60.4 \leq \overline{x} \leq 68.4) = P(64.40) = 1.0$

   If we take a random sample of five salaries, there is a 100% chance that the mean of the sample selected will be within four (i.e., $4,000) of the population mean.

7.9   (a) $\mu = \Sigma x / N = 90/6 = 15$ cm

(b)

Sample	Lengths	$\overline{x}$
A,B	19, 14	16.5
A,C	19, 15	17.0
A,D	19, 9	14.0
A,E	19, 16	17.5
A,F	19, 17	18.0
B,C	14, 15	14.5
B,D	14, 9	11.5
B,E	14, 16	15.0
B,F	14, 17	15.5
C,D	15, 9	12.0
C,E	15, 16	15.5
C,F	15, 17	16.0
D,E	9, 16	12.5
D,F	9, 17	13.0
E,F	16, 17	16.5

(c)

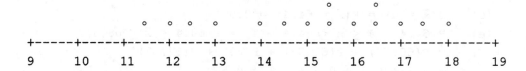

```
                                    o           o
              o   o   o   o       o   o   o   o   o   o   o
        +-----+-----+-----+-----+-----+-----+-----+-----+-----+-----+
        9    10    11    12    13    14    15    16    17    18    19
```

(d) $P(\overline{x} = \mu) = P(\overline{x} = 15) = 1/15 = 0.067$

(e) $P(15 - 1 \le \overline{x} \le 15 + 1) = P(14 \le \overline{x} \le 16)$

$$= P(14) + P(14.5) + P(15) + P(15.5) + P(16)$$
$$= 1/15 + 1/15 + 1/15 + 2/15 + 1/15 = 6/15$$
$$= 0.4$$

If we take a random sample of two bullfrogs, there is a 40% chance
that their mean length will be within one cm of the population
mean length.

7.11 (b)

Sample	Lengths	$\overline{x}$
A,B,C	19, 14, 15	16.0
A,B,D	19, 14, 9	14.0
A,B,E	19, 14, 16	16.3
A,B,F	19, 14, 17	16.7
A,C,D	19, 15, 9	14.3
A,C,E	19, 15, 16	16.7
A,C,F	19, 15, 17	17.0
A,D,E	19, 9, 16	14.7
A,D,F	19, 9, 17	15.0
A,E,F	19, 16, 17	17.3
B,C,D	14, 15, 9	12.7
B,C,E	14, 15, 16	15.0
B,C,F	14, 15, 17	15.3
B,D,E	14, 9, 16	13.0
B,D,F	14, 9, 17	13.3
B,E,F	14, 16, 17	15.7
C,D,E	15, 9, 16	13.3
C,D,F	15, 9, 17	13.7
C,E,F	15, 16, 17	16.0
D,E,F	9, 16, 17	14.0

(c)

```
                         o     o       o      o   o
               o  o  o  o  o  o  o  o  o  o  o  o  o  o
      +-----+-----+-----+-----+-----+-----+-----+-----+-----+-----+
      9    10    11    12    13    14    15    16    17    18    19
```

(d) $P(\overline{x} = \mu) = P(\overline{x} = 15) = 2/20 = 0.10$

(e) $P(15 - 1 \le \overline{x} \le 15 + 1) = P(14 \le \overline{x} \le 16)$

$= P(14) + P(14.3) + P(14.7) + P(15) + P(15.3) + P(15.7) + P(16)$
$= 2/20 + 1/20 + 1/20 + 2/20 + 1/20 + 1/20 + 2/20 = 10/20$

$= 0.50$

If we take a random sample of three bullfrogs, there is a 50% chance that their mean length will be within one cm of the population mean length.

7.13 (b)

Sample	Lengths	$\overline{x}$
A,B,C,D,E	19, 14, 15, 9, 16	14.6
A,B,C,D,F	19, 14, 15, 9, 17	14.8
A,B,C,E,F	19, 14, 15, 16, 17	16.2
A,B,D,E,F	19, 14, 9, 16, 17	15.0
A,C,D,E,F	19, 15, 9, 16, 17	15.2
B,C,D,E,F	14, 15, 9, 16, 17	14.2

(c)

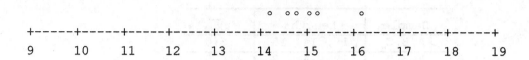

```
                                     o  oo oo        o
        +-----+-----+-----+-----+-----+-----+-----+-----+-----+-----+
        9    10    11    12    13    14    15    16    17    18    19
```

(d) $P(\overline{x} = \mu) = P(\overline{x} = 15) = 1/6 = 0.167$

(e) $P(15 - 1 \leq \overline{x} \leq 15 + 1) = P(14 \leq \overline{x} \leq 16)$

$= P(14.2) + P(14.6) + P(14.8) + P(15) + P(15.2)$

$= 1/6 + 1/6 + 1/6 + 1/6 + 1/6 = 5/6 = 0.833$

If we take a random sample of five bullfrogs, there is an 83.3% chance that their mean length will be within one cm of the population mean length.

7.15   Increasing the sample size tends to reduce the sampling error.

7.17   (a)   If a sample of size n = 1 is taken from a population of size N, there are N possible samples.

(b)   Since each sample mean is based upon a single observation, the possible $\overline{x}$-values and the population values are the same.

(c)   There is no difference between taking a random sample of size n = 1 from a population and selecting a member at random from the population.

Exercises 7.2

7.19   Obtaining the mean and standard deviation of $\overline{x}$ is a first step in approximating the sampling distribution of the mean by a normal distribution because the normal distribution is completely determined by its mean and standard deviation.

7.21   Yes. The spread of the distribution of $\overline{x}$ gets smaller as the sample size increases. Since that spread is measured by the standard deviation of $\overline{x}$, the standard deviation also gets smaller.

7.23   Standard error of the mean. The standard deviation of $\overline{x}$ determines the amount of sampling error to be expected when a population mean is estimated by a sample mean.

7.25   (a) $\mu = \Sigma x/N = 322/5 = 64.4$

(b) $\mu_{\overline{x}} = \dfrac{\sum \overline{x}}{N} = \dfrac{66.5 + 57.0 + \ldots + 72.5}{10} = 64.4$

(c) $\mu_{\overline{x}} = \mu = 64.4$

7.27   (b) $\mu_{\overline{x}} = \dfrac{\sum \overline{x}}{N} = \dfrac{59.00 + 69.33 + \ldots + 63.00}{10} = 64.4$

(c) $\mu_{\overline{x}} = \mu = 64.4$

7.29   (b) $\mu_{\bar{x}} = \dfrac{\sum \bar{x}}{N} = \dfrac{64.4}{1} = 64.4$

(c) $\mu_{\bar{x}} = \mu = 64.4$

7.31   (a) $\mu_{\bar{x}} = \mu = 5.8 \; days; \quad \sigma_{\bar{x}} = \sigma/\sqrt{n} = 4.3/\sqrt{75} = 0.50 \; day$

(b) $\mu_{\bar{x}} = 5.8 \; days; \quad \sigma_{\bar{x}} = 4.3/\sqrt{500} = 0.19 \; day$

7.33   (a) $\mu_{\bar{x}} = \mu = 8.5 \; years; \quad \sigma_{\bar{x}} = \sigma/\sqrt{n} = 2.6/\sqrt{50} = 0.37 \; year$

(b) $\mu_{\bar{x}} = 8.5 \; years; \quad \sigma_{\bar{x}} = 2.6/\sqrt{200} = 0.18 \; year$

7.35   (a)

$$\sigma_x = \sqrt{\dfrac{\sum x^2}{N} - \mu_x^2}$$

$$= \sqrt{\dfrac{70^2 + 63^2 + \ldots + 70^2}{5} - 64.4^2} = \sqrt{4266.0 - 64.4^2} = 10.89$$

(b)

$$\sigma_{\bar{x}} = \sqrt{\dfrac{\sum \bar{x}^2}{N} - \mu_x^2}$$

$$= \sqrt{\dfrac{66.5^2 + 57.0^2 + \ldots + 72.5^2}{10} - 64.4^2} = \sqrt{4191.85 - 64.4^2} = 6.67$$

(c)   Formula (1) is appropriate here because the sample size is not small relative to the population size. The sample size is 40% of the population size. By rule of thumb, the correction factor cannot be ignored if the sample size is larger than 5% of the population size.

(d) $\sigma_{\bar{x}} = \sqrt{\dfrac{5-2}{5-1}} \cdot \dfrac{10.89}{\sqrt{2}} = 6.67$

This is identical to the result in part (b).

(e) $\sigma_{\bar{x}} = 10.89/\sqrt{2} = 7.70$

Formula (2) yields such a poor approximation because the sample size is not small relative to the population size.

7.37   (b)

$$\sigma_{\bar{x}} = \sqrt{\frac{\sum \bar{x}^2}{N} - \mu_x^2}$$

$$= \sqrt{\frac{59.00^2 + 69.33^2 + \ldots + 63.00^2}{10} - 64.4^2} = \sqrt{4167.134 - 64.4^2} = 4.447$$

(c)   Formula (1) is appropriate here because the sample size is not small relative to the population size. The sample size is 60% of the population size. By rule of thumb, the correction factor cannot be ignored if the sample size is larger than 5% of the population size.

(d) $\sigma_{\bar{x}} = \sqrt{\frac{5-3}{5-1} \cdot \frac{10.89}{\sqrt{3}}} = 4.446$

This matches the result in part (b). They are slightly different due to rounding error.

(e) $\sigma_{\bar{x}} = 10.89/\sqrt{3} = 6.29$

Formula (2) yields such a poor approximation because the sample size is not small relative to the population size.

7.39   (b)

$$\sigma_{\bar{x}} = \sqrt{\frac{\sum \bar{x}^2}{N} - \mu_x^2}$$

$$= \sqrt{\frac{64.4^2}{1} - 64.4^2} = \sqrt{4147.36 - 64.4^2} = 0$$

(c)   Formula (1) is appropriate here because the sample size is not small relative to the population size. The sample size is 100% of the population size. By rule of thumb, the correction factor cannot be ignored if the sample size is larger than 5% of the population size.

(d) $\sigma_{\bar{x}} = \sqrt{\dfrac{5-5}{5-1} \cdot \dfrac{10.89}{\sqrt{5}}} = 0$

(e) $\sigma_{\bar{x}} = 10.89/\sqrt{5} = 4.87$

Formula (2) yields such a poor approximation because the sample size is not small relative to the population size.

7.41 (a)   Since n $\geq$ 1, N $-$ n $\leq$ N $-$ 1. Also, if n $\leq$ 0.05N, then N $-$ n $\geq$ N $-$ 0.05N = 0.95N. Therefore,

$$\sqrt{\frac{N-n}{N-1}} \leq \sqrt{\frac{N-1}{N-1}} = 1$$

and if n $\leq$ 0.05N,

$$\sqrt{\frac{N-n}{N-1}} \geq \sqrt{\frac{0.95N}{N-1}} \geq \sqrt{\frac{0.95N}{N}} = \sqrt{0.95} = 0.97 \quad .$$

Consequently, if n $\leq$ 0.05N,

$$0.97 \leq \sqrt{\frac{N-n}{N-1}} \leq 1 \quad .$$

(b)   If n $\leq$ 0.05N, then by part (a),

$$0.97 \cdot \frac{\sigma}{\sqrt{n}} \leq \sqrt{\frac{N-n}{N-1}} \cdot \frac{\sigma}{\sqrt{n}} \leq \frac{\sigma}{\sqrt{n}}$$

This shows that there is very little difference in the values given by Formulas (1) and (2) when the sample size is no larger than 5% of the population size.

7.43 No. For example, in Example 7.2, the population consists of five observations (76, 78, 79, 81, and 86). The population median is 79. For samples of size 2, the median and mean are identical and therefore have the same sampling distribution. The mean of the sampling distribution of the median equals the mean of the sampling distribution of the mean which is shown to be 80 in the example. Since this is not equal to the population median, it is clear that, in general, the sample median is not an unbiased estimator or the population median.

7.45 (a)   Theoretically, the mean of possible sample means is 266 and the standard deviation is $16/\sqrt{9}$ = 5.33.

(b)   Follow instructions.

(c)   Follow instructions.

(d)   We would expect the mean of the 2000 sample means to be roughly 266 and the standard deviation of the sample means to be about 5.33 since we are taking a sample of 2000 means from a theoretical sampling distribution with mean and standard deviation given in part (a).

(e)   The mean of our 2000 means was 266.84 and the standard deviation was 5.37.

(f)   The answers in part (d) differ from the theoretical values given in part (c) as a result of random error or sampling variability.

Exercises 7.3

7.47 (a)   The sampling distribution of the mean is approximately normally distributed with mean

$\mu_{\bar{x}} = 100$ and standard deviation $\sigma_{\bar{x}} = 28/\sqrt{49} = 4$.

(b)   No assumptions were made about the distribution of the population.

(c)   Part (a) cannot be answered if the sample size is n = 16. Since the distribution of the population is not specified, we need a sample size of at least 30 to apply Key Fact 7.6.

7.49 (a)   The probability distribution of $\bar{x}$ is normal with mean μ and standard deviation $\sigma/\sqrt{n}$.

(b)   The answer to part (a) does not depend on how large the sample size is because the population being sampled is normally distributed.

(c)   The mean of $\bar{x}$ is $\mu_{\bar{x}} = \mu$; its standard deviation is $\sigma_{\bar{x}} = \sigma/\sqrt{n}$.

(d)   No.

7.51 (a)   All four graphs are centered at the same place because $\mu_{\bar{x}} = \mu$ and because normal curves are centered at their μ-parameter.

(b)   Since $\sigma_{\bar{x}} = \sigma/\sqrt{n}$, we see that $\sigma_{\bar{x}}$ decreases as n increases. This results in a diminishing of the spread, because the spread of a distribution is determined by its σ-parameter. As a consequence, we see that the larger the sample size, the greater the likelihood for small sampling error.

(c)   The graphs in Figure 7.6(a) are bell-shaped because, for normally distributed populations, the random variable $\bar{x}$ is normally distributed (regardless of the sample size).

(d)   The graphs in Figures 7.6(b) and 7.6(c) become bell-shaped as the sample size increases because of the central limit theorem; the probability distribution of $\bar{x}$ tends to a normal distribution as the sample size increases.

7.53 (a)   Because the times themselves are normally distributed, the sampling distribution for means of samples of size 4 will also be normal and will have mean $\mu_{\bar{x}} = \mu = 61$ and standard deviation

$\sigma_{\bar{x}} = \sigma/\sqrt{n} = 9/\sqrt{4} = 4.5$. Thus the distribution of the possible sample means for samples of four finishing times will be normal with mean 61 minutes and standard deviation 4.5 minutes.

(b)   Because the times themselves are normally distributed, the sampling distribution for means of samples of size 9 will also be normal and will have mean $\mu_{\bar{x}} = \mu = 61$ and standard deviation

$\sigma_{\bar{x}} = \sigma/\sqrt{n} = 9/\sqrt{9} = 3.0$. Thus the distribution of the possible sample means for samples of nine finishing times will be normal with mean 61 minutes and standard deviation 3.0 minutes.

(c)   To facilitate the comparison of the three graphs, we have overlaid them on one set of axes.

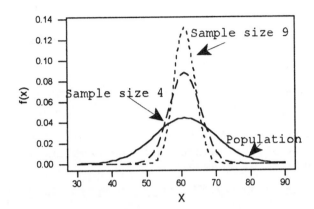

7.55 (a) Because the sample size is large, the sampling distribution for means of samples of size 100 will be approximately normal and will have mean $\mu_{\bar{x}} = \mu = \$8657$ and standard deviation

$\sigma_{\bar{x}} = \sigma/\sqrt{n} = 7500/\sqrt{100} = 750$. Thus the distribution of the

possible sample means for samples of 100 alimony incomes received by women will be approximately normal with mean $8657 and standard deviation $750.

(b) Because the sample size is large, the sampling distribution for means of samples of size 1000 will be approximately normal and will have mean $\mu_{\bar{x}} = \mu = \$8657$ and standard deviation

$\sigma_{\bar{x}} = \sigma/\sqrt{n} = 7500/\sqrt{1000} = \$237.17$. Thus the distribution of the

possible sample means for samples of 1000 alimony incomes received by women will be approximately normal with mean $8657 and standard deviation $237.17.

(c) Since the sample size is at least 30, it is not necessary to assume that the population of annual alimony payments is normally distributed.

7.57 (a)   (a) $\sigma_{\bar{x}} = 9/\sqrt{4} = 4.5;$ $P(56 \leq \bar{x} \leq 66):$

z-score computations: Area less than z:

$\bar{x} = 56 \rightarrow z = \dfrac{56-61}{4.5} = -1.11$ 0.1335

$\bar{x} = 66 \rightarrow z = \dfrac{66-61}{4.5} = 1.11$ 0.8665

Total area = 0.8665 - 1335 = 0.7330

There is a 73.30% chance that the sampling error will be less than 5 for samples of size 4.

(b)   (a) $\sigma_{\overline{x}} = 9/\sqrt{9} = 3.0$; $P(56 \leq \overline{x} \leq 66)$:

z-score computations: Area less than z:

$$\overline{x} = 56 \rightarrow z = \frac{56-61}{3.0} = -1.67 \qquad\qquad 0.0475$$

$$\overline{x} = 66 \rightarrow z = \frac{66-61}{3.0} = 1.67 \qquad\qquad 0.9525$$

Total area = 0.9525 - 0.0475 = 0.9050

There is a 90.50% chance that the sampling error will be less than 5 for samples of size 9.

7.59   (a) $\sigma_{\overline{x}} = 7500/\sqrt{100} = 750$; $P(8157 \leq \overline{x} \leq 9157)$:

z-score computations: Area less than z:

$$\overline{x} = 8,157 \rightarrow z = \frac{8157-8657}{750} = -0.67 \qquad\qquad 0.2514$$

$$\overline{x} = 9,157 \rightarrow z = \frac{9157-8657}{750} = 0.67 \qquad\qquad 0.7486$$

Total area = 0.7486 - 0.2514 = 0.4972

(b) $n = 1000$; $\sigma_{\overline{x}} = 7500/\sqrt{1000} = 237.17$; $P(8157 \leq \overline{x} \leq 9157)$:

z-score computations: Area less than z:

$$\overline{x} = 8,157 \rightarrow z = \frac{8157-8657}{237.17} = -2.11 \qquad\qquad 0.0174$$

$$\overline{x} = 9,157 \rightarrow z = \frac{9157-8657}{237.17} = 2.11 \qquad\qquad 0.9826$$

Total area = 0.9826 - 0.0174 = 0.9652

7.61 $\sigma_{\overline{x}} = 17.20/\sqrt{500} = 0.7692$; $P(\mu - 1 \leq \overline{x} \leq \mu + 1)$:

z-score computations: Area less than z:

$$\overline{x} = \mu - 1 \rightarrow z = \frac{(\mu - 1) - \mu}{0.7692} = -1.30 \qquad\qquad 0.0968$$

$$\overline{x} = \mu + 1 \rightarrow z = \frac{(\mu + 1) - \mu}{0.7692} = 1.30 \qquad\qquad 0.9032$$

Total area = 0.9032 - 0.0968 = 0.8064

7.63 (a) $\sigma/\sqrt{n}$ (b) $2\sigma/\sqrt{n}$ (c) $3\sigma/\sqrt{n}$

(d) $z_{\alpha/2} \cdot \sigma/\sqrt{n}$

7.65 $\mu = 40;$ $\sigma = 1.5;$ $\sigma_{\overline{x}} = 1.5/\sqrt{10} = 0.474$

(a) P(x $\leq$ 39):
z-score computation: Area less than z:

$$x = 39 \rightarrow z = \frac{39-40}{1.5} = -0.67$$ 0.2514

Total area = 0.2514

(b) P($\overline{x} \leq$ 39):
z-score computation: Area less than z:

$$\overline{x} = 39 \rightarrow z = \frac{39-40}{0.474} = -2.11$$ 0.0174

Total area = 0.0174

(c) If I bought an *individual* bag of water-softener salt and it weighed 39 lb, I *would not* consider this to be enough evidence to refute the company's claim that the bags contain an average of 40 lb of salt. The result in part (a) tells me that I can expect about a (fairly high) 25% chance that an individual bag will weigh 39 lb or less.

(d) If I bought 10 bags of water-softener salt and their *mean* weight was 39 lb, I *would* consider this to be enough evidence to refute the company's claim that the bags contain an average of 40 lb of salt. The result in part (b) tells me that there is only a 1.74% chance that 10 randomly selected bags will have a mean weight of 39 lb or less. Perhaps the true mean is not 40 lb, but something less than 40 lb.

7.67 (a) We have Minitab take 2000 random samples of size n = 9 from a normally distributed population with mean 266 and standard

deviation 16 by choosing **Calc ▶ Random Data ▶ Normal...**, typing 2000 in the **Generate rows of data** text box, typing C1-C9 in the **Store in column(s)** text box, typing 266 in the **Mean** text box and 16 in the **Standard deviation** text box, and clicking **OK**. These commands tell Minitab to place a total of 18000 observations from the population into columns C1-C9, with 2000 observations in each column. Then, our first random sample of size n = 9 is the first row of columns C1-C9, our second random sample of size n = 9 is the second row of columns C1-C9, and so on.

(b) We compute the sample mean of each of the 2000 samples by choosing

Calc ▶ Row statistics..., clicking on the **Mean** button, typing C1-C9 in the **Input variables** text box and XBAR in the **Store result in** text box, and clicking **OK**. This command instructs Minitab to compute the means of the 2000 rows of C1-C9 and to place those means in a column named XBAR. Thus, the 2000 $\overline{x}$-values are now in XBAR. (We will not print these values.)

(c) The mean is obtained by choosing **Calc ▶ Column statistics...**, clicking on the **Mean** button, selecting XBAR in the **Input variable**

text box, and clicking **OK**.   The result is
Mean of XBAR = 266.03

Similarly, the standard deviation is obtained by choosing **Calc ▶ Column statistics...**, clicking on the **Standard deviation** button, selecting XBAR in the **Input variable** text box, and clicking **OK**. The result is
Standard deviation of XBAR = 5.2753
To get a histogram of the 2000 sample means stored in XBAR, choose

Graph ▶ Histogram..., select XBAR for the **X** variable for **Graph 1**, and click **OK**.   The result is

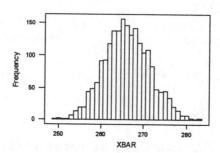

(d)   Theoretically, the distribution of all possible sample means for samples of size nine from a normal population should have mean 266 days, standard deviation $\sigma / \sqrt{n} = 16 / \sqrt{9} = 5.33$, and a normal distribution.

(e)   The histogram is close to bell-shaped, is centered near 266, and most of the data lies within three standard deviations (3 x 5.33 = 16) of 266, i.e. between 250 and 282. The mean of the 2000 sample means is 266.03, very close to 266, and the standard deviation of the sample means is 5.27, very close to 5.33.

REVIEW TEST FOR CHAPTER 7

1.   Errors that result from sampling because sampling provides information for only a portion of a population are called sampling errors.

2.   The sampling distribution of a statistic is the set of all possible observations of the statistic for a sample of a given size.

3.   Two other terms are 'sampling distribution of the mean' and 'the distribution of the variable $\overline{x}$.'

4.   The set of possible means exhibits less and less variability as the sample size increases, that is, the set becomes more and more clustered about the population mean. This means that as the sample size increases, there is a greater chance that the value of the sample mean from any sample is close to the value of the population mean.

5.   (a)   The error results from using the mean income tax, $\overline{x}$, of the 125,000 tax returns sampled as an estimate of the mean income tax, μ, of all 1994 tax returns.

(b)   The sampling error is $6104 - $6192 = -$88.

(c)   No, not necessarily. However, increasing the sample size from 125,000 to 250,000 would increase the likelihood for small sampling error.

(d)   Increase the sample size.

6.   (a) $\mu = \dfrac{108}{6} = \$18$ (thousands)

(b)

Sample	Salaries	$\bar{x}$
A,B,C,D	8, 12, 16, 20	14
A,B,C,E	8, 12, 16, 24	15
A,B,C,F	8, 12, 16, 28	16
A,B,D,E	8, 12, 20, 24	16
A,B,D,F	8, 12, 20, 28	17
A,B,E,F	8, 12, 24, 28	18
A,C,D,E	8, 16, 20, 24	17
A,C,D,F	8, 16, 20, 28	18
A,C,E,F	8, 16, 24, 28	19
A,D,E,F	8, 20, 24, 28	20
B,C,D,E	12, 16, 20, 24	18
B,C,D,F	12, 16, 20, 28	19
B,C,E,F	12, 16, 24, 28	20
B,D,E,F	12, 20, 24, 28	21
C,D,E,F	16, 20, 24, 28	22

(c)

```
                            •
               •     •      •     •     •
      •     •     •     •      •     •     •      •
  •   •     •     •     •      •     •     •      •
  +----+----+----+----+----+----+----+----+----+  x̄
  14   15   16   17   18   19   20   21   22
```

(d) $P(|\bar{x} - \mu| \le 1) = P(\bar{x} = 17) + P(\bar{x} = 18) + P(\bar{x} = 19)$

(e) $\qquad\qquad = 2/15 + 3/15 + 2/15 = 7/15 = 0.4666$

$$\mu_{\bar{x}} = \frac{\sum \bar{x}}{N} = \frac{270}{15} = 18.0$$

(g)   Yes. The mean of the sampling distribution of $\bar{x}$ is the same as the mean of the population, which is 18.

7.   (a) $\mu_{\bar{x}} = 506; \quad \sigma_{\bar{x}} = 237/\sqrt{25} = 47.4$

(b) $\mu_{\bar{x}} = 506; \quad \sigma_{\bar{x}} = 237/\sqrt{200} = 16.76$

(c)   The value of $\sigma_{\bar{x}}$ will be smaller than 16.76 because $\sigma_{\bar{x}} = \sigma/\sqrt{n}$.

Thus, the larger the sample size, the smaller the value of $\sigma_{\bar{x}}$.

8.   (a)   False. By the central limit theorem, the random variable $\bar{x}$ is approximately normally distributed.   Furthermore, $\mu_{\bar{x}} = \mu = 40$ and

$\sigma_{\bar{x}} = \sigma/\sqrt{n} = 10/\sqrt{100} = 1$.   Thus, $P(30 \leq \bar{x} \leq 50)$ equals the area

under the normal curve with parameters $\mu_{\bar{x}} = 40$

and $\sigma_{\bar{x}} = 1$ that lies between 30 and 50.   Applying the usual
techniques, we find that area to be 1.0000 to four decimal places.
Hence, there is almost a 100% chance that the mean of the sample
will be between 30 and 50.

(b)   This is not possible to tell, since we do not know the
distribution of the population.

(c)   True.   Referring to part (a), we see that $P(39 \leq \bar{x} \leq 41)$ equals

the area under the normal curve with parameters $\mu_{\bar{x}} = 40$

and $\sigma_{\bar{x}} = 1$ that lies between 39 and 41.   Applying the usual

techniques, we find that area to be 0.6826.   Hence, there is about
a 68.26% chance that the mean of the sample will be between 39 and
41.

9.   (a)   False.   Since the population is normally distributed, so is the
random variable $\bar{x}$.

Furthermore, $\mu_{\bar{x}} = \mu = 40$ and $\sigma_{\bar{x}} = \sigma/\sqrt{n} = 10/\sqrt{100} = 1$.   Hence, as in

Problem 4(a), we find that there is almost a 100% chance that the
mean of the sample will be between 30 and 50.

(b)   True.   Since the population is normally distributed, percentages
for the population are equal to areas under the normal curve with
parameters $\mu = 40$ and $\sigma = 10$.   Applying the usual techniques, we
find that the area under that normal curve between 30 and 50 is
0.6826.

(c)   True.   From part (a), we see that the random variable $\bar{x}$ is

normally distributed with $\mu_{\bar{x}} = 40$ and $\sigma_{\bar{x}} = 1$.   Hence, as in

Problem 8(c), we find that there is about a 68.26% chance that the
mean of the sample will be between 39 and 41.

10.   (a) $\mu = 585$ and $\sigma = 45$:

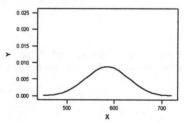

(b) $\mu_{\bar{x}} = 585$; $\sigma_{\bar{x}} = 45/\sqrt{3} = 25.98$:

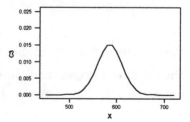

(c) $\mu_{\overline{x}} = 585;$ $\sigma_{\overline{x}} = 45/\sqrt{9} = 15:$

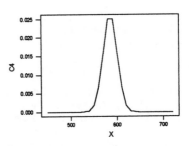

11. $n = 3;$ $\mu_{\overline{x}} = 585;$ $\sigma_{\overline{x}} = 45/\sqrt{3} = 25.98$

(a) $P(575 \le \overline{x} \le 595):$
   z-score computations: Area less than z:

   $\overline{x} = 575 \rightarrow z = \dfrac{575 - 585}{25.98} = -0.38$ 0.3520

   $\overline{x} = 595 \rightarrow z = \dfrac{595 - 585}{25.98} = 0.38$ 0.6480

   Total area = 0.6480 - 0.3520 = 0.2960, or 29.60%.

(b) The probability is 0.2960 that a sample of size three will have a mean within $10 of the population mean of $585.

(c) There is a 29.6% chance that the mean monthly rent $\overline{x}$ of the three studio apartments obtained will be within $10 of the population mean monthly rent of $585.

(d) $n = 75;$ $\sigma_{\overline{x}} = 45/\sqrt{75} = 5.2;$ $P(575 \le \overline{x} \le 595):$

   z-score computations: Area less than z:

   $\overline{x} = 575 \rightarrow z = \dfrac{575 - 585}{5.2} = -1.92$ 0.0274

   $\overline{x} = 595 \rightarrow z = \dfrac{595 - 585}{5.2} = 1.92$ 0.9726

   Total area = 0.9726 - 0.0274 = 0.9452 or 94.52%

   The probability is 0.9452 that a sample of size 75 will have a mean within $10 of the population mean of $585.

   There is a 94.52% chance that the mean monthly rent $\overline{x}$ of the 75 studio apartments obtained will be within $10 of the population mean monthly rent of $585.

12. (a) For a normally distributed population, the random variable $\overline{x}$ is normally distributed, regardless of the sample size. Also, we know that $\mu_{\bar{x}} = \mu$.   Consequently, since the normal curve for a

normally distributed population or random variable is centered at its μ-parameter, all three curves are centered at the same place.

(b)   Curve B corresponds to the larger sample size.

Since $\sigma_{\bar{x}} = \sigma/\sqrt{n}$, the larger the sample size, the smaller the value

of $\sigma_{\bar{x}}$ and, hence, the smaller the spread of the normal curve for $\bar{x}$.   Thus, Curve B, which has the smaller spread, corresponds to the larger sample size.

(c)   The spread of each curve is different because $\sigma_{\bar{x}} = \sigma/\sqrt{n}$, and the

spread of a normal curve is determined by $\sigma_{\bar{x}}$.   Thus, different sample sizes result in normal curves with different spreads.

(d)   Curve B corresponds to the sample size that will tend to produce less sampling error.   The smaller the value of $\sigma_{\bar{x}}$, the smaller the sampling error tends to be.

(e)   When x is normally distributed, $\bar{x}$ always has a normal distribution as well.

13. $n = 500$; $\mu_{\bar{x}} = \mu$; $\sigma_{\bar{x}} = 50{,}900/\sqrt{500} = 2276.32$

(a) $P(\mu - 2{,}000 \leq \bar{x} \leq \mu + 2{,}000)$:
z-score computations: Area less than z:

$\bar{x} = \mu - 2{,}000 \rightarrow z = \dfrac{(\mu - 2000) - \mu}{2276.32} = -0.88$ 0.1894

$\bar{x} = \mu + 2{,}000 \rightarrow z = \dfrac{(\mu + 2{,}000) - \mu}{2276.32} = 0.88$ 0.8106

Total area = 0.8106 − 0.1894 = 0.6212

(b)   To answer part (a), it is not necessary to assume that the population is normally distributed because the sample size is large and, therefore, $\bar{x}$ is approximately normally distributed, regardless of the distribution of the population of life-insurance amounts.

If the sample size were 20 instead of 500, it would be necessary to assume normality because the sample size is small.

(c) $n = 5000$; $\sigma_{\bar{x}} = 50{,}900/\sqrt{5000} = 719.83$; $P(119{,}400 \leq \bar{x} \leq 123{,}400)$:

z-score computations: Area less than z:

$\bar{x} = \mu - 2{,}000 \rightarrow z = \dfrac{(\mu - 2{,}000) - \mu}{719.83} = -2.78$ 0.0027

$\bar{x} = \mu + 2{,}000 \rightarrow z = \dfrac{(\mu + 2{,}000) - \mu}{719.83} = 2.78$ 0.9973

Total area = 0.9973 − 0.0027 = 0.9946

15. $\mu = 5$; $\sigma = 0.5$; $\sigma_{\overline{x}} = 0.5/\sqrt{10} = 0.158$

(a) $P(x \leq 4.5)$:

z-score computation: Area less than z:

$$x = 4.5 \rightarrow z = \frac{4.5-5}{0.5} = -1$$ 0.1587

Total area = 0.1587

If the paint lasts 4.5 years, I would not consider this to be substantial evidence against the manufacturer's claim that the paint will last an average of five years.

Assuming the manufacturer's claim is correct, the probability is 0.1587 that the paint will last 4.5 years or less on a (randomly selected) house painted with the paint. In other words, there is a (fairly high) 15.87% chance that the paint would last 4.5 years or less, if the manufacturer's claim is correct.

(b) $P(\overline{x} \leq 4.5)$:

z-score computation: Area less than z:

$$\overline{x} = 4.5 \rightarrow z = \frac{4.5-5}{0.158} = -3.16$$ 0.0008

Total area = 0.0008

For 10 houses, if the paint lasts an average of 4.5 years, I would consider this to be substantial evidence against the manufacturer's claim that the paint will last an average of five years.

Assuming the manufacturer's claim is correct, the probability is 0.0008 that the paint will last an average of 4.5 years or less for 10 (randomly selected) houses painted with the paint. In other words, there is less than a 0.1% chance that that would occur, if the manufacturer's claim is correct.

(c) $P(\overline{x} \leq 4.9)$:

z-score computation: Area less than z:

$$\overline{x} = 4.9 \rightarrow z = \frac{4.9-5}{0.158} = -0.63$$ 0.2643

Total area = 0.2643

For 10 houses, if the paint lasts an average of 4.9 years, I would not consider this to be substantial evidence against the manufacturer's claim that the paint will last an average of five years.

Assuming the manufacturer's claim is correct, the probability is 0.2643 that the paint will last an average of 4.9 years or less for 10 (randomly selected) houses painted with the paint. In other words, there is a (fairly high) 26.43% chance that that would occur, if the manufacturer's claim is correct.

15.   (a)   We have Minitab take 1000 random samples of size n = 4 from a normally distributed population with mean 500 and standard

deviation 100 by choosing **Calc ▶ Random Data ▶ Normal...**, typing <u>1000</u> in the **Generate rows of data** text box, typing <u>C1-C4</u> in the **Store in column(s)** text box, typing <u>500</u> in the **Mean** text box and <u>100</u> in the **Standard deviation** text box, and clicking **OK**.

(b)   We compute the sample mean of each of the 1000 samples by choosing

Calc ▶ Row statistics...,clicking on the **Mean** button, typing <u>C1-C9</u> in the **Input variables** text box and <u>XBAR</u> in the **Store result in** text box, and clicking **OK**. (We will not print these values.)

(c)   The mean is obtained by choosing Calc ▶ Column statistics..., clicking on the **Mean** button, selecting XBAR in the **Input variable** text box, and clicking **OK**. Our result is
Mean of XBAR = 500.59

Similarly, the standard deviation is obtained by choosing Calc ▶ **Column statistics...**, clicking on the **Standard deviation** button, selecting XBAR in the **Input variable** text box, and clicking **OK**. Our result is
Standard deviation of XBAR = 48.672

To get a histogram of the 1000 sample means stored in XBAR, choose

Graph ▶ Histogram..., select XBAR for the **X** variable for **Graph 1**, and click **OK**. The result is shown below.

(d)   Theoretically, the distribution of all possible sample means for samples of size four from this normal population should have mean 500 , standard deviation
$\sigma / \sqrt{n} = 100 / \sqrt{4} = 50$, and a normal distribution.

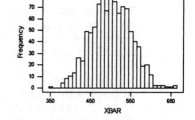

(e)   The histogram is close to bell-shaped, is centered near 500, and most of the data lies within three standard deviations (3 x 50 = 150) of 500, i.e. between 350 and 650. The mean of the 1000 sample means is 500.59, very close to 500, and the standard deviation of the sample means is 48.672, very close to 50.

16.   (a)   A uniform distribution between 0 and 1:

(b)   We have Minitab take 2000 random samples of size n = 2 from a uniformly distributed population between 0 and 1 by choosing **Calc**

▶ **Random Data** ▶ **Normal...**, typing <u>2000</u> in the **Generate rows of data** text box, typing <u>C1-C2</u> in the **Store in column(s)** text box, typing <u>0.0</u> in the **Lower endpoint** text box and <u>1.0</u> in the **Upper endpoint** text box, and clicking **OK**.

(c)   We compute the sample mean of each of the 1000 samples by choosing **Calc ▶ Row statistics...**, clicking on the **Mean** button, typing <u>C1-C2</u> in the **Input variables** text box and <u>XBAR</u> in the **Store result in** text box, and clicking **OK**. (We will not print these values.)

(d)   The mean of the sample means is obtained by choosing **Calc ▶ Column statistics...**, clicking on the **Mean** button, selecting XBAR in the **Input variable** text box, and clicking **OK**. Our result is
Mean of XBAR = 0.50454

Similarly, the standard deviation is obtained by choosing **Calc ▶ Column statistics...**, clicking on the **Standard deviation** button, selecting XBAR in the **Input variable** text box, and clicking **OK**. Our result is
Standard deviation of XBAR = 0.20545

(e)   Theoretically, the distribution of all possible sample means for samples of size two from this uniform normal population should have mean $(0 + 1)/2 = .5$ and standard deviation
$\sigma / \sqrt{n} = ((1-0)/\sqrt{12})/\sqrt{2} = 0.2041$. The Minitab simulation results are very close to these theoretical values.

(f)   To get a histogram of the 1000 sample means stored in XBAR, choose **Graph ▶ Histogram...**, select XBAR for the **X** variable for **Graph 1**, and click **OK**. The result is shown below.

The histogram is more triangle shaped than bell-shaped. Since the population is far from normal and the sample size is only two, we would not expect the sampling distribution to be bell-shaped.

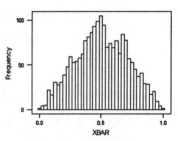

(g)   Repeating the process above, but using C1-C35 for the data in each sample, we obtained

Mean of XBAR = 0.49932
Standard deviation of XBAR = 0.049240

The theoretical distribution of the means has mean 0.5 and standard deviation
$\sigma / \sqrt{n} = ((1-0)/\sqrt{12})/\sqrt{35} = 0.0488$.
The Minitab simulation values are very close to these. The histogram for the means of the samples of size 35 is shown at the right. It is much more bell-shaped than for samples of size two, as we would expect since n > 30.

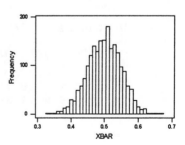

CHAPTER 8 ANSWERS

Exercises 8.1

8.1 The value of a statistic that is used to estimate a parameter is called a <u>point estimate</u> of the parameter.

8.3 (a) $\bar{x} = 1640/20 = \$82.00$

(b) Since some sampling error is expected, it is unlikely that the sample mean will exactly equal the population mean μ.

8.5 (a) $\bar{x} = 240.3/35 = 6.87$ lb

8.7 $n = 20$; $\bar{x} = 82.00$; $\sigma = 16$

(a)

$$\bar{x} - 2 \cdot (\sigma/\sqrt{n}) \quad to \quad \bar{x} + 2 \cdot (\sigma/\sqrt{n})$$

$$82.00 - 2 \cdot (16/\sqrt{20}) \quad to \quad 82.00 + 2 \cdot (16/\sqrt{20})$$

$$\$74.84 \quad to \quad \$89.16$$

(b) We can be 95.44% confident that the mean price μ of all science books is somewhere between $74.84 and $89.16.

(c) The true population price may or may not lie in the confidence interval, but we can be 95.44% confident that it does.

8.9 $n = 35$; $\bar{x} = 6.87$; $\sigma = 1.9$

(a)

$$\bar{x} - 2 \cdot (\sigma/\sqrt{n}) \quad to \quad \bar{x} + 2 \cdot (\sigma/\sqrt{n})$$

$$6.87 - 2 \cdot (1.9/\sqrt{35}) \quad to \quad 6.87 + 2 \cdot (1.9/\sqrt{35})$$

$$6.22 \; lbs \quad to \quad 7.51 \; lbs$$

(b) We can be 95.44% confident that the mean weight μ of all newborns is somewhere between 6.22 lbs and 7.51 lbs.

(c) This confidence interval is not exact because the sampling distribution of the mean $\bar{x}$ is not exactly normal.

8.11

$$P\left(\bar{x} - 3 \cdot \frac{\sigma}{\sqrt{n}} < \mu < \bar{x} + 3 \cdot \frac{\sigma}{\sqrt{n}}\right) = 0.997$$

We can be 99.74% confident that the mean μ is somewhere between $\bar{x} - 3 \cdot (\sigma/\sqrt{n})$ and $\bar{x} + 3 \cdot (\sigma/\sqrt{n})$.

$n = 36$; $\bar{x} = 38.28$; We compute s to be 7.940. Since the sample is large, we will assume that $\sigma = 7.940$.

$$\bar{x} - 3 \cdot (\sigma/\sqrt{n}) \quad to \quad \bar{x} + 3 \cdot (\sigma/\sqrt{n})$$

$$38.28 - 3 \cdot (7.940/\sqrt{36}) \quad to \quad 38.28 + 3 \cdot (7.940/\sqrt{36})$$

$$34.31 \quad to \quad 42.25$$

We can be 99.74% confident that the mean μ is between \$34,310 and \$42,250.

Exercises 8.2

8.13 (a) Confidence level = 0.90; $\alpha = 0.10$

(b) Confidence level = 0.99; $\alpha = 0.01$

8.15 (a) By saying that a 1-α confidence interval is exact, we mean that the true confidence level is equal to 1-α.

(b) By saying that a 1-α confidence interval is approximately correct, we mean that the true confidence level is only approximately equal to 1-α.

8.17 When we use the abbreviation "normal population," we mean that the variable under consideration is normally distributed on the population of interest.

8.19 A statistical procedure is robust if it is insensitive to departures from the assumptions on which it is based.

8.21 Before performing a statistical inference procedure, we should always examine the data to see whether any of the conditions required for using the procedure have been violated.

8.23 n = 25; $\bar{x}$ = 643; σ = 247

(a) Step 1: $\alpha = 0.10$; $z_{\alpha/2} = z_{0.05} = 1.645$

Step 2:

$$\bar{x} - z_{\alpha/2} \cdot (\sigma/\sqrt{n}) \quad to \quad \bar{x} + z_{\alpha/2} \cdot (\sigma/\sqrt{n})$$

$$643 - 1.645 \cdot (247/\sqrt{25}) \quad to \quad 643 + 1.645 \cdot (247/\sqrt{25})$$

$$561.7 \quad to \quad 724.3 \text{ square feet}$$

(b) We can be 90% confident that the mean size μ of household vegetable gardens in the U.S. is somewhere between 561.7 and 724.3 square feet.

8.25 n = 40; $\bar{x}$ = 136.88; σ = 12.0

(a) Step 1: $\alpha = 0.10$; $z_{\alpha/2} = z_{0.05} = 1.645$

Step 2:

$$\bar{x} - z_{\alpha/2} \cdot (\sigma/\sqrt{n}) \quad to \quad \bar{x} + z_{\alpha/2} \cdot (\sigma/\sqrt{n})$$

$$136.88 - 1.645 \cdot (12.0/\sqrt{40}) \quad to \quad 136.88 + 1.645 \cdot (12.0/\sqrt{40})$$

$$133.8 \quad to \quad 140.0 \text{ }lb$$

(b)   We can be 90% confident that the mean weight μ of all American women 5 feet 4 inches tall and in the age group 18-24 years is somewhere between 133.8 and 140.0 lb.

8.27 $n = 150$; $\overline{x} = \$16,107$; $s = \$4241$

Step 1: $\alpha = 0.10$; $z_{\alpha/2} = z_{0.05} = 1.645$

Step 2:

$$\overline{x} - z_{\alpha/2} \cdot (\sigma/\sqrt{n}) \quad to \quad \overline{x} + z_{\alpha/2} \cdot (\sigma/\sqrt{n})$$

$$16,107 - 1.96 \cdot (4241/\sqrt{150}) \quad to \quad 16,107 + 1.96 \cdot (4241/\sqrt{150})$$

$$\$15,428.30 \quad to \quad \$16,785.70$$

8.29 $n = 40$; $\overline{x} = 136.88$; $\sigma = 12.0$

(a)   Step 1: $\alpha = 0.01$; $z_{\alpha/2} = z_{0.005} = 2.575$

Step 2:

$$\overline{x} - z_{\alpha/2} \cdot (\sigma/\sqrt{n}) \quad to \quad \overline{x} + z_{\alpha/2} \cdot (\sigma/\sqrt{n})$$

$$136.88 - 2.575 \cdot (12.0/\sqrt{40}) \quad to \quad 136.88 + 2.575 \cdot (12.0/\sqrt{40})$$

$$132.0 \quad to \quad 141.8 \; lb$$

(b)   The confidence interval in part (a) is longer than the one in Exercise 8.25 because we have changed the confidence level from 90% in Exercise 8.25 to 99% in this exercise. Notice that increasing the confidence level from 90% to 99% increases the $z_{\alpha/2}$-value from 1.645 to 2.575. The larger z-value, in turn, results in a longer interval.

(c)

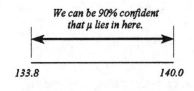

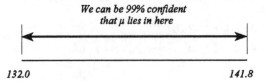

(d)   The 90% confidence interval is shorter and therefore provides a more concise estimate of μ.

8.31

x	f	xf	$x^2 f$
2	198	396	792
3	118	354	1062
4	101	404	1616
5	59	295	1475
6	12	72	432
7	3	21	147
8	8	64	512
9	1	9	81
	500	1615	6117

(a)   Step 1: $\alpha = 0.05$; $z_{\alpha/2} = z_{0.025} = 1.96$

Step 2:

$$\overline{x} - z_{\alpha/2} \cdot (\sigma/\sqrt{n}) \quad to \quad \overline{x} + z_{\alpha/2} \cdot (\sigma/\sqrt{n})$$

$$3.23 - 1.96 \cdot (1.3/\sqrt{500}) \quad to \quad 3.23 + 1.96 \cdot (1.3/\sqrt{500})$$

$$3.12 \quad to \quad 3.34$$

(b)   We can be 95% confident that the mean size μ of all American families is somewhere between 3.12 and 3.34 members.

8.33   (a) Using Minitab, with the data in a column named WEIGHTS, we first create a column of normal scores. To do this, choose **Calc ▶ Calculator...**, type <u>NORMAL</u> in the **Store result in variable** text box, select **normal scores** from the function list, select WEIGHTS to replace **number** in the **Expression:** text box. Click **OK**. To create the Normal Probability Plot, choose **Graph ▶ Plot..** and select NORMAL for the **Y** variable for **Graph 1** and WEIGHTS for the **X** variable. Click **OK**. For the boxplot, choose **Graph ▶ Boxplot...** and select WEIGHTS for the **Y** variable for **Graph 1**. Click **OK**. For the histogram, choose **Graph ▶ Histogram...** and select WEIGHTS for the **X** variable for **Graph 1**. Click **OK**. For the stem-and-leaf plot, choose **Graph ▶ Character Graphs ▶ Stem-and-leaf...** and select WEIGHTS in the **Variables** text box. Click **OK**. The results are shown below.

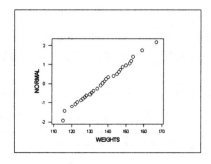

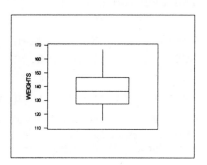

```
Stem-and-leaf of WEIGHTS    N = 40
Leaf Unit = 1.0


      4    11 5566
      7    12 023
     11    12 5678
     17    13 012244
    (8)    13 66678899
     15    14 03
     13    14 556778
      7    15 02344
      2    15 9
      1    16
      1    16 7
```

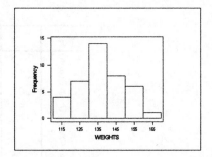

(b)   Choose **Stat ▶ Basic statistics ▶ 1-Sample z...**, specify WEIGHTS
      in the **Variables** text box, select the **Confidence interval** option
      button, click in the **Level** text box and type <u>90</u>, click in the Σ
      text box and type <u>12.0</u>, and click **OK**. The result is

      The assumed σ = 12.0

```
Variable      N      Mean    StDev   SE Mean      90.0 % C.I.
WEIGHTS      40    136.88    12.77      1.90   ( 133.75,  140.00)
```

(c)   There were no potential outliers and the sample size was large.
      Therefore there should be no problem with using the z-procedure.

8.35 (a)   With the data in a columns named AGES, choose **Stat ▶ Basic**

      statistics ▶ 1-Sample z..., specify AGES in the **Variables** text
      box, select the **Confidence interval** option button, click in the
      Level text box and type <u>95</u>, click in the Σ text box and type <u>21.2</u>,
      and click **OK**. The result is
      The assumed σ = 21.2

```
Variable      N      Mean    StDev   SE Mean      95.0 % C.I.
AGES         35     55.40    19.88      3.58   (  48.37,   62.43)
```

(b)   We first create a column of normal scores. To do this, choose

      Calc ▶ Mathematical Expressions..., type <u>NORMAL</u> in the **Variable
      (New or Modified):** text box and <u>NSCORES('AGES')</u> in the **Expression:**
      text box. Click **OK**. To create the Normal Probability Plot, choose

      Graph ▶ Plot.. and select NORMAL for the **Y** variable for **Graph 1**
      and AGES for the **X** variable. Click **OK**. For the boxplot, choose

Graph ▶ Boxplot... and select AGES for the **Y** variable for **Graph**

1. Click **OK**. For the histogram, choose **Graph** ▶ **Histogram...** and select AGES for the **X** variable for **Graph 1**. Click **OK**. For the

stem-and-leaf plot, choose **Graph** ▶ **Character Graphs** ▶ **Stem-and-leaf...** and select AGES in the **Variables** text box. Click **OK**. The results are shown below.

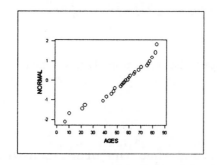

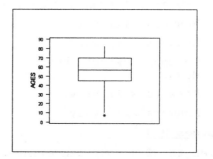

```
Stem-and-leaf of AGES        N  = 35
Leaf Unit = 1.0


     1     0  7
     2     1  0
     4     2  13
     6     3  88
    13     4  1157788
    (6)    5  345679
    16     6  0114588
     9     7  056799
     3     8  233
```

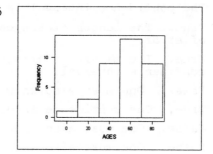

(c)   One observation, 7, is noted as a potential outlier. After removing it from the data and repeating part (a), the results are The assumed σ = 21.2

Variable	N	Mean	StDev	SE Mean	95.0 % C.I.	
AGES	34	56.82	18.28	3.64	(49.70,	63.95)

(d)   The mean has changed from 55.40 to 56.82, the standard deviation from 19.88 to 18.28, and the confidence interval from (48.37,62.43) to (49.70,63.95). Considering the width of the interval, the change in the confidence interval is quite small and the effect of the potential outlier is not great. The use of the z-interval procedure is reasonable.

8.37 (a) The standard deviation is $4261 for all 4-year colleges and universities.

(b) The standard deviation is $4169 for all 4-year colleges and universities in the sample.

(c) There were 150 colleges and universities in the sample.

(d) The mean tuition and fees for the institutions in the sample was $16,107.

(e) The 95% confidence interval for the mean tuition and fees for all 4-year colleges and universities is ($15,428, $16,786).

8.39 (a) Answers will vary.

(b) Answers will vary.

(c) We would expect about 95 to contain the population mean of 266 days.

(d) Answers will vary.

(e) Answers will vary.

Exercises 8.3

8.41 The length of a confidence interval, and thus the precision with which $\overline{x}$ estimates μ, is determined by the margin of error.

8.43 (a) The length of the confidence interval is twice the margin of error; i.e., 2 x 3.4 = 6.8,

(b) The confidence interval is 52.8 $\pm$ 3.4 = (49.4, 56.2).

8.45 (a) True. The length of the confidence interval is twice the margin of error.

(b) True. The margin of error is one-half the length of the confidence interval.

(c) False. One must also know the sample mean.

(d) True. The confidence interval is mean $\pm$ margin of error.

(e) False. One must also know n and σ.

(f) False. One must also know n and σ.

(g) True. $E = z_{\alpha/2} \cdot (\sigma/\sqrt{n})$

(h) True. From part (g), $E = z_{\alpha/2} \cdot (\sigma/\sqrt{n})$. Therefore

$z_{\alpha/2} = E \cdot (\sqrt{n}/\sigma)$. Once $z_{\alpha/2}$ has been determined, the confidence level can be determined using the standard normal table.

8.47 If σ is unknown, one could apply Formula 8.1, by first taking a preliminary sample of size 30 or more to obtain s as an estimate of σ.

8.49 (a) E = (724.3 - 561.7)/2 = 81.3

(b) $E = 1.645 \left(\dfrac{247}{\sqrt{25}} \right) = 81.3$

8.51 (a) E = (140.0 - 133.8)/2 = 3.1

(b) We are 90% confident that the maximum error made in using $\overline{x}$ to estimate μ is 3.1.

(c) The margin of error of the estimate is specified to be E = 2.0 lb. Also, for a 99% confidence interval, $z_{\alpha/2} = z_{0.005} = 2.575$.

$$n = \left[\frac{z_{\alpha/2}\cdot\sigma}{E}\right]^2 = \left[\frac{2.575\cdot 12.0}{2.0}\right]^2 = 238.7 \text{ or } 239$$

$$\bar{x} - z_{\alpha/2}\cdot (s/\sqrt{n}) \quad to \quad \bar{x} + z_{\alpha/2}\cdot (s/\sqrt{n})$$

(d)
$$134.2 - 2.575\cdot (12.0/\sqrt{239}) \quad to \quad 134.2 + 2.575\cdot (12.0/\sqrt{239})$$
$$132.2 \quad to \quad 136.2 \text{ lb}$$

8.53 (a) The margin of error of the estimate is specified to be E = 2 years.

$$n = \left[\frac{z_{\alpha/2}\cdot\sigma}{E}\right]^2 = \left[\frac{1.96\cdot 13.36}{2.0}\right]^2 = 171.4 \text{ or } 172$$

(b) We used s in place of σ because σ was unknown. We can do this because the sample of size 36 is large enough to provide an estimate of σ and the variation is not likely to change much from one year to the next.

8.55 (a) $\alpha = 0.05$; $z_{\alpha/2} = z_{0.025} = 1.96$; $\sigma = 10$

$$E = 1.96\left(\frac{10}{\sqrt{4}}\right) = 9.80$$

(b) $\alpha = 0.05$; $z_{\alpha/2} = z_{0.025} = 1.96$; $\sigma = 10$

$$E = 1.96\left(\frac{10}{\sqrt{16}}\right) = 4.90$$

(c) It appears that quadrupling the sample size will halve the margin of error. Therefore increasing n from 16 to 64 will decrease the margin of error from 4.90 to 2.45.

Exercises 8.4

8.57 The formula for the standardized version of $\bar{x}$ uses σ, whereas the studentized version uses s.

8.59 (a) The standardized version of $\bar{x}$

is $(\bar{x} - \mu)/(\sigma/\sqrt{n}) = (108 - 100)/(16/\sqrt{4}) = 1.00$

(b) The studentized version of $\bar{x}$ is

$(\bar{x} - \mu)/(s/\sqrt{n}) = (108 - 100)/(12/\sqrt{4}) = 1.333$

8.61 (a) The distribution of $(\overline{x} - 61)/(9/\sqrt{12})$ is standard normal.

   (b) The distribution of $(\overline{x} - 61)/(s/\sqrt{12})$ is a t distribution with 11 degrees of freedom.

8.63 The variation in the possible values of the standardized version of $\overline{x}$ is due only to the variation in $\overline{x}$ while the variation in the studentized version results not only from the variation in $\overline{x}$, but also from the variation in the sample standard deviation.

8.65 For df = 6:

   (a) $t_{0.10} = 1.440$ (b) $t_{0.025} = 2.447$ (c) $t_{0.01} = 3.143$

8.67 (a) $t_{0.10} = 1.323$ (b) $t_{0.01} = 2.518$

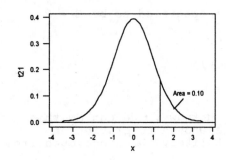

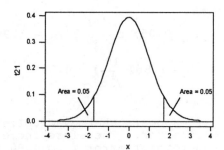

(c) $-t_{0.025} = -2.080$ (d) $\pm t_{0.05} = \pm 1.721$

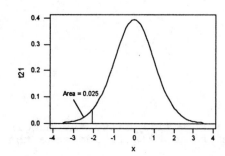

8.69 It is reasonable to use the t-interval procedure since the sample size is large and for large degrees of freedom (99), the t-distribution is very similar to the standard normal distribution. Another way of expressing this is that the sampling distribution of $\overline{x}$ is approximately normal when n is large so the standardized and studentized versions of $\overline{x}$ are essentially the same.

8.71 $n = 51$; $df = 51$; $t_{\alpha/2} = t_{0.005} = 2.678$; $\bar{x} = 61.49$; $s = 3.754$

(a)

$$\bar{x} - t_{\alpha/2} \cdot (s/\sqrt{n}) \quad to \quad \bar{x} + t_{\alpha/2} \cdot (s/\sqrt{n})$$

$$61.49 - 2.678 \cdot (3.754/\sqrt{51}) \quad to \quad 61.47 + 2.678 \cdot (3.754/\sqrt{51})$$

$$60.1 \quad to \quad 62.9 \; bushels$$

(b) It does appear that the farmer can get a better mean yield than the national average by using the new fertilizer because the confidence interval does not contain, and is to the right of, the national average of 58.4 bushels per acre.

8.73 $n = 20$; $df = 19$; $t_{\alpha/2} = t_{0.025} = 2.093$; $\bar{x} = 44.85$; $s = 3.392$

(a)

$$\bar{x} - t_{\alpha/2} \cdot (s/\sqrt{n}) \quad to \quad \bar{x} + t_{\alpha/2} \cdot (s/\sqrt{n})$$

$$44.85 - 2.093 \cdot (3.392/\sqrt{20}) \quad to \quad 44.85 + 2.093 \cdot (3.392/\sqrt{20})$$

$$43.3 \quad to \quad 46.4 \; inches$$

(b) We can be 95% confident that the mean height μ of all six-year-old girls is somewhere between 43.3 and 46.4 inches.

8.75 $n = 63$; $df = 62$; $t_{\alpha/2} = t_{0.05} = 1.671$; $\bar{x} = \$58.90$; $s = \$12.75$

(a)

$$\bar{x} - t_{\alpha/2} \cdot (s/\sqrt{n}) \quad to \quad \bar{x} + t_{\alpha/2} \cdot (s/\sqrt{n})$$

$$\$58.90 - 1.671 \cdot (12.75/\sqrt{63}) \quad to \quad \$58.90 + 1.671 \cdot (12.75/\sqrt{63})$$

$$\$56.22 \quad to \quad \$61.58$$

(b) We can be 90% confident that the mean μ for monthly fuel expenditure for all household vehicles is somewhere between $56.22 and $61.58.

8.77 $n = 22$; $\bar{x} = 25.82$; $s = 7.71$; $df = 21$; $t_{\alpha/2} = t_{0.05} = 1.721$

(a)

$$\bar{x} - t_{\alpha/2} \cdot (s/\sqrt{n}) \quad to \quad \bar{x} + t_{\alpha/2} \cdot (s/\sqrt{n})$$

$$25.82 - 1.721 \cdot (7.71/\sqrt{22}) \quad to \quad 25.82 + 1.721 \cdot (7.71/\sqrt{22})$$

$$23.0 \quad to \quad 28.6 \; minutes$$

(b) We can be 90% confident that the mean commuting time μ for local bicycle commuters in the city is somewhere between 23.0 and 28.6 minutes.

(c)   n=21; $\overline{x}$ = 24.76; s = 6.05; df = 20; $t_{\alpha/2}$ = $t_{0.05}$ = 1.725

$$\overline{x} - t_{\alpha/2} \cdot (s/\sqrt{n}) \quad to \quad \overline{x} + t_{\alpha/2} \cdot (s/\sqrt{n})$$

$$24.76 - 1.725 \cdot (6.05/\sqrt{21}) \quad to \quad 25.82 + 1.725 \cdot (6.05/\sqrt{21})$$

22.5   to   27.0 minutes

(d)   As expected, the mean and standard deviation both decrease when the potential outlier is eliminated from the data. However, the confidence interval has not changed dramatically. If you obtain a normal probability plot of the original data, you will see that the possible non-normality caused by the potential outlier is not great. While the sample size is such that one must be alert to the possible influence of outliers, in this instance the use of the t-interval procedure is reasonable.

8.79 (a)   We could not provide entries for every possible degrees of freedom because the number of possibilities is infinite.

(b)   As the degrees of freedom increase, the difference in consecutive entries becomes very small, making it unnecessary to list every possibility.

(c)   Anytime the actual degrees of freedom lies between two consecutive entries in the table, we use the table value for the lower number of degrees of freedom. Thus for $t_{0.05}$ with 85 df, we use the value for 80 df, 1.664; for 52 df, use 1.676; and for 78 df, use 1.667. This is a conservative approach, resulting in margins of error which are never smaller than we are entitled to have.

8.81 The observed values of the studentized and standardized versions of $\overline{x}$ are the same for any sample size n whenever the sample standard deviation s is identical to the population standard deviation σ.

8.83 (a)   With the data in a column named BUSHELS, first create a column of normal scores. To do this, choose **Calc ▶ Calculator...**, type NORMAL in the **Store result in variable:** text box, select **normal scores** from the function list, select BUSHELS to replace **number** in the **Expression:** text box, and click **OK**. To create the Normal Probability Plot, choose **Graph ▶ Plot..** and select NORMAL for the **Y** variable for **Graph 1** and BUSHELS for the **X** variable. Click **OK**.

For the boxplot, choose **Graph ▶ Boxplot...** and select BUSHELS for the **Y** variable for **Graph 1**. Click **OK**. For the histogram, choose **Graph ▶ Histogram...** and select BUSHELS for the **X** variable for **Graph 1**. Click **OK**. For the stem-and-leaf plot, choose **Graph ▶ Character Graphs ▶ Stem-and-leaf...** and select BUSHELS in the **Variables** text box. Click **OK**. The results are shown below.

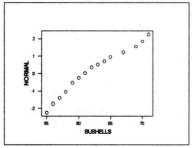
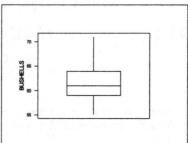

Stem-and-leaf of BUSHELS   N = 51
Leaf Unit = 0.10

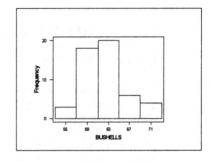

```
  1    55  0
  3    56  00
  5    57  00
 10    58  00000
 20    59  0000000000
 21    60  0
(10)    61  0000000000
 20    62  0000
 16    63  00
 14    64  0000
 10    65  000
  7    66
  7    67  000
  4    68
  4    69  00
  2    70  0
  1    71  0
```

(b)   Choose **Stat ▶ Basic statistics ▶ 1-Sample t...**, specify BUSHELS in the
      Variables text box, select the **Confidence interval** option button, click
      in the **Level** text box and type <u>99</u>, and click **OK**. The result is

Variable	N	Mean	StDev	SE Mean	99.0 % C.I.
BUSHELS	51	61.490	3.754	0.526	(60.082, 62.898)

(c)   Although the graphs indicate a slight skewness to the right, the large
      sample size and lack of outliers justifies the use of the t-interval
      procedure.

8.85 (a)   The sample standard deviation is s = 3.392 inches.

     (b)   There were n = 20 girls in the sample.

     (c)   The sample mean height is 44.850 inches.

     (d)   The 95% confidence interval for the population mean height μ is
           (43.263, 46.437) inches.

8.87 (a)   Your results will vary. To obtain the 2000 samples using Minitab,

           Choose **Calc ▶ Random Data ▶ Normal...**, type <u>2000</u> in the **Generate
           rows of data** text box, type <u>C1-C5</u> in the **Store in Column(s):** text
           box, type <u>.270</u> in the **Mean** text box, and type <u>.031</u> in the **Standard
           deviation:** text box. Click **OK**.

     (b)   To find the mean and sample standard deviation in each row

           (sample), choose **Calc ▶ Row statistics...** and click on **Mean**. Type
           <u>C1-C5</u> in the **Input variable(s):** text box and type <u>C6</u> in the **Store**

result in: text box. Repeat this last process, selecting **Standard Deviation** instead of **Mean** and put the results in C7.

(c)   To obtain the Standardized version of each $\overline{x}$ in the sample, choose **Calc ▶ Calculator...,** type STANDARD in the **Store results in variable:** text box, type (C6-.270)/(.031/SQRT(5)) in the **Expression:** text box and click **OK.**

(d)   Choose **Graph ▶ Histogram...,** select STANDARD as the **X** variable for **Graph 1.** To facilitate a comparison in part (k),. Select the **Frame** button, then **Min and Max... .** Click in the **X minimum** box and type -10 into the text box. Click in the **X maximum** box and type 10 in the text box. Click **OK** and click **OK.** Our graph is shown below, but yours will differ, yet look similar.

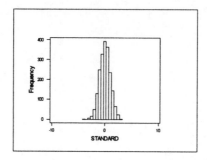

(e)   In theory, the distribution of the standardized version of $\overline{x}$ is standard normal.

(f)   The histogram in (d) appears to be very close to standard normal. Recall that the distribution is centered at zero and 99.74% of the data should be within 3 standard deviations of the mean. It does appear that this is so for this simulated data.

(g)   To obtain the Studentized version of each $\overline{x}$ in the sample, choose **Calc ▶ Calculator...,** type STUDENT in the **Store results in variable:** text box, type (C6-.270)/(C7/SQRT(5) in the **Expression:** text box and click **OK.**

(h)   Choose **Graph ▶ Histogram...,** select STUDENT as the **X** variable for **Graph 1** and click **OK.** To facilitate a comparison in part (k),. Select the **Frame** button, then **Min and Max... .** Click in the **X minimum** box and type -10 into the text box. Click in the **X maximum** box and type 10 in the text box. Click **OK** and click **OK.** Our graph follows, but yours will differ, yet look similar.

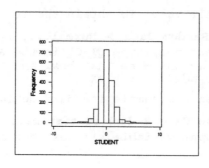

(i)   In theory, the distribution of the studentized version of $\overline{x}$ has a t-distribution with 4 degrees of freedom.

(j)   The distribution shown is symmetric about zero as is a t-distribution.

(k)   The histogram of the studentized version in (h) is more spread out than that of the standardized version in (d). The reason is that there is more variability in the t-distribution due to the extra uncertainty arising from the use of s instead of σ.

8.89 (a)   If *DataDisk* is in floppy drive a, invoke the macro by typing in the Session window %a:\macros\tcurves.mac. Type 5 in response to the question *How many different t-curves do you want?* Then type 1 2 5 10 20 in response to *Enter the degrees of freedom for the t-curves*. All five t-curves and a standard normal curve are overlaid in the resulting plot which follows. The lowest curve in the center of the plot is the t-curve with 1 degree of freedom; as the degrees of freedom increases, the curves become higher in the center and lower at the edges. The highest curve in the center of the plot is the standard normal curve. When seen on the computer monitor, the distinct curves will be shown in color.

Standard Normal Curve and Selected t-Curves

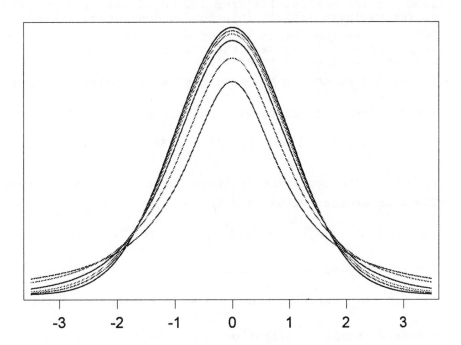

(b)   The graph illustrates that the curves become more and more like the standard normal as the degrees of freedom increase.

REVIEW TEST FOR CHAPTER 8

1. A point estimate of a parameter consists of a single value with no indication of the accuracy of the estimate. A confidence interval consists of an interval of numbers obtained from a point estimate of the parameter together with a percentage that specifies how confident we are that the parameter lies in the interval.

2. False. We are 95% confident that the mean lies in the interval from 33.8 to 39.0, but about 5% of the time, the procedure will produce an interval that does not contain the population mean. Therefore, we cannot say that the mean must lie in the interval.

3. The z-interval procedure can be used almost anytime with large samples because the sampling distribution of $\bar{x}$ is approximately normal for large n. The same is true for t-interval procedure because when n is large, the t distribution is very similar to the normal distribution. However, when n is small, especially when n is 15 or less, the z-interval and t-interval procedures will not provide reliable estimates if the distribution of the underlying variable is not normal. For sample sizes in the range of 15 to 30, both procedures can be used if the data is roughly normal and has no outliers.

4. Approximately 950 of 1000 95% confidence intervals for a population mean would actually contain the true value of the mean.

5. Before applying a particular statistical inference procedure, we should look at the sample data to see if there appear to be any violations of the conditions required for the use of the procedure.

6. (a) Reducing the sample size from 100 to 50 will reduce the precision of the estimate (result in a longer confidence interval).

    (b) Reducing the confidence level from .95 to .90 will maintaining the sample size will increase the precision of the estimate (result in a shorter confidence interval).

7. (a) The length of the confidence interval is twice the margin of error or 2 x 10.7 = 21.4.

    (b) The confidence interval will be 75.2 $\pm$ 10.7 = (64.5, 85.9)

8. (a) $E = z_{\alpha/2} \cdot (\sigma/\sqrt{n}) = 1.645 \cdot (12/\sqrt{9}) = 6.58$

    (b) To obtain the confidence interval, you also need to know $\bar{x}$.

9. (a) The standardized value of $\bar{x}$ is

    $z = (\bar{x} - \mu) / (\sigma/\sqrt{n}) = (262.1 - 266) / (16/\sqrt{10}) = -0.77$

    (b) The studentized value of $\bar{x}$ is

    $z = (\bar{x} - \mu) / (s/\sqrt{n}) = (262.1 - 266) / (20.4/\sqrt{10}) = -0.60$

10. (a) standard normal distribution

    (b) t distribution with 2 degrees of freedom

11. The curve that looks more like the standard normal curve has the larger degrees of freedom because, as the number of degrees of freedom gets larger, t-curves look increasingly like the standard normal curve.

12. (a) The t-interval procedure should be used.

    (b) The z-interval procedure should be used.

(c)   The z-interval procedure should be used.

(d)   Neither procedure should be used.

(e)   The z-interval procedure should be used.

(f)   Neither procedure should be used.

13. $n = 36$, $\overline{x} = 58.53$, $\sigma = 13.0$, $z_{\alpha/2} = z_{0.025} = 1.96$

$$\overline{x} - z_{\alpha/2} \cdot (\sigma/\sqrt{n}) \quad to \quad \overline{x} + z_{\alpha/2} \cdot (\sigma/\sqrt{n})$$

$$58.53 - 1.96 \cdot (13.0/\sqrt{36}) \quad to \quad 58.53 + 1.96 \cdot (13.0/\sqrt{36})$$

$$54.3 \quad to \quad 62.8 \; years$$

14.   A confidence-interval estimate specifies how confident we are that an (unknown) parameter lies in the interval. This interpretation is presented correctly by (c). A *specific* confidence interval either will or will not contain the true value of the population mean μ; the *specific* interval is either sure to contain μ or sure not to contain μ. This interpretation is *not* presented correctly by (a), (b), and (d).

15. $n = 23$, $\overline{x} = 60.1$, $\sigma = 4.3$, $z_{\alpha/2} = z_{0.005} = 2.575$

(a)
$$\overline{x} - z_{\alpha/2} \cdot (\sigma/\sqrt{n}) \quad to \quad \overline{x} + z_{\alpha/2} \cdot (\sigma/\sqrt{n})$$

$$60.1 - 2.575 \cdot (4.3/\sqrt{23}) \quad to \quad 60.1 + 2.575 \cdot (4.3/\sqrt{23})$$

$$57.8 \quad to \quad 62.4 \; hours$$

(b)   We can be 99% confident that the mean battery life μ is somewhere between 57.8 and 62.4 hours.

(c)   Since the sample size is not particularly large, the normal probability plot should be roughly linear and not indicate the presence of any outliers.

16.   (a) $E = z_{\alpha/2} \cdot (\sigma/\sqrt{n}) = 2.575 \cdot (4.3/\sqrt{23}) = 2.3 \; hours$

(b)   We can be 99% confident that the maximum error made in using $\overline{x}$ to estimate μ is 2.3 hours.

(c)   The margin of error of the estimate is specified to be $E = 0.5$ hours.

$$n = \left[\frac{z_{\alpha/2} \cdot \sigma}{E} \right]^2 = \left[\frac{2.575 \cdot 4.3}{0.5} \right]^2 = 490.4 \; or \; 491$$

(d)
$$\overline{x} - z_{\alpha/2} \cdot (\sigma/\sqrt{n}) \quad to \quad \overline{x} + z_{\alpha/2} \cdot (\sigma/\sqrt{n})$$

$$59.8 - 2.575 \cdot (4.3/\sqrt{491}) \quad to \quad 59.8 + 2.575 \cdot (4.3/\sqrt{491})$$

$$59.3 \quad to \quad 60.3 \; hours$$

17. (a) $t_{0.025} = 2.101$ (b) $t_{0.05} = 1.734$

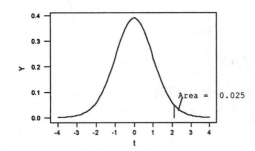   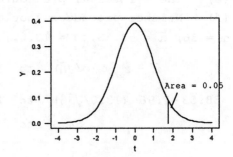

(c) $-t_{0.10} = -1.330$ (d) $\pm t_{0.005} = \pm\,2.878$

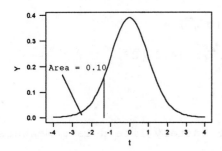   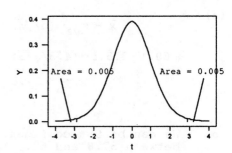

18. $n = 20$; $df = 19$; $t_{\alpha/2} = t_{0.025} = 2.093$; $\overline{x} = 7.71$; $s = 1.78$

(a)

$$\overline{x} - t_{\alpha/2}\cdot (s/\sqrt{n}) \quad to \quad \overline{x} + t_{\alpha/2}\cdot (s/\sqrt{n})$$

$$7.71 - 2.093\cdot (1.780/\sqrt{20}) \quad to \quad 7.71 + 2.093\cdot (1.780/\sqrt{20})$$

$$6.88 \quad to \quad 8.54 \; hours$$

(b)   We can be 95% confident that the mean daily viewing time μ of all American households is somewhere between 6.88 and 8.54 hours.

(c)   The result in part (a) does not provide evidence of an increase in average daily viewing time because the 1992 average of seven hours and four minutes does not lie completely to the left of the confidence interval in part (a).

19.   Choose **Stat ▶ Basic statistics ▶ 1-Sample z...,** specify AGES in the **Variables** text box, select the **Confidence interval** option button, click in the **Level** text box and type <u>95</u>, click in the Σ text box and type <u>13.0,</u> and click **OK**.   The result is

The assumed σ = 13.0

Variable	N	Mean	StDev	SE Mean	95.0 % C.I.
AGES	36	58.53	13.36	2.17	(54.28, 62.78)

20. (a) THE ASSUMED Σ = 13.0 (d) MEAN = 58.53

(b) STDEV = 13.36 (e) 95.0 PERCENT C.I. = (54.28, 62.78)

(c) n = 36

21. (a) With the data in a column named VIEW and a second blank column named NORMAL, we choose **Calc ▶ Calculator...**, specify NORMAL in the **Store results in variables** text box, select **normal scores** from the function list, select VIEW to replace **numbers** in the

Expression text box, and click **OK**. Then choose **Graph ▶ Plot...**, specify NORMAL in the **Y** variable text box for **Graph 1** and VIEW in the **X** variable for **Graph 1**, and click **OK**. The result is

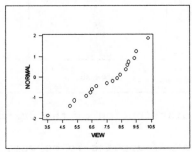

The plot is roughly linear without outliers. With n = 20, it is reasonable to use the t interval procedure.

(b) Choose **Stat ▶ Basic statistics ▶ 1-Sample t...**, specify VIEW in the **Variables** text box, select the **Confidence interval** option button, click in the **Level** text box and type 95, and click **OK**. The result is

Variable	N	Mean	StDev	SE Mean	95.0 % C.I.
VIEW	20	7.710	1.780	0.398	(6.877, 8.543)

22. (a) MEAN = 7.710 (d) n = 20

(b) STDEV = 1.780 (e) 95.0 PERCENT C.I. = (6.877, 8.543)

(c) SE Mean = 0.398   hours

23. (a) Your results will vary. To obtain the 3000 samples using Minitab,

Choose **Calc ▶ Random Data ▶ Normal...**, type 3000 in the **Generate rows of data** text box, type C1-C4 in the **Store in Column(s):** text box, type 500 in the **Mean** text box, and type 100 in the **Standard deviation:** text box. Click **OK**.

(b) To find the mean and sample standard deviation in each row

(sample), choose **Calc ▶ Row statistics...** and click on **Mean**. Type C1-C4 in the **Input variable(s):** text box and type C5 in the **Store**

result in: text box. Repeat this last process, selecting **Standard Deviation** instead of **Mean** and put the results in C6.

(c)   To obtain the Standardized version of each $\overline{x}$ in the sample, choose

    Calc ▶ Calculator..., type STANDARD in the **Store results in variable:** text box, type (C5-500)/(100/SQRT(4)) in the **Expression:** text box and click **OK**.

(d)   Choose **Graph ▶ Histogram...**, select STANDARD as the **X** variable for **Graph 1**. To facilitate a comparison in part (k),. Select the **Frame** button, then **Min and Max...** . Click in the **X minimum** box and type -20 into the text box. Click in the **X maximum** box and type 20 in the text box. Click **OK** and click **OK**. Our graph is shown below. Yours will differ, yet look similar.

)

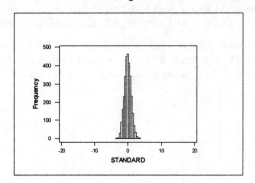

(e)   In theory, the distribution of the standardized version of $\overline{x}$ is standard normal.

(f)   The histogram in (e) appears to be very close to standard normal. Recall that the distribution is centered at zero and 99.74% of the data should be within 3 standard deviations of the mean. It does appear that this is so for this simulated data.

(g)   To obtain the Studentized version of each $\overline{x}$ in the sample, choose

    Calc ▶ Calculator..., type STUDENT in the **Store results in variable** text box, type (C5-500)/(C6/SQRT(4) in the **Expression:** text box and click **OK**.

(h)   Choose **Graph ▶ Histogram...**, select STUDENT as the **X** variable for **Graph 1** and click **OK**. To facilitate a comparison in part (k),. Select the **Frame** button, then **Min and Max...** . Click in the **X minimum** box and type -20 into the text box. Click in the **X maximum** box and type 20 in the text box. Click **OK** and click **OK**. Our graph is shown below, but yours will differ, yet look similar.

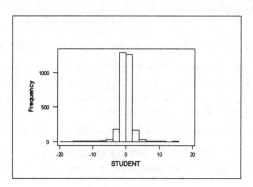

(i) In theory, the distribution of the studentized version of $\bar{x}$ is a t-distribution with 3 degrees of freedom.

(j) The distribution shown is symmetric about zero as is a t-distribution.

(k) The histogram of the studentized version in (h) is much more spread out than that of the standardized version in (d). The reason is that there is more variability in the t-distribution due to the extra uncertainty arising from the use of s instead of σ.

Exercises 9.1

9.1   A hypothesis is a statement that something is true.

9.3   (a)   The population mean μ is equal to some fixed amount; i.e., $\mu = \mu_0$.

        (b)   The population mean μ is greater than μ_0; i.e., $\mu > \mu_0$.

                The population mean μ is less than μ_0; i.e., $\mu < \mu_0$.

                The population mean μ is unequal to μ_0; i.e., $\mu \neq \mu_0$.

9.5   Let μ denote last year's mean amount telephone expenditure per consumer.

        (a) H_0: $\mu = \$690$ (b) H_a: $\mu > \$690$ (c)   right-tailed test

9.7   Let μ denote the mean daily iron intake of all adult females under the age of 51.

        (a) H_0: $\mu = 18$ mg   (b) H_a: $\mu < 18$ mg   (c)   left-tailed test

9.9   Let μ denote the mean body temperature.

        (a) H_0: $\mu = 98.6°$ F   (b) H_a: $\mu \neq 98.6°$ F   (c)   two-tailed test

9.11   (a) H_0: $\mu = 5.6$ (mean number of radios per U.S. household in 1990)

                H_a: $\mu \neq 5.6$ (mean number of radios has changed from 1990)

        (b)   If the sample mean number of radios $\bar{x}$ differs by too much from 5.6, then we should be inclined to reject H_0 and conclude that H_a is true. From the data, we compute $\bar{x} = 5.889$. The question is whether the difference of 0.289 between the sample mean of 5.889 and the hypothesized population mean of 5.6 can be attributed to sampling error or whether the difference is large enough to indicate that the population mean is not 5.6.

        (c)   Since n is large, $\bar{x}$ will have approximately a normal distribution with mean μ and standard deviation $\sigma / \sqrt{n} = \sigma / \sqrt{45}$.

        (d)   It is quite unlikely that the sample mean $\bar{x}$ will be more than two standard deviations away from the population mean μ. If $\bar{x}$ is more than two standard deviations away from μ, then reject H_0 and conclude that H_a is true. Otherwise, do not reject H_0.

        (e)   We have $\sigma = 1.9$, $n = 45$, $\bar{x} = 5.889$, and $\mu = 5.6$ under H_0 true.

                Thus, $z = (5.889 - 5.6)/(1.9/\sqrt{45}) = 1.02$. Since $\bar{x}$ is less than two standard deviations away from 5.6, we do not reject H_0.

9.13   (a)   If the mean weight $\bar{x}$ of the 50 bags of pretzels sampled is more than one standard deviation away from 454 grams, then reject the null hypothesis that $\mu = 454$ grams and conclude that the alternative hypothesis, which is $\mu \neq 454$ grams, is true. Otherwise, do not reject the null hypothesis.

            Graphically, the decision criterion looks like:

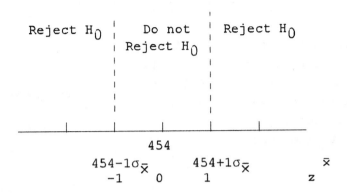

(b)

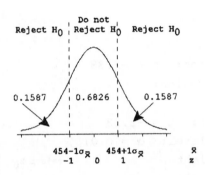

The lower figure shows that, using our decision criterion, the probability is 0.3174 (= 1 - 0.6826 = 0.1587 + 0.1587) of rejecting the null hypothesis if it is in fact true.

(c)   We have $\sigma = 7.9$, $n = 25$, $\bar{x} = 450$, and $\mu = 454$ if H_0 is true. Thus, $z = (450 - 454)/(7.8/\sqrt{25}) = -2.56$. The sample mean $\bar{x}$ is 2.56 standard deviations below the null hypothesis mean of 454 grams. Since the mean weight $\bar{x}$ of 25 bags of pretzels sampled is more than one standard deviation away from 454 grams, we reject the null hypothesis that $\mu = 454$ grams and conclude that the alternative hypothesis, which is $\mu \neq 454$ grams, is true. In other words, the data provide sufficient evidence to conclude that the packaging machine is not working properly.

9.15   If the null hypothesis is true, the chance of incorrectly rejecting it is .0456 when using the 95.44% part of the 68.26-95.44-99.74 rule.

Exercises 9.2

9.17   (a)   This statement is true: If it is important not to reject a true null hypothesis, i.e., not to make a Type I error, then the hypothesis test should be performed at a small significance level. This can be appreciated by considering the meaning of the significance level. The significance level is equal to the probability of making a Type I error. The smaller the significance level, the smaller the probability of rejecting a true null hypothesis.

       (b)   This statement is true: Decreasing the significance level results in an increase in the probability of making a Type II error. This can be appreciated by considering the relation between Type I and

Type II error probabilities. For a fixed sample size, the smaller the Type I error probability (which is equal to the significance level), the larger the Type II error probability.

9.19 (a) Rejection region: $z \geq 1.645$

(b) Nonrejection region: $z < 1.645$

(c) Critical value: $z = 1.645$

(d) Significance level: $\alpha = 0.05$

(e) Graph: see right

(f) Right-tailed test

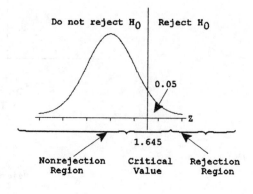

9.21 (a) Rejection region: $z \leq -2.33$

(b) Nonrejection region:
$z > -2.33$

(c) Critical value: $z = -2.33$

(d) Significance level: $\alpha = 0.01$

(e) Graph: see right

(f) Left-tailed test

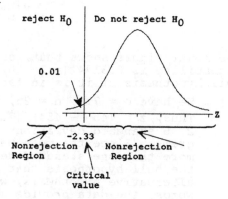

9.23 (a) Rejection region: $z \leq -1.645$
or $z \geq 1.645$

(b) Nonrejection region: $-1.645 < z < 1.645$

(c) Critical values: $z = \pm 1.645$

(d) Significance level: $\alpha = 0.10$

(e) Graph: see right

(f) Two-tailed test

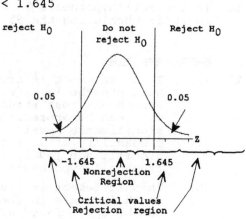

9.25 (a) A Type I error would occur if, in fact, μ = \$690, but the results of the sampling lead to the conclusion that μ > \$690.

(b) A Type II error would occur if, in fact, μ > \$690, but the results of the sampling fail to lead to that conclusion.

(c) A correct decision would occur if, in fact, μ = \$690 and the results of the sampling do not lead to the rejection of that fact; or if, in fact, μ > \$690 and the results of the sampling lead to that conclusion.

(d) If, in fact, last year's mean telephone expenditure per consumer is equal to the 1994 mean of \$690, and we do not reject the null hypothesis that μ = \$690, we made a correct decision.

(e) If, in fact, last year's mean telephone expenditure per consumer is greater than the 1994 mean of \$690, and we do not reject the null hypothesis that μ = \$690, we made a Type II error.

9.27 (a) A Type I error would occur if, in fact, μ = 18 mg, but the results of the sampling lead to the conclusion that μ < 18 mg.

(b) A Type II error would occur if, in fact, μ < 18 mg, but the results of the sampling fail to lead to that conclusion.

(c) A correct decision would occur if, in fact, μ = 18 mg and the results of the sampling do not lead to the rejection of that fact; or if, in fact, μ < 18 mg and the results of the sampling lead to that conclusion.

(d) If, in fact, the mean iron intake is not less than 18 mg, and we reject the null hypothesis that μ = 18 mg, we made a Type I error.

(e) If, in fact, the mean iron intake is less than 18 mg, and we reject the null hypothesis that μ = 18 mg, we made a correct decision.

9.29 (a) A Type I error would occur if, in fact, μ = 98.6° F, but the results of the sampling lead to the conclusion that $\mu \neq 98.6°$ F.

(b) A Type II error would occur if, in fact, $\mu \neq 98.6°$ F, but the results of the sampling fail to lead to that conclusion.

(c) A correct decision would occur if, in fact, $\mu = 98.6°$ F and the results of the sampling do not lead to the rejection of that fact; or if, in fact, $\mu \neq 98.6°$ F and the results of the sampling lead to that conclusion.

(d) If, in fact, the mean temperature of all healthy humans equals the 98.6° F, and we do not reject the null hypothesis that μ = 98.6° F, we made a correct decision.

(e) If, in fact, the mean temperature all healthy humans does not equal 98.6° F, and we do not reject the null hypothesis that $\mu = 98.6°$ F, we made a Type II error.

9.31 (a) P(Type I error) = α = 0.

(b) If α = 0, P(Type II error) = β = 1.

9.33 In this exercise, we are told that failing to reject the null hypothesis corresponds to approving the nuclear reactor for use. This action—approving the nuclear reactor—suggests that the null hypothesis must be something like: "The nuclear reactor is safe." This further suggests that the alternative hypothesis is something like: "The nuclear reactor is unsafe." Putting things together, the Type II error in this situation is: "Approving the nuclear reactor for use when, in fact, it is unsafe." This type of error has consequences that are catastrophic. Thus, the property that we want the Type II error probability to exhibit is that it be small.

9.35 (a) The probability of a Type I error is the same as the significance level.

(b)   If the mean net weight being packaged is 447 g, then the distribution of $\bar{x}$ is a normal distribution with mean 447 g and standard deviation $\sigma/\sqrt{n} = 7.8/\overline{2} = 1.56$ g.

(c) β is the probability of not rejecting null hypothesis when it is actually false. In this case, β is the probability that $\bar{x}$ falls between 450.88 g and 457.12 g when μ = 447 grams and $\sigma_{\bar{x}}$ = 1.56 g. Thus

$$\beta = P(450.88 < \bar{x} < 457.12) = P(\frac{450.88 - 447}{1.56} < z < \frac{457.12 - 447}{1.56})$$

$$= P(2.49 < z < 6.49) = 1.0000 - 0.9936 = 0.0064$$

(d)   The probability of a Type II error is an area between the two critical values of $\bar{x}$ computed above (i.e., between $\bar{x}_1 = 450.88$ g and $\bar{x}_r = 457.12$ g) **assuming** that the true mean is any one of the thirteen values of μ presented in this part of the exercise. As a probability statement, this is written $P(450.88 < \bar{x} < 457.12)$.

Since the sample size in this exercise is large (i.e., n = 30), the random variable $\bar{x}$ is approximately normally distributed with mean $\mu_{\bar{x}} = \mu$ and standard deviation $\sigma_{\bar{x}} = \sigma/\sqrt{n}$. Thus, in order to calculate $P(450.88 < \bar{x} < 457.12)$, we implement the z-score formulas

$$z = \frac{\bar{x}_1 - \mu_a}{\sigma/\sqrt{n}} \quad \text{and} \quad z = \frac{\bar{x}_r - \mu_a}{\sigma/\sqrt{n}} ,$$

insert the necessary elements into the right-hand side of each formula itself, and proceed with using Table II to find the appropriate areas.

Notice that $\bar{x}_1 = 450.88$ and $\bar{x}_r = 457.12$ and that the standard deviation to be inserted into each formula has already been presented; i.e., $\sigma/\sqrt{n} = 7.8/\sqrt{25} = 1.56$. Most importantly, the value of the population mean to be inserted into each formula is not the value of μ **assuming** that the null hypothesis is true; i.e., it is not $\mu_0 = 454$. It is, instead, an alternative value of μ, as indicated by the symbol μ_a in each of the formulas.

For this part of the exercise, we are given thirteen alternative "true mean" values for μ. This translates into 26 z-scores that need to be computed (i.e., two for each value of the "true mean"). In turn, we calculate the area associated with each pair of z-scores and then use this information to compute β, defined as the probability of a Type II error.

The appropriate calculations are:

True mean μ	z-score computation	P(Type II error) β
448	$z = \dfrac{450.88 - 448}{7.8/\sqrt{25}} = 1.85$ $z = \dfrac{457.12 - 448}{7.8/\sqrt{25}} = 5.85$	$1.0000 - 0.9678 = 0.0322$
449	$z = \dfrac{450.88 - 449}{7.8/\sqrt{25}} = 1.21$ $z = \dfrac{457.12 - 449}{7.8/\sqrt{25}} = 5.21$	$1.0000 - 0.8869 = 0.1131$
450	$z = \dfrac{450.12 - 450}{7.8/\sqrt{25}} = 0.56$ $z = \dfrac{457.12 - 450}{7.8/\sqrt{25}} = 4.56$	$1.0000 - 0.7123 = 0.2977$
451	$z = \dfrac{450.88 - 451}{7.8/\sqrt{25}} = -0.08$ $z = \dfrac{457.12 - 451}{7.8/\sqrt{25}} = 3.92$	$1.0000 - 0.4681 = 0.5319$
452	$z = \dfrac{450.88 - 452}{7.8/\sqrt{25}} = -0.72$ $z = \dfrac{457.12 - 452}{7.8/\sqrt{25}} = 3.28$	$0.9995 - 0.2358 = 0.7637$
453	$z = \dfrac{450.88 - 453}{7.8/\sqrt{25}} = -1.36$ $z = \dfrac{457.12 - 453}{7.8/\sqrt{25}} = 2.64$	$0.9959 - 0.0869 = 0.9090$

True mean μ	z-score computation	P(Type II error) β
455	$z = \dfrac{450.88 - 455}{7.8/\sqrt{25}} = -2.64$ $z = \dfrac{457.12 - 455}{7.8/\sqrt{25}} = 1.36$	$0.9131 - 0.0041 = 0.9090$
456	$z = \dfrac{450.88 - 456}{7.8/\sqrt{25}} = -3.28$ $z = \dfrac{457.12 - 456}{7.8/\sqrt{25}} = 0.72$	$0.7642 - 0.0005 = 0.7637$
457	$z = \dfrac{450.88 - 457}{7.8/\sqrt{25}} = -3.92$ $z = \dfrac{457.12 - 457}{7.8/\sqrt{25}} = 0.08$	$0.5319 - 0.0000 = 0.5319$
458	$z = \dfrac{450.88 - 458}{7.8/\sqrt{25}} = -4.56$ $z = \dfrac{457.12 - 458}{7.8/\sqrt{25}} = -0.56$	$0.2877 - 0.0000 = 0.2877$
459	$z = \dfrac{450.88 - 459}{7.8/\sqrt{25}} = -5.21$ $z = \dfrac{457.12 - 459}{7.8/\sqrt{25}} = -1.21$	$0.1131 - 0.0000 = 0.1131$
460	$z = \dfrac{450.88 - 460}{7.8/\sqrt{25}} = -5.85$ $z = \dfrac{457.12 - 460}{7.8/\sqrt{25}} = -1.85$	$0.0322 - 0.0000 = 0.0322$

True mean μ	z-score computation	P(Type II error) β
461	$z = \dfrac{450.88 - 461}{7.8/\sqrt{25}} = -6.49$ $z = \dfrac{457.12 - 461}{7.8/\sqrt{25}} = -2.49$	$0.0064 - 0.0000 = 0.0064$

To summarize this part of the exercise, notice that the answer for each β value is presented in the third column of the previous table.

(e) Consider columns 1 and 3 of the table in part (d). Also consider a graph whose vertical axis is labeled β and whose horizontal axis is labeled μ. Plot the points of β in column 3 of the table versus the respective values of μ in column 1 and then connect the points with a smooth curve. This curve is presented below.

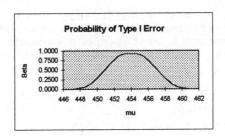

Recall that the value of μ, assuming that the null hypothesis is true, is $\mu_0 = 454$ g. The previous graph tells us that the farther the true value of μ is from the null hypothesis value of 454 g, the smaller is the probability of making a Type II error; i.e., the smaller is β.

All of this is reasonable. We would expect it to be more likely for a false null hypothesis to be detected--and hence β to be small--when the true value of μ is far from the null hypothesis value than when it is close.

Exercises 9.3

9.37 Critical value: $z_{0.01} = 2.33$ **9.39** Critical values: $\pm z_{0.05} = \pm 1.645$

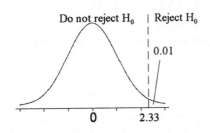

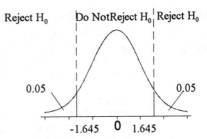

9.41   Critical value: $-z_{0.05} = -1.645$

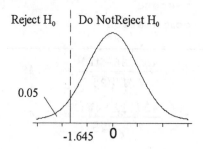

9.43 $n = 40$, $\sigma = 350$, $\overline{x} = 28{,}175/40 = 704.375$

Step 1: H_0: $\mu = \$690$, H_a: $\mu > \$690$

Step 2: $\alpha = 0.05$

Step 3:   Critical value = 1.645

Step 4: $z = (704.375 - 690)/(350/\sqrt{40}) = 0.26$

Step 5:   Since $0.26 < 1.645$, do not reject H_0.

Step 6:   At the 5% significance level, the data do not provide sufficient evidence to conclude that last year's mean amount μ of telephone expenditures per consumer unit has increased over the 1994 mean of \$690.

9.45 $n = 45$, $\overline{x} = 14.68$, $\sigma = 4.2$

Step 1: H_0: $\mu = 18$ mg, H_a: $\mu < 18$ mg

Step 2: $\alpha = 0.01$

Step 3:   Critical value = -2.33

Step 4: $z = (14.68 - 18)/(4.2/\sqrt{45}) = -5.30$

Step 5:   Since $-5.30 < -2.33$, reject H_0.

Step 6:   At the 1% significance level, the data provide sufficient evidence to conclude that adult females under the age of 51 are, on the average, getting less than the RDA of 18 mg of iron. Considering that iron deficiency causes anemia and that iron is required for transporting oxygen in the blood, this result could have practical significance as well.

9.47 $n = 93$, $\overline{x} = 98.12°F$, $\sigma = 0.63°F$

Step 1: H_0: $\mu = 98.6°F$, H_a: $\mu \neq 98.6°F$

Step 2: $\alpha = 0.01$

Step 3:   Critical values = ±2.575

Step 4: $z = (98.12 - 98.60)/(0.63/\sqrt{93}) = -7.35$

Step 5:   Since $-7.35 < -2.575$, reject H_0.

Step 6:   At the 5% significance level, the data provide sufficient evidence to conclude that the mean body temperature μ of all healthy humans differs from the widely accepted value of 98.6°F. The practical significance of this result is not great since temperatures lower than normal do not usually give rise to any concern and when a person is considered to have a dangerously high temperature, it is so high that whether normal is 98.6 or 98.1 makes no difference.

9.49 $n = 30$, $\bar{x} = 25.23$, $\sigma = 1.4$

Step 1: $H_0: \mu = 26$ mpg, $H_a: \mu < 26$ mpg

Step 2: $\alpha = 0.05$

Step 3: Critical value = -1.645

Step 4: $z = (25.23 - 26)/(1.4/\sqrt{30}) = -3.01$

Step 5: Since $-3.01 < -1.645$, reject H_0.

Step 6: At the 5% significance level, the data support the consumer group's conjecture. While the 0.77 mpg difference may not be great enough to dissuade a car buyer, it is enough to question the accuracy of the manufacturer's claim.

9.51 (a) The z-test is not appropriate. The sample size is moderate and the data are highly skewed.

(b) The z-test is not appropriate. The sample size is small and an outlier is present.

(c) The z-test is appropriate. The sample size is large. Thus mild skewness is not a problem.

9.53 (a) The following expressions are equivalent:

$$\bar{x} - z_{a/2} \cdot \frac{\sigma}{\sqrt{n}} \leq \mu_0 \leq \bar{x} + z_{a/2} \cdot \frac{\sigma}{\sqrt{n}}$$

$$- z_{a/2} \cdot \frac{\sigma}{\sqrt{n}} \leq \mu_0 - \bar{x} \leq z_{a/2} \cdot \frac{\sigma}{\sqrt{n}}$$

$$- z_{a/2} \cdot \frac{\sigma}{\sqrt{n}} \leq \bar{x} - \mu_0 \leq z_{a/2} \cdot \frac{\sigma}{\sqrt{n}}$$

$$- z_{a/2} \leq \frac{\bar{x} - \mu_0}{\sigma/\sqrt{n}} \leq z_{a/2}$$

(b) From part (a) we see that μ_0 lies in the $(1 - \alpha)$-level confidence interval for μ if and only if the test statistic

$$z = \frac{\bar{x} - \mu_0}{\sigma/\sqrt{n}}$$

lies in the nonrejection region.

Exercises 9.4

9.55 (a) A Type I error occurs if the data leads to rejecting the null hypothesis when it is, in fact, true.

(b) A Type II error occurs if the data leads to not rejecting the null hypothesis when it is, in fact, false.

(c) The significance level is the probability associated with the test procedure of rejecting the null hypothesis when it is actually true, i.e., it is the probability of making a Type I error.

9.57 Since μ is unknown, the power curve enables one to evaluate the effectiveness of a hypothesis test for a variety of values of μ.

9.59   If the significance level is decreased without changing the sample size, the rejection region is made smaller (in probability terms). This makes the non-rejection region larger, i.e., β gets larger. This, in turn, makes the power $1 - \beta$ smaller.

9.61   (a)   Note: $z = \dfrac{\overline{x} - \mu_0}{\sigma/\sqrt{n}}$ $\Rightarrow$ $\overline{x} = \mu_0 + z \cdot \sigma/\sqrt{n}$

Since this is a right-tailed test, we would reject H_0 if $z \geq 1.645$; or equivalently if $\overline{x} \geq 690 + 1.645(350)/\sqrt{40} = 781.0$.

So we reject H_0 if $\overline{x} \geq 781.0$; otherwise do not reject H_0.

(b) $\alpha = 0.05$

(c)

True mean μ	z-score computation	P(Type II error) β	Power $1 - \beta$
720	$z = \dfrac{781.0 - 720}{350/\sqrt{40}} = 1.10$	0.8643	0.1357
750	$z = \dfrac{781.0 - 750}{350/\sqrt{40}} = 0.56$	0.7123	0.2877
780	$z = \dfrac{781.0 - 780}{350/\sqrt{40}} = 0.02$	0.5080	0.4920
810	$z = \dfrac{781.0 - 810}{350/\sqrt{40}} = -0.52$	0.3015	0.6985
840	$z = \dfrac{781.0 - 840}{350/\sqrt{40}} = -1.07$	0.1423	0.8577
870	$z = \dfrac{781.0 - 870}{350/\sqrt{40}} = -1.61$	0.0537	0.9463
900	$z = \dfrac{781.0 - 900}{350/\sqrt{40}} = -2.15$	0.0158	0.9842
930	$z = \dfrac{781.0 - 930}{350/\sqrt{40}} = -2.69$	0.0036	0.9964

(d)

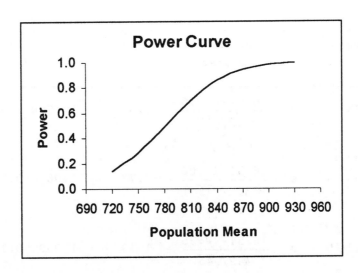

9.63   (a)   Note: $z = \dfrac{\overline{x} - \mu_0}{\sigma/\sqrt{n}} \Rightarrow \overline{x} = \mu_0 + z \cdot \sigma/\sqrt{n}$

Since this is a left-tailed test, we would reject H_0 if $z \leq -2.33$; or equivalently if $\overline{x} \leq 18 - 2.33(4.2)/\sqrt{45} = 16.54$.

So reject H_0 if $\overline{x} \leq 16.54$; otherwise do not reject H_0.

(b) $\alpha = 0.01$

(c)

True mean μ	z-score computation	P(Type II error) β	Power $1 - \beta$
15.50	$z = \dfrac{16.54 - 15.50}{4.2/\sqrt{45}} = 1.66$	1.0000−0.9515=0.0485	0.9515
15.75	$z = \dfrac{16.54 - 15.75}{4.2/\sqrt{45}} = 1.26$	1.0000−0.8962=0.1038	0.8962
16.00	$z = \dfrac{16.54 - 16.00}{4.2/\sqrt{45}} = 0.86$	1.0000−0.8051=0.1949	0.8051
16.25	$z = \dfrac{16.54 - 16.25}{4.2/\sqrt{45}} = 0.46$	1.0000−0.6772=0.3228	0.6772

True mean μ	z-score computation	P(Type II error) β	Power $1 - \beta$
16.75	$z = \dfrac{16.54 - 16.75}{4.2/\sqrt{45}} = -0.34$	$1.0000 - 0.3669 = 0.6331$	0.3669
17.00	$z = \dfrac{16.54 - 17.00}{4.2/\sqrt{45}} = -0.73$	$1.0000 - 0.2327 = 0.7673$	0.2327
17.25	$z = \dfrac{16.54 - 17.25}{4.2/\sqrt{45}} = -1.13$	$1.0000 - 0.1292 = 0.8708$	0.1292
17.50	$z = \dfrac{16.54 - 17.50}{4.2/\sqrt{45}} = -1.53$	$1.0000 - 0.0630 = 0.9370$	0.0630
17.75	$z = \dfrac{16.54 - 17.75}{4.2/\sqrt{45}} = -1.93$	$1.0000 - 0.0268 = 0.9732$	0.0268

(d)

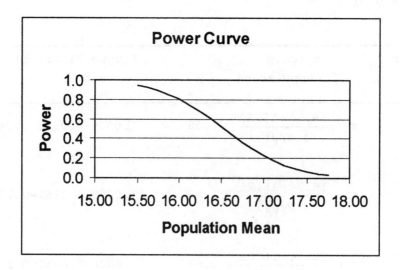

9.65   (a)   Note: $z = \dfrac{\bar{x} - \mu_0}{\sigma/\sqrt{n}} \Rightarrow \bar{x} = \mu_0 + z \cdot \sigma/\sqrt{n}$

Since this is a two-tailed test, we would reject H_0 if $|z| \geq 2.575$;

equivalently if $\bar{x} \leq 98.6 - 2.575(0.63)/\sqrt{93} = 98.43.$ or

$$\bar{x} \geq 98.6 + 2.575(0.63)/\sqrt{93} = 98.77$$

So reject H_0 if $\bar{x} \leq 98.43$ or $\bar{x} \geq 98.77$; else do not reject H_0.

(b) $\alpha = 0.01$

(c)

True mean μ	z-score computation	P(Type II error) β	Power $1 - \beta$
98.30	$z = \dfrac{98.43-98.30}{0.63/\sqrt{93}} = 1.99$ $z = \dfrac{98.77-98.30}{0.63/\sqrt{93}} = 7.19$	$1.0000 - 0.9767 = 0.0233$	0.9767
98.35	$z = \dfrac{98.43-98.35}{0.63/\sqrt{93}} = 1.22$ $z = \dfrac{98.77-98.35}{0.63/\sqrt{93}} = 6.43$	$1.0000 - 0.8888 = 0.1112$	0.8888
98.40	$z = \dfrac{98.43-98.40}{0.63/\sqrt{93}} = 0.46$ $z = \dfrac{98.77-98.40}{0.63/\sqrt{93}} = 5.66$	$1.0000 - 0.6772 = 0.3228$	0.6772
98.45	$z = \dfrac{98.43-98.45}{0.63/\sqrt{93}} = -0.31$ $z = \dfrac{98.77-98.45}{0.63/\sqrt{93}} = 4.90$	$1.0000 - 0.3783 = 0.6217$	0.3783

True mean μ	z-score computation	P(Type II error) β	Power $1 - \beta$
98.50	$z = \dfrac{98.43 - 98.50}{0.63/\sqrt{93}} = -1.07$ $z = \dfrac{98.77 - 98.50}{0.63/\sqrt{93}} = 4.13$	$1.0000 - 0.1423 = 0.8577$	0.1423
98.55	$z = \dfrac{98.43 - 98.55}{0.63/\sqrt{93}} = -1.84$ $z = \dfrac{98.77 - 98.55}{0.63/\sqrt{93}} = 3.37$	$0.9996 - 0.0329 = 0.9667$	0.0333
98.65	$z = \dfrac{98.43 - 98.65}{0.63/\sqrt{93}} = -3.37$ $z = \dfrac{98.77 - 98.65}{0.63/\sqrt{93}} = 1.84$	$0.9671 - 0.0004 = 0.9667$	0.0333
98.70	$z = \dfrac{98.43 - 98.70}{0.63/\sqrt{93}} = -4.13$ $z = \dfrac{98.77 - 98.70}{0.63/\sqrt{93}} = 1.07$	$0.8577 - 0.0000 = 0.8577$	0.1423
98.75	$z = \dfrac{98.43 - 98.75}{0.63/\sqrt{93}} = -4.90$ $z = \dfrac{98.77 - 98.75}{0.63/\sqrt{93}} = 0.31$	$0.6217 - 0.0000 = 0.6217$	0.3783
98.80	$z = \dfrac{98.43 - 98.80}{0.63/\sqrt{93}} = -5.66$ $z = \dfrac{98.77 - 98.80}{0.63/\sqrt{93}} = -0.46$	$0.3228 - 0.0000 = 0.3228$	0.6772

True mean μ	z-score computation	P(Type II error) β	Power $1 - \beta$
98.85	$z = \dfrac{98.43 - 98.85}{0.63/\sqrt{93}} = -6.43$ $z = \dfrac{98.77 - 98.85}{0.63/\sqrt{93}} = -1.22$	$0.1112 - 0.0000 = 0.1112$	0.8888
98.90	$z = \dfrac{98.43 - 98.90}{0.63/\sqrt{93}} = -7.19$ $z = \dfrac{98.77 - 98.90}{0.63/\sqrt{93}} = -1.99$	$0.0233 - 0.0000 = 0.0233$	0.9767

(d)

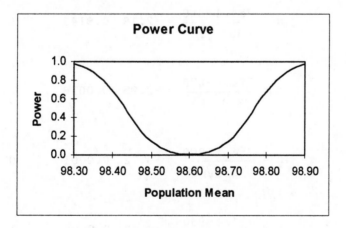

9.67 From Exercise 9.61

(a) Note: $z = \dfrac{\overline{x} - \mu_0}{\sigma/\sqrt{n}} \Rightarrow \overline{x} = \mu_0 + z \cdot \sigma/\sqrt{n}$

Since this is a right-tailed test, we would reject H_0 if $z \geq$ 1.645; or equivalently if $\overline{x} \geq 690 + 1.645(350)/\sqrt{80} = 754.4$.

So reject H_0 if $\overline{x} \geq 754.4$; otherwise do not reject H_0.

(b) $\alpha = 0.05$

(c)

True mean μ	z-score computation		P(Type II error) β	Power $1 - \beta$
720	$z = \dfrac{754.4 - 720}{350/\sqrt{80}}$	$= 0.88$	0.8106	0.1894
750	$z = \dfrac{754.4 - 750}{350/\sqrt{80}}$	$= 0.11$	0.5438	0.4562
780	$z = \dfrac{754.4 - 780}{350/\sqrt{80}}$	$= -0.65$	0.2578	0.7422
810	$z = \dfrac{754.4 - 810}{350/\sqrt{80}}$	$= -1.42$	0.0778	0.9222
840	$z = \dfrac{754.4 - 840}{350/\sqrt{80}}$	$= -2.19$	0.0143	0.9857
870	$z = \dfrac{754.4 - 870}{350/\sqrt{80}}$	$= -2.95$	0.0016	0.9984
900	$z = \dfrac{754.4 - 900}{350/\sqrt{80}}$	$= -3.72$	0.0001	0.9999
930	$z = \dfrac{754.4 - 930}{350/\sqrt{80}}$	$= -4.48$	0.0000	1.0000

(d)

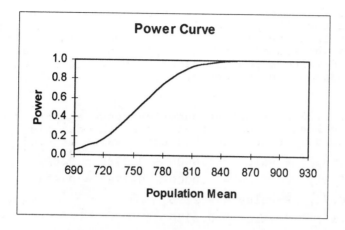

Comparing the Power curves, we can see that the principle which seems to be illustrated is that increasing the sample size for a hypothesis test without changing the significance level increases the power of the test.

9.69 (a)

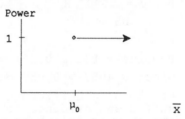

(b)   The curve in (a) portrays that, ideally, one desires the value for the power for any given μ_a to be as close to 1 as possible.

9.71 (a)

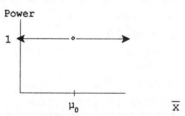

(b)   The curve in (a) portrays that, ideally, one desires the value for the power for any given μ_a to be as close to 1 as possible.

Exercises 9.5

9.73   Two reasons why it is a good idea to include the P-value are: (1) it permits people reading only the conclusions of a study to make their own evaluation; and (2) it provides the reader with the actual significance of the test.

9.75   In the **critical value** approach, we determine critical values based on the significance level. The critical values determine where the rejection and nonrejection regions lie for the test statistic. If the value of the test statistic falls in the rejection region, the null hypothesis is rejected. In the **P-value** approach, the test statistic is computed and then the probability of observing a value as extreme or more extreme than the value obtained is determined. If the P-value is smaller than the significance level, the null hypothesis is rejected. Reporting a P-value allows a reader to draw his/her own conclusion based on the strength of the evidence.

9.77   True

9.79   A P-value of 0.02 provides stronger evidence against the null hypothesis than does a value of 0.03. It says that the data is less likely if the null hypothesis is true than when the P-value is 0.03.

9.81   (a) $z = 2.03$,   P-value = $1.0000 - 0.9788 = 0.0212$

  (b) $z = -0.31$,   P-value = $1.0000 - 0.3783 = 0.6217$

9.83   (a) $z = -0.74$,   P-value = 0.2296

  (b) $z = 1.16$,   P-value = 0.8770

9.85   (a) $z = -1.66$,   Left-tail probability = 0.0485

      P-value = $0.0485 \times 2 = 0.0970$

  (b) $z = 0.52$,   Right-tail probability = $1.0000 - 0.6985 = 0.3015$

      P-value = $0.3015 \times 2 = 0.6030$

9.87   (See Exercise 9.43 for classical approach results.)

  Step 1: H_0: $\mu = \$690$, H_a: $\mu > \$690$

  Step 2: $\alpha = 0.05$

  Step 3: $z = 0.26$

  Step 4:   P-value = $P(z \geq 0.26) = 1.0000 - 0.6026 = 0.3974$

  Step 5:   Since $0.4052 > 0.05$, do not reject H_0.

  Step 6:   At the 5% significance level, the data do not provide sufficient evidence to conclude that last year's mean amount μ of telephone expenditure per consumer unit has increased over the 1994 mean of \$690.

  Using Table 9.12, we classify the strength of evidence against the null hypothesis as weak or none because $P > 0.10$.

9.89   (See Exercise 9.45 for classical approach results.)

  Step 1: H_0: $\mu = 18$ mg, H_a: $\mu < 18$ mg

  Step 2: $\alpha = 0.01$

  Step 3: $z = -5.30$

  Step 4:   P-value = $P(z < -5.30) = 0.0000$ (to four decimal places)

  Step 5:   Since $0.0000 < 0.01$, reject H_0.

  Step 6:   At the 1% significance level, the data provide sufficient evidence to conclude that adult females under the age of 51 are, on the average, getting less than the RDA of 18 mg of iron.

  Using Table 9.12, we classify the strength of evidence against the null hypothesis as very strong because $P < 0.01$.

9.91   (See Exercise 9.47 for classical approach results.)

  (a)   Step 1: H_0: $\mu = 98.6$, H_a: $\mu \neq 98.6$

    Step 2: $\alpha = 0.01$

Step 3: $z = -7.35$

Step 4:   P-value $= P(z \le -7.35$ or $z \ge 7.35) = 2(0.0000) = 0.0000$

Step 5:   Since $0.0000 < 0.05$, reject H_0.

Step 6:   At the 1% significance level, the data provide sufficient evidence to conclude that the mean body temperature of healthy humans differs from the value of 98.6 degrees Fahrenheit.

Using Table 9.12, we classify the strength of evidence against the null hypothesis as very strong P < 0.01.

9.93 (a)   We used Minitab to produce the boxplot at the right. With the data in a column named MPG, **choose Graph ▶ Boxplot...**, select MPG for the **Y** variable for **Graph 1**. Click **OK**. One observation, 29.6, is identified as a potential outlier.

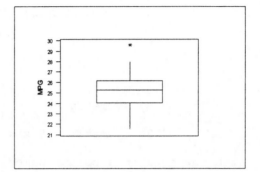

(b)   See Exercise 9.49 for classical approach results.)

Step 1: H_0: $\mu = 26$ mpg, H_a: $\mu < 26$ mpg

Step 2: $\alpha = 0.05$

Step 3: $z = -3.01$

Step 4:   P-value $= P(z \le -3.01) = 0.0013$

Step 5:   Since $0.0013 < 0.05$, reject H_0.

Step 6:   At the 5% significance level, the data support the consumer group's conjecture.

Using Table 9.12, we classify the strength of evidence against the null hypothesis as very strong because P < 0.01.

(c)   After removing 29.6 from the data, we conduct the hypothesis again:

$n = 29$, $\overline{x} = 25.079$, $\sigma = 1.4$

Step 1: H_0: $\mu = 26$ mpg, H_a: $\mu < 26$ mpg

Step 2: $\alpha = 0.05$

Step 3:   Critical value $= -1.645$

Step 4: $z = (25.079 - 26)/(1.4/\sqrt{29}) = -3.54$

Step 4:   P-value $= P(z \le -3.54) = 0.0002$

Step 5:   Since $0.0002 < 0.05$, reject H_0.

Step 6:   At the 5% significance level, the data support the consumer group's conjecture.

Using Table 9.12, we classify the strength of evidence against the null hypothesis as very strong because P < 0.01.

(d)   Removing the outlier at the high end of the data had the effect of lowering the sample mean and making the z-value for the test even more negative.

(e)   The conclusion regarding the hypothesis test is the same whether or not the potential outlier is removed. The data support the consumer group's conjecture.

9.95 (a) The P-value is expressed as $P(z \le z_0)$ if the hypothesis test is left-tailed.

(b) The P-value is expressed as $P(|z| \ge |z_0|)$ if the test is two-tailed.

9.97 Given that x can be transformed to z (and x_0 to z_0), we have:

1. $P(x \ge x_0) = P(z \ge z_0)$, for a right-tailed test

2. $P(x \le x_0) = P(z \le z_0)$, for a left-tailed test

3. $2 \cdot \min \{P(x \le x_0), P(x \ge x_0)\} = 2 \cdot \min \{P(z \le z_0), P(z \ge z_0)\}$

$$= \begin{cases} 2 \cdot P(z \le z_0) & \text{if } z_0 < 0 \\ 2 \cdot P(z \ge z_0) & \text{if } z_0 \ge 0 \end{cases}$$

$$\text{By symmetry} = \left. \begin{cases} 2 \cdot P(z \ge -z_0) & \text{if } z_0 < 0 \\ 2 \cdot P(z \ge z_0) & \text{if } z_0 \ge 0 \end{cases} \right\} = 2 \cdot P(z \ge |z_0|)$$

$$\text{By symmetry} = P(z \le -|z_0|) + P(z \ge |z_0|) = P(|z| \ge |z_0|)$$

9.99 With the data in a column named EXPENDIT, we choose **Stat ▶ Basic Statistics ▶ 1-Sample z...**, select EXPENDIT in the **Variables** text box, click on the **Test mean** button, type <u>690</u> in the **Test mean** text box, click on the arrow in the **Alternative** text box and select 'greater than,' type <u>350</u> in the Σ text box, and click **OK**. The result is

Test of μ = 690.0 vs μ > 690.0
The assumed σ = 350

Variable	N	Mean	StDev	SE Mean	Z	P-Value
EXPENDIT	40	704.4	384.4	55.3	0.26	0.40

The P-value of 0.40 is greater than the significance level of .05, so we conclude that the data do not provide sufficient evidence that the mean value has increased over the 1994 mean of $690.

9.101 With the data in a column named MPG, we choose **Stat ▶ Basic Statistics ▶ 1-Sample z...**, select MPG in the **Variables** text box, click on the **Test mean** button, type <u>26</u> in the **Test mean** text box, click on the arrow in the **Alternative** text box and select 'less than,' type <u>1.4</u> in the Σ text box, and click **OK**. The result is

Test of μ = 26.000 vs μ < 26.000
The assumed σ = 1.40

Variable	N	Mean	StDev	SE Mean	Z	P-Value
MPG	30	25.230	1.592	0.256	-3.01	0.0013

The P-value of 0.0013 is less than the significance level of .05, so we conclude that the data provide sufficient evidence that the mean value is less than the manufacturer's claim of 26 mpg.

9.103 (a) H_0: $\mu = 6.7$ days, H_a: $\mu < 6.5$ days

(b)   THE ASSUMED SIGMA = 7.70

(c)   STDEV = 7.01

(d)   N = 40

(e)   MEAN = 6.450 days

(f)   Z = -0.21

(g)   P-VALUE = 0.42

(h)   The smallest significance level at which H_0 can be rejected is 0.42.

(i)   Since 0.42 > 0.05, do not reject H_0. Thus, at the 5% significance level, the data do not provide sufficient evidence to conclude that the mean hospital stay μ for this year will be less than the 1994 mean of 6.7 days.

Exercises 9.6

9.105 In the z-test, it is assumed that σ is known. In the t-test, σ is unknown.

9.107 n = 36, df = 35, $\bar{x} = 1582.11$, s = 351.69

Step 1: H_0: $\mu = \$1644$, H_a: $\mu \neq \$1644$

Step 2: $\alpha = 0.05$

Step 3:   Critical values = ±2.030

Step 4: $t = (1582.11 - 1644)/(351.69/\sqrt{36}) = -1.056$

Step 5:   Since -2.030 < -1.056 < 2.030, do not reject H_0. Note: For the p-value approach, p-value > 0.20. So, since the p-value > α, do not reject H_0.

Step 6:   At the 5% significance level, the data do not provide sufficient evidence to conclude that the 1994 mean apparel and services expenditure μ for households in the Midwest differed from the national average of \$1567.

9.109 n = 25, df = 24, $\bar{x} = 436.4$, s = 85.5

Step 1: H_0: $\mu = 428$, H_a: $\mu > 428$

Step 2: $\alpha = 0.10$

Step 3:   Critical value = 1.318

Step 4: $t = (436.4 - 428)/(85.5/\sqrt{25}) = 0.491$

Step 5:   Since 0.491 < 1.318, do not reject H_0. Note: For the P-value approach, p-value > 0.10. So, since the p-value > α, do not reject H_0.

Step 6:   At the 10% significance level, the data do not provide sufficient evidence to conclude that the mean SAT verbal score μ for last year is greater than the mean of 428 points for 1995.

9.111 n = 15, df = 14, $\bar{x} = 40.79$, s = 3.501

Step 1: H_0: $\mu = 46.0$¢/lb, H_a: $\mu \neq 46.0$¢/lb

Step 2: $\alpha = 0.05$

Step 3:   Critical values = ±2.145

Step 4: $t = (48.4 - 46.0)/(3.501/\sqrt{15}) = 2.656$

Step 5: Since 2.656 > 2.145, reject H_0. Note: For the p-value approach, $0.01 < P < 0.02$. So, since the p-value $< \alpha$, reject H_0.

Step 6: At the 5% significance level, the data provide sufficient evidence to conclude that the mean retail price μ for bananas now is different from the 1994 mean of 46.0 cents per pound. The practical significance of this result is minimal since the difference is only 2.4¢/lb.

9.113 $n = 40$, df $= 39$, $\overline{x} = 58.40$, $s = 20.42$

(a) Step 1: H_0: $\mu = 64$ lb, H_a: $\mu < 64$ lb

Step 2: $\alpha = 0.05$

Step 3: Critical value $= -1.690$

Step 4: $t = (58.40 - 64)/(20.42/\sqrt{40}) = -1.734$

Step 5: Since $-1.734 < -1.690$, reject H_0. Note: For the p-value approach, $0.025 <$ p-value < 0.05. So, since p-value $< \alpha$, reject H_0.

Step 6: At the 5% significance level, it appears that last year's mean beef consumption is less than the 1990 mean of 64 lbs. This result has would have little practical significance to the individual beef eater, but to beef producers, it would have great practical significance since the apparent decline in consumption by the total population is considerable.

(b) After removing the four potential outliers,

$n = 36$, df $= 35$, $\overline{x} = 64.11$, $s = 11.02$

Step 1: H_0: $\mu = 64$ lb, H_a: $\mu < 64$ lb

Step 2: $\alpha = 0.05$

Step 3: Critical value $= -1.690$

Step 4: $t = (64.11 - 64)/(11.02/\sqrt{36}) = 0.060$

Step 5: Since $0.060 > -1.690$, do not reject H_0. Note: For the p-value approach, p-value > 0.10. So, since p-value $> \alpha$, do not reject H_0.

Step 6: At the 5% significance level, there is not sufficient evidence to conclude that last year's mean beef consumption is less than the 1990 mean of 64 pounds.

(c) The results in parts (a) and (b) are very different. With the potential outliers included in the data, there appears to be a decrease in the mean beef consumption. With the outliers removed, no such decrease is apparent.

(d) Since the Department of Agriculture's 1990 data would have included the beef consumption of people who ate little or no beef, it is entirely appropriate to include data from such people in last year's sample.

(e) Since the results of the analysis change considerably when the potential outliers are removed, it would be best to increase the size of the sample to reduce the effect of the potential outliers on the results or to use a non-parametric method of analysis which does not assume normality.

9.115 (a) $n = 10$, $df = 9$, $\bar{x} = 1050.2$ hours, $s = 65.8$ hours

Step 1: H_0: $\mu = 1000$ hours, H_a: $\mu > 1000$ hours

Step 2: $\alpha = 0.01$

Step 3:   Critical value = 2.821

Step 4: $z = (1050.2 - 1000)/(65.8/\sqrt{10}) = 2.413$

Step 5:   Since 2.413 < 2.821, do not reject H_0. Note: For the P-value approach, 0.010 < P-value < 0.025. So, since P-value > α, do not reject H_0.

Step 6:   At the 1% significance level, the data do not provide sufficient evidence to conclude that the new bulbs will outlast the old bulbs.

(b) (i)   If z were used instead of t, the critical value of z would be 2.33.

(ii)   The actual critical value of t used was 2.821.

(iii)   The mistaken use of a z critical value, when a t critical value should be used, makes it more likely that the null hypothesis will be rejected. Particularly for small sample sizes, the critical values of t are much larger in absolute value (for given α-levels) than critical values of z, thus making rejection of H_0 less probable. We see this occurring directly above. Using z, we have 2.41 > 2.33, so we reject H_0. Using t, we have 2.41 < 2.821, so we do not reject H_0.

9.117 (a)   If t were mistakenly used instead of z, the critical value of t would be 1.761. The critical value of z that should have been used is 1.645. The desired significance level corresponding to the correct z-value of 1.645 is 0.05. The significance level that corresponds with the critical value actually used is found by consulting the z table. A z-value of 1.761 results in a tail area (i.e., a significance level) of 0.0392. This is lower than the desired significance level of 0.05.

(b)   In the situation described, if the t-table is mistakenly used instead of the z-table to obtain the critical value(s), the actual significance level of the resulting test will be lower than α. This is demonstrated in part (a).

9.119 (a)   With the data in a column named EXPENDIT, to obtain the normal probability plot, we choose **Calc ▶ Calculator...**, type <u>NSCORE</u> in the **Store result in variable** text box, select **Normal scores** from the function list, select EXPENDIT to replace **numbers** in the **Expression** text box, and click **OK**. We now choose **Graph ▶ Plot...**, select NSCORE for the **Y** variable for **Graph 1** and EXPENDIT for the **X** variable, and click **OK**. To obtain the boxplot, we choose **Graph ▶ Boxplot...**, select EXPENDIT for the **Y** variable for **Graph1**, and click **OK**. To obtain the histogram, we choose **Graph ▶ Histogram...**, select EXPENDIT for the **Y** variable for **Graph1**, and click **OK**. To obtain the stem-and-leaf plot, we choose **Graph ▶ Stem-and-leaf...**, select EXPENDIT in the **Variables** text box, and click **OK**. The four resulting plots are

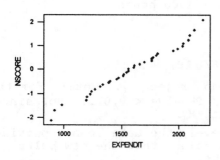

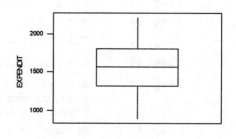

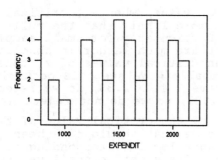

```
Stem-and-leaf of EXPENDIT   N  = 36
Leaf Unit = 10

    1      8  8
    3      9  18
    3     10
    5     11  99
    8     12  236
   12     13  1389
   14     14  57
   (5)    15  11356
   17     16  0159
   13     17  5569
    9     18  1
    8     19  58
    6     20  3489
    2     21  3
    1     22  1
```

(b)   To perform the one-sample t-test, we choose **Stat ▶ Basic**
       statistics ▶ 1-Sample t..., select EXPENDIT in the **Variables** text
       box, click the **Test Mean** button and type <u>1644</u> in the **Test mean**
       text box, select **not equal** in the **Alternative** box, and click **OK**.
       The results in the Session Window are

       Test of μ = 1644.0 vs μ not = 1644.0

Variable	N	Mean	StDev	SE Mean	T	P
EXPENDIT	36	1582.1	351.7	58.6	-1.06	0.30

(c)   None of the plots indicate any serious departure from the
       assumption of normality required for the t-test, so the procedure
       in part (b) is justified.

9.121 (a) H_0: μ = 711, H_a: μ < 711

       (b)   STDEV = 237.5

       (c)   N = 37

       (d)   MEAN = 517.4

       (e)   T = -4.96

(f)   P-VALUE = 0.0000

(g)   The smallest significance level at which H_0 can be rejected is 0.0000 to four decimal places.

(h)   Since 0.0000 < 0.05, reject H_0. At the 5% significance level, the data provide sufficient evidence to conclude that, on the average, the fuel consumed per vehicle in Hawaii in 1995 is less than in the United States as a whole. The practical significance of this result is that Hawaiian drivers get better mileage due to the constantly warm temperatures and they are, of course, somewhat more limited in the distances which they can drive as a result of living on one of the Hawaiian islands. As a result, their cars probably last longer and dealers probably sell fewer new cars.

9.123 (a)   With the data in a column named BEEF, choose **Graph ▶ Boxplot...**, select BEEF as the **Y** variable for **Graph 1**, and click **OK**. The plot is shown at the right. Although it appears that there are three potential outliers, there are two observations of zero, and thus there are four potential outliers (0, 0, 8, and 20).

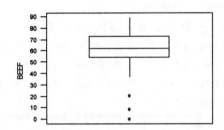

(b)   To perform the t-test with the original data, choose **Stat ▶ Basic Statistics ▶ 1-sample t...** and select Beef in the **Variables** text box, click on **Test mean**, and enter <u>64</u> in the **Test mean** text box. Choose **less than** from the **Alternative** text box, and click **OK**. The results are

Test of $\mu = 64.00$ vs $\mu < 64.00$

Variable	N	Mean	StDev	SE Mean	T	P-Value
BEEF	40	58.40	20.42	3.23	-1.73	0.045

(c)   Enter the Data window and delete the four data values 0, 0, 8, and 20. Then repeat the process in part (b) above. The results are

Test of $\mu = 64.00$ vs $\mu < 64.00$

Variable	N	Mean	StDev	SE Mean	T	P-Value
BEEF	36	64.11	11.02	1.84	0.06	0.52

(d)   Deleting the four low outliers raises the mean, decreases the standard deviation, and changes the P-value from 0.045 to 0.52 from strong evidence against the null hypothesis to weak or no evidence against the null hypothesis. Since the four outliers are presumably legitimate observations, they can not simply be deleted and ignored. Therefore, either the sample size should be increased to reduce the effect of outliers on the result or a nonparametric method of analysis which is not based on the normality assumption should be used.

Exercises 9.7

9.125 The advantages of nonparametric methods are that they do not require normality, they make use of fewer and simpler calculations than do parametric methods, and they are resistant to outliers. The disadvantage of nonparametric methods is that they tend to give less accurate results than parametric methods when the assumptions underlying the parametric methods are actually met.

9.127 Because the D-value for such a data value equals 0 and so we cannot attach a sign to the rank of $|D|$.

9.129 (a)   Wilcoxon signed-rank test   (b)   Wilcoxon signed-rank test

(c)   Neither

9.131 $\eta_0 = 34.3$, $n = 10$, $\alpha = 0.05$

Step 1: H_0: $\eta = 34.3$, H_a: $\eta > 34.3$

Step 2: $\alpha = 0.05$

Step 3:   Critical value = 44

Step 4:

x	$x-\eta_0=D$	$\|D\|$	Rank of $\|D\|$	Signed Rank R
40	5.7	5.7	3	3
43	8.7	8.7	4	4
60	25.7	25.7	10	10
47	12.7	12.7	6	6
12	-22.3	22.3	8	-8
37	2.7	2.7	2	2
55	20.7	20.7	7	7
9	-25.3	25.3	9	-9
34	-0.3	0.3	1	-1
24	-10.3	10.3	5	-5

Step 5:   W = sum of the + ranks = 32

Step 6:   Since W < 44, do not reject H_0.

Step 7:   At the 5% significance level, the data do not provide sufficient evidence to conclude that the median age has increased over the 1995 median age of 34.3 years.

9.133 $\mu_0 = 12$, $n = 14$, $\alpha = 0.05$ (The value of 12.0 was deleted from the original sample)

Step 1: H_0: $\mu = 12$, H_a: $\mu < 12$

Step 2: $\alpha = 0.05$

Step 3:   Critical value = 26

Step 4:

x	$x - \mu_0 = D$	$\lvert D \rvert$	Rank of $\lvert D \rvert$	Signed Rank R
10.9	−1.1	1.1	2	−2
15.0	3.0	3.0	9	9
14.2	2.2	2.2	4.5	4.5
11.4	−0.6	0.6	1	−1
7.1	−4.9	4.9	12	−12
9.2	−2.8	2.8	7	−7
10.1	−1.9	1.9	3	−3
9.2	−2.8	2.8	7	−7
8.8	−3.2	3.2	10	−10
9.8	−2.2	2.2	4.5	−4.5
6.6	−5.4	5.4	13	−13
4.4	−7.6	7.6	14	−14
14.8	2.8	2.8	7	7
8.0	−4.0	4.0	11	−11

Step 5:   W = sum of the + ranks = 20.5

Step 6:   Since W < 26, reject H_0.

Step 7:   At the 5% significance level, the data do provide sufficient evidence to conclude that the new antacid tablet works faster.

9.135 $\mu_0 = 46.0$, $n = 15$, $\alpha = 0.05$

(a)   Step 1: H_0: $\mu = 46.0$, H_a: $\mu \neq 46.0$

Step 2: $\alpha = 0.05$

Step 3:   Critical values = 21, 84

Step 4:

x	$x - \mu_0 = D$	$\lvert D \rvert$	Rank of $\lvert D \rvert$	Signed Rank R
48	2	2	3.5	3.5
50	4	4	8	8
48	2	2	3.5	3.5
45	−1	1	1.5	−1.5
52	6	6	12	12
53	7	7	14	14
49	3	3	5.5	5.5
43	−3	3	5.5	−5.5
42	−4	4	8	−8
45	−1	1	1.5	−1.5
52	6	6	12	12
52	6	6	12	12
46	0	0	---	---
50	4	4	8	8
51	5	5	10	10

Step 5:   W = sum of the + ranks = 88.5

Step 6:   Since W > 84, reject H₀.

Step 7:   At the 5% significance level, the data provide sufficient evidence to conclude that the mean retail price for bananas now is different from the 1994 mean of 46.0 cents per pound.

(b)   A Wilcoxon signed-rank test is permissible because a normally distribution population is symmetric.

9.137 n = 16, $\bar{x}$ = 306, s = 8.671, α = 0.05

(a)   Step 1:   H₀: μ = 310, Hₐ: μ < 310

    Step 2:   α = 0.05

    Step 3:   Critical value = -1.753

    Step 4: $t = (306 - 310)/(8.671/\sqrt{16}) = -1.845$

    Step 5:   Since -1.845 < -1.753, reject H₀. Note: For the p-value approach, 0.025 < p-value < 0.05. So, since p-value < α, reject H₀.

    Step 6:   At the 5% significance level, the data do provide sufficient evidence to conclude that the mean content, μ, is less than the advertised content of 310 ml.

(b)   Step 1:   H₀: μ = 310, Hₐ: μ < 310

    Step 2:   α = 0.05

    Step 3:   Critical value = 36

    Step 4:

x	x-μ₀=D	|D|	Rank of|D|	Signed Rank R
297	-13	13	14	-14
311	1	1	2	2
322	12	12	12.5	12.5
315	5	5	7	7
318	8	8	9	9
303	-7	7	8	-8
307	-3	3	5	-5
296	-14	14	15	-15
306	-4	4	6	-6
291	-19	19	16	-16
312	2	2	4	4
309	-1	1	2	-2
300	-10	10	10.5	-10.5
298	-12	12	12.5	-12.5
300	-10	10	10.5	-10.5
311	1	1	2	2

Step 5:   W = sum of the + ranks = 36.5

Step 6:   Since W > 36, do not reject H₀.

Step 7:   At the 5% significance level, the data do not provide sufficient evidence to conclude that the mean content, μ, is less than the advertised content of 310 ml.

(c)   Since the population is normally distributed, the t-test is more powerful than the Wilcoxon signed-rank test; that is, the t-test is more likely to detect a false null hypothesis.

9.139 (a)   If John is not unlucky, he should expect to wait 15 minutes for the train, on the average.

(b)   If John is not unlucky, the distribution of the times he waits for the trains should be a uniform distribution over the interval from 0 to 30 minutes.

(c)   Step 1: H_0: $\eta = 15$, H_a: $\eta \neq 15$

Step 2: $\alpha = 0.10$

Step 3:   Critical value = 56

Step 4:

x	$x - \eta_0 = D$	|D|	Rank of |D|	Signed Rank R
24	9	9	6.5	6.5
26	11	11	9.5	8.5
20	5	5	3.5	3.5
4	-11	11	9.5	-9.5
20	5	5	3.5	3.5
3	-12	12	11	-11
19	4	4	2	4
5	-10	10	8	-8
28	13	13	12	12
16	1	1	1	1
22	7	7	5	5
24	9	9	6.5	6.5

Step 5:   W = sum of the + ranks = 50.5

Step 6:   Since W = 50.5 , do not reject H_0.

Step 7:   At the 10% significance level, the data do not provide sufficient evidence to conclude that John waits more than 15 minutes for the train, on the average.

(d)   Since the population is uniform (which is symmetric), the Wilcoxon test is appropriate.

(e)   Since the population is symmetric and non-normal, the Wilcoxon signed-rank test is more powerful than the t-test and more appropriate than the t-test which assumes normality.

9.141 (a)   Step 1: H_0: $\eta = 12$ minutes, H_a: $\eta < 12$ minutes

Step 2: $\alpha = 0.05$

Step 3:   Critical value = -1.645

Step 4:   See Step 4 in the solution to Exercise 9.133.

Step 5:   From Step 5 in the solution to Exercise 9.13, W = 20.5. Now:

$$z = \frac{W - n(n+1)/4}{\sqrt{n(n+1)(2n+1)/24}} = \frac{20.5 - 14(14+1)/4}{\sqrt{14(14+1)(2\cdot14+1)/24}} = -2.01.$$

Step 6:   Since $-2.01 < -1.645$, reject H_0.

Step 7:   It appears that the new antacid works faster.

(b)   Both the Wilcoxon signed-rank test and the normal approximation led to rejection of the null hypothesis.

9.143 (a)   Summing the ranks corresponding to the "+" signs in each row results in a value for W. All eight possible values for W are presented in the last column.

Rank			
1	2	3	W
+	+	+	6
+	+	−	3
+	−	+	4
+	−	−	1
−	+	+	5
−	+	−	2
−	−	+	3
−	−	−	0

(b)   Since there are eight equally likely outcomes, the probability that a sample will match any particular row of the table is 1/8 = 0.125.

(c)   The probability distribution of the random variable W when n = 3 is:

(d) A histogram for the probability distribution of W for n=3 is:

W	P(W)
0	0.125
1	0.125
2	0.125
3	0.250
4	0.125
5	0.125
6	0.125

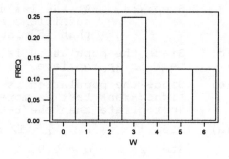

(e)   For a left-tailed test with n = 3 and α = 0.125, the critical value W_ℓ equals 0.

9.145   With the data in a column named AGES, we choose **Stat** ▶

Nonparametrics ▶ **1-Sample Wilcoxon...**, select AGES in the **Variables** text box, click on the **Test median** button and type <u>34.3</u> in its text box, select 'greater than' in the **Alternative** text box, and click **OK**. The result is

Test of median = 34.30 versus median > 34.30

	N	N for test	Wilcoxon Statistic	P	Estimated Median
AGES	10	10	32.0	0.342	37.00

9.147 (a) H_0: η = 7.2 yr, H_a: η < 7.2 yr

(b) 50 (c) 0 (d) 6.882 yr

(e) 0.450 (f) 0.450

(g) Do not reject H_0. At the 5% significance level, the data do not provide sufficient evidence to conclude that last year's median marriage duration is less than the 1995 median of 7.2 years.

(h) The Wilcoxon signed-rank procedure is based on the assumption that the underlying distribution of observations is symmetric. Marriages cannot last less than 0 years, and there are many marriages that last 25 years, sometimes 50 years and more. If the median is somewhere around 7 years, then there is a good chance that the distribution of marriage lengths is somewhat skewed to the right, making the Wilcoxon signed-rank test inappropriate.

Exercises 9.8

9.149 (a) One-sample z-test, One-sample t-test, Wilcoxon signed-rank test

(b) The z-test assumes that σ is known, and that the population is normal or the sample is large. The t-test assumes that σ is unknown, and that the population is normal or the sample is large. The signed-rank test assumes only that the population is symmetric.

(c) $= (\overline{x} - \mu_0)/(\sigma/\sqrt{n})$ $t = (\overline{x} - \mu_0)/(s/\sqrt{n}$

W = the sum of the positive ranks

9.151 (a) Yes. The t-test can be used when the sample size is large. It is almost equivalent to the z-test in this situation.

(b) Yes. The Wilcoxon signed-rank test can be used when the population distribution is symmetric.

(c) The Wilcoxon signed-rank test is preferable in this situation since it is more powerful (more likely to detect a false null hypothesis) when the population is symmetric, but non-normal.

9.153 Since we have normality and σ is known, use the z-test.

9.155 Since we have a large sample with no outliers and σ is unknown, use the t-test.

9.157 Since we have a symmetric non-normal distribution, use the Wilcoxon signed-rank test.

9.159 The distribution looks skewed and the sample size is not large. Consult a statistician.

REVIEW TEST FOR CHAPTER 9

1. (a) A null hypothesis always specifies a single value for the parameter of a population which is of interest.

(b) The alternative hypothesis reflects the purpose of the hypothesis test which can be to determine that the parameter of interest is greater than, less than, or different from the single value specified in the null hypothesis.

(c) The test statistic is a quantity calculated from the sample, under the assumption that the null hypothesis is true, which is used as a basis for deciding whether or not to reject the null hypothesis.

(d)   The rejection region is a set of values of the test statistic which lead to rejection of the null hypothesis.

(e)   The nonrejection region is a set of values of the test statistic which lead to not rejecting the null hypothesis.

(f)   The critical values are values of the test statistic that separate the rejection region from the nonrejection region.

2.   (a)   The statement is expressing the fact that there is variability in the net weights of the boxes' content and some boxes may actually contain less than the printed weight on the box. However, the net weights for each day's production will average slightly more than the printed weight.

(b)   To test the truth of this statement, we would use a null hypothesis that stated that the population mean net weight of the boxes was <u>equal</u> to the printed weight and an alternative hypothesis that stated that the population mean net weight of the boxes was <u>greater than</u> the printed weight.

(c)   Null hypothesis: Population mean net weight = 76 oz
Alternative hypothesis: Population mean net weight > 76 oz
or
H_0: $\mu = 76$
H_a: $\mu > 76$

3.   (a)   Roughly speaking, there is a range of values of the test statistic which one could reasonably expect to occur if the null hypothesis were true. If the value of the test statistic is one which would not be expected to occur when the null hypothesis is true, then we reject the null hypothesis.

(b)   To make this procedure objective and precise, we specify the probability with which we are willing to reject the null hypothesis when it is actually true. This is called the significance level of the test and is usually some small number like 0.05 or 0.01. Specifying the significance level allows us to determine the range of values of the test statistic that will lead to rejection of the null hypothesis. If the computed value of the test statistic falls in this "rejection region," then the null hypothesis is rejected. If it does not fall in the rejection region, then the null hypothesis is not rejected.

4.   We would use the alternative hypothesis $\mu \neq \mu_0$ if we wanted to determine whether the population mean were <u>different from</u> the value μ_0 specified in the null hypothesis. We would use the alternative hypothesis $\mu > \mu_0$ if we wanted to determine whether the population mean were <u>greater than</u> from the value μ_0 specified in the null hypothesis. We would use the alternative hypothesis $\mu < \mu_0$ if we wanted to determine whether the population mean were <u>less than</u> the value μ_0 specified in the null hypothesis.

5.   (a)   A Type I error is made whenever the null hypothesis is true, but the value of the test statistic leads us to reject the null hypothesis. A Type II error is made whenever the null hypothesis is false, but the value of the test statistic leads us to not reject the null hypothesis.

(b)   The probability of a Type I error is represented by α and that of a Type II error by β.

(c)   If the null hypothesis is true, the test statistic can lead us to either reject or not reject the null hypothesis. The first is the correct decision, while the latter constitutes a Type I error. Thus a Type I error is the only type of error possible when the null hypothesis is true.

(d)   If the null hypothesis is not rejected, a correct decision has been made if the null hypothesis is, in fact, true. But if the null hypothesis is false, we have made a Type II error. Thus a Type II error is the only type of error possible when the null hypothesis is not rejected.

6.   Assuming that the null hypothesis is true, find the value of the test statistic for which the probability of obtaining a value greater than the specified value is 0.05.

7.   (a)   If the population standard deviation is known, and the population is normal or the sample size is large, we can use the one-sample

z-statistic, $z = (\bar{x} - \mu_0) / (\sigma / \sqrt{n})$.

(b)   If the population standard deviation is unknown, and the population is normal or the sample size is large, we can use the

one-sample t-statistic, $t = (\bar{x} - \mu_0) / (s / \sqrt{n})$.

(c)   If the population is symmetric, we can use the Wilcoxon signed-rank statistic.

8.   (a)   A hypothesis test is exact if the actual significance level is the same as the one that is stated.

(b)   A hypothesis test is approximately correct if the significance level is only approximately equal to the one that is stated.

9.   A statistically significant result occurs when the value of the test statistic falls in the rejection region. A result has practical significance when it is statistically significant and the result also is different enough from results expected under the null hypothesis to be important to the consumer of the results. By taking large enough sample sizes, almost any result can be made statistically significant due to the increased ability of the test to detect a false null hypothesis, but small differences from the conditions expressed by the null hypothesis may not be important, that is, they may not have practical significance.

10.   The probability of a Type II error is increased when the significance level is decreased for a fixed sample size.

11.   (a)   The power of a hypothesis test is the probability of rejecting the null hypothesis when the null hypothesis is false.

(b)   The power of a test increases when the sample size is increased while keeping the significance level constant.

12.   (a)   The P-value of a hypothesis test is the probability, assuming that the null hypothesis is true, of getting a value of the test statistic that is as extreme or more extreme than the one actually obtained.

(b)   True. If the null hypothesis were true, a value of the test statistic with a P-value of 0.02 would be more extreme than one with a P-value of 0.03.

(c)   True. If the P-value is 0.74, this means that 74% of the time when the null hypothesis is true, the value of the test statistic would be more extreme than the one actually obtained.

(d)   The P-value of a hypothesis test is also called the observed significance level since it represents the lowest possible significance level at which the null hypothesis could have been rejected.

13.   In the critical-value approach, the null hypothesis is rejected if the value of the test statistic falls in the rejection region which is determined by the chosen significance level. In the P-value approach, the test statistic is computed and then the probability of obtaining a

value as extreme or more extreme than the one actually obtained is found. This is the P-value. The advantages of providing the P-value are that the observed significance level of the of the test is given and the reader of the results can determine for him/herself whether the results are strong enough evidence against the null hypothesis to reject it.

14.   Non-parametric methods have the advantages of involving fewer and simpler calculations than parametric methods and are more resistant to outliers and other extreme values. Parametric methods are preferred when the population is normal or the sample size is large since they are more powerful than non-parametric methods and thus tend to give more accurate results than non-parametric methods under those conditions.

15.   Let μ denote last year's mean cheese consumption by Americans.

   (a) H_0: $\mu = 27.3$ lb

   (b) H_a: $\mu > 27.3$ lb

   (c)   This is a right-tailed test.

16.   (a)   Rejection region: $z \geq 1.28$

   (b)   Nonrejection region: $z < 1.28$

   (c)   Critical value: $z = 1.28$

   (d)   Significance level: $\alpha = 0.10$

   (e)   Graph: see right

   (f)   Right-tailed test

17.   (a)   A Type I error would occur if, in fact, $\mu = 27.3$ lb, but the results of the sampling lead to the conclusion that $\mu > 27.3$ lb.

   (b)   A Type II error would occur if, in fact, $\mu > 27.3$ lb, but the results of the sampling fail to lead to that conclusion.

   (c)   A correct decision would occur if, in fact, $\mu = 27.3$ lb and the results of the sampling do not lead to the rejection of that fact; or if, in fact, $\mu > 27.3$ lb and the results of the sampling lead to that conclusion.

   (d)   If, in fact, last year's mean consumption of cheese for all Americans has not increased over the 1995 mean of 27.3 lb, and we do not reject the null hypothesis that $\mu = 27.3$ lb, we made a correct decision.

   (e)   If, in fact, last year's mean consumption of cheese for all Americans has increased over the 1995 mean of 27.3 lb, and we fail to reject the null hypothesis that $\mu = 27.3$ lb, we made a Type II error.

18.   (a) $\alpha = 0.10$

   (b)   The distribution of $\bar{x}$ will be approximately normal with a mean of 27.5 and a standard deviation of $6.9/\sqrt{35} = 1.2$.

   (c)   Note: $z = \dfrac{\bar{x} - \mu_0}{\sigma/\sqrt{n}} \;\Rightarrow\; \bar{x} = \mu_0 + z \cdot \sigma/\sqrt{n}$

Since this is a right-tailed test, we would reject H_0 if $z \geq 1.28$; or equivalently if $\bar{x} \geq 27.3 + 1.28(6.9)/\sqrt{35} = 28.79$.

So reject H_0 if $\bar{x} \geq 28.79$; otherwise do not reject H_0.

If $\mu = 27.5$, then P(Type II error) $= P(\bar{x} \leq 28.79)$

$= P(z \leq (28.79 - 27.5)/(6.9/\sqrt{35})) = P(z \leq 1.11) = 0.8665$

(d-e) Assuming that the true mean μ is one of the values listed, the distribution of $\bar{x}$ will be approximately normal with that mean and with a standard deviation of 1.166. The computations of β and the power 1 −β are shown in the table below.

True mean μ	z-score computation	P(Type II error) β	Power 1 − β
27.5	$z = \dfrac{28.79-27.5}{6.9/\sqrt{35}} = 1.11$	0.8665	0.1335
28.0	$z = \dfrac{28.79-28.0}{6.9/\sqrt{35}} = 0.68$	0.7517	0.2483
28.5	$z = \dfrac{28.79-28.5}{6.9/\sqrt{35}} = 0.25$	0.5987	0.4013
29.0	$z = \dfrac{28.79-29.0}{6.9/\sqrt{35}} = -0.18$	0.4286	0.5714
29.5	$z = \dfrac{28.79-29.5}{6.9/\sqrt{35}} = -0.61$	0.2709	0.7291
30.0	$z = \dfrac{28.79-30.0}{6.9/\sqrt{35}} = -1.04$	0.1492	0.8508
30.5	$z = \dfrac{28.79-30.5}{6.9/\sqrt{35}} = -1.47$	0.0708	0.9292
31.0	$z = \dfrac{28.79-31.0}{6.9/\sqrt{35}} = -1.89$	0.0294	0.9706
31.5	$z = \dfrac{28.79-31.5}{6.9/\sqrt{35}} = -2.32$	0.0102	0.9898
32.0	$z = \dfrac{28.79-32.0}{6.9/\sqrt{35}} = -2.75$	0.0030	0.9970

(f)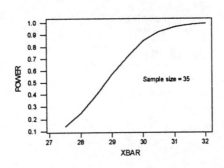

(g)   The distribution of $\bar{x}$ will be approximately normal with a mean of 27.5 and a standard deviation of $6.9/\sqrt{100} = 0.69$.

(h)   Note: $z = \dfrac{\bar{x}-\mu_0}{\sigma/\sqrt{n}} \;\Rightarrow\; \bar{x} = \mu_0 + z\cdot\sigma/\sqrt{n}$

Since this is a right-tailed test, we would reject H_0 if $z \geq 1.28$; or equivalently if $\bar{x} \geq 27.3 + 1.28(6.9)/\sqrt{100} = 28.18$.

So reject H_0 if $\bar{x} \geq 28.18$; otherwise do not reject H_0.

If $\mu = 27.5$, then

$$
\begin{aligned}
P(\text{Type II error}) &= P(\bar{x} \leq 28.18) \\
&= P(z \leq (28.18 - 27.5)/(6.9/\sqrt{10})) \\
&= P(z \leq 0.99) = 0.8389
\end{aligned}
$$

(i-j) Assuming that the true mean μ is one of the values listed, the distribution of $\bar{x}$ will be approximately normal with that mean and with a standard deviation of 0.69. The computations of β and the power $1 -\beta$ are shown in the following table.

True mean μ	z-score computation	P(Type II error) β	Power $1 - \beta$
27.5	$z = \dfrac{28.18-27.5}{6.9/\sqrt{100}} = 0.99$	0.8389	0.1511
28.0	$z = \dfrac{28.18-28.0}{6.9/\sqrt{100}} = 0.26$	0.6026	0.3974
28.5	$z = \dfrac{28.18-28.5}{6.9/\sqrt{100}} = -0.46$	0.3228	0.6772
29.0	$z = \dfrac{28.18-29.0}{6.9/\sqrt{100}} = -1.19$	0.1170	0.8830
29.5	$z = \dfrac{28.18-29.5}{6.9/\sqrt{100}} = -1.91$	0.0281	0.9719
30.0	$z = \dfrac{28.18-30.0}{6.9/\sqrt{100}} = -2.64$	0.0041	0.9959
30.5	$z = \dfrac{28.18-30.5}{6.9/\sqrt{100}} = -3.36$	0.0004	0.9996
31.0	$z = \dfrac{28.18-31.0}{6.9/\sqrt{100}} = -4.09$	0.0000	1.0000

True mean μ	z-score computation	P(Type II error) β	Power $1 - \beta$
31.5	$z = \dfrac{28.18 - 31.5}{6.9/\sqrt{100}} = -4.81$	0.0000	1.0000
32.0	$z = \dfrac{28.18 - 32.0}{6.9/\sqrt{100}} = -5.54$	0.0000	1.0000

(k)

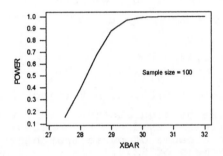

(l)   The principle being illustrated is that increasing the sample size for a hypothesis test without changing the significance level increases the power.

19.   (a) $n = 35$, $\bar{x} = 973/35 = 27.8$, $\sigma = 6.9$

Step 1: H_0: $\mu = 27.3$ lb, H_a: $\mu > 27.3$ lb

Step 2: $\alpha = 0.10$

Step 3:   Critical value = 1.28

Step 4: $z = (27.8 - 27.3)/(6.9/\sqrt{35}) = 0.43$

Step 5:   Since $0.43 < 1.28$, do not reject H_0.

Step 6:   At the 10% significance level, the data do not provide sufficient evidence to conclude that last year's mean cheese consumption μ for all Americans has increased over the 1995 mean of 27.3 lb.

(b)   Given the conclusion in part (a), if an error has been made, it must be a Type II error. This is because, given that the null hypothesis was not rejected, the only error that could be made is the error of not rejecting a false null hypothesis.

20.   (a)   Step 1: H_0: $\mu = 27.3$, H_a: $\mu > 27.3$

Step 2: $\alpha = 0.10$

Step 3: $z = 0.43$

Step 4: $P = 1 - 0.6664 = 0.3336$

Step 5:   Since $0.3336 > 0.10$, do not reject H_0.

Step 6:   At the 10% significance level, the data do not provide sufficient evidence to conclude that last year's mean cheese consumption μ for all Americans has increased over the 1995 mean of 27.3 lb.

(b)   Using Table 9.12, we classify the strength of evidence against the null hypothesis as weak or none because P > 0.10.

21.   n = 41, $\bar{x}$ = 260, s = 84

Step 1: H_0: μ = \$279, H_a: μ < \$279

Step 2: α = 0.05

Step 3:   Critical value = -1.684

Step 4:   t = (260 - 279)/(84/$\sqrt{41}$) = -1.448

Step 5:   Since -1.448 > -1.684, do not reject H_0.

Step 6:   At the 5% significance level, the data do not provide sufficient evidence to conclude that the mean value lost because of purse snatching has decreased from the 1994 mean of \$279.

22.   (a)   Since -1.684 < -1.448 -1.303, we have 0.05 < P < 0.010.

(b)   Step 1: H_0: μ = \$279, H_a: μ < \$279

Step 2: α = 0.05

Step 3:   t = -1.448

Step 4:   0.05 < P < 0.10

Step 5:   Since P > 0.05, do not reject H_0.

Step 6:   At the 5% significance level, the data do not provide sufficient evidence to conclude that the mean value lost because of purse snatching has decreased from the 1994 mean of \$279.

(c)   Using Table 9.12 , we classify the strength of evidence against the null hypothesis as moderate because 0.05 < P < 0.10.

23.   n = 15, df = 14, $\bar{x}$ = 28.753, s = 1.595

(a)   Step 1: H_0: μ = 29 mpg, H_a: μ $\neq$ 29 mpg

Step 2: α = 0.05

Step 3:   Critical values = ±2.145

Step 4:   t = (28.753 - 29)/(1.595/) = -0.600

Step 5:   Since -2.145 < -0.600 < 2.145, do not reject H_0.

Step 6:   At the 5% significance level, the data do not provide sufficient evidence to conclude the company's report was incorrect.

(b)   Step 1: H_0: μ = 29 mpg, H_a: μ $\neq$ 29 mpg

Step 2: α = 0.05

Step 3:   t = -0.600

Step 4:   P > 0.20

Step 5:   Since the P-value > 0.05, do not reject H_0.

Step 6:   At the 5% significance level, the data do not provide sufficient evidence to conclude that the company's report was incorrect.

(c)   Weak or none.

24.   (a)   Step 1: H_0: μ = 29 mpg, H_a: μ $\neq$ 29 mpg

Step 2: α = 0.05

Step 3:   Critical values = 25 ,95

Step 4:

Mileage x	Difference D = x − 29	|D|	Rank of |D|	Signed Rank R
27.3	−1.7	1.7	10.5	−10.5
30.9	1.9	1.9	12	12
25.9	−3.1	3.1	15	−15
31.2	2.2	2.2	13	13
29.7	0.7	0.7	6	6
28.8	−0.2	0.2	2	−2
29.4	0.4	0.4	3.5	3.5
28.5	−0.5	0.5	5	−5
28.9	−0.1	0.1	1	−1
31.6	2.6	2.6	14	14
27.8	−1.2	1.2	7.5	−7.5
27.8	−1.2	1.2	7.5	−7.5
28.6	−0.4	0.4	3.5	−3.5
27.3	−1.7	1.7	10.5	−10.5
27.6	−1.4	1.4	9	−9

Step 5: $W =$ sum of + ranks = 48.5

Step 6:   Since $25 < W < 95$, do not reject H_0.

Step 7:   At the 5.0% significance level, the data do not provide sufficient evidence to conclude that the company's report was incorrect.

(b)   In performing the hypothesis test of part (a), the assumption made is that the distribution of the gas mileages is symmetric.

(c)   Under the assumption that gas mileages are normally distributed, it is permissible to perform a Wilcoxon signed-rank test for the mean gas mileage because a normally distributed population is symmetric.

25.   The t-test, because that hypothesis-testing procedure is designed specifically for a normally distributed population; the t-test is more powerful than the Wilcoxon signed-rank test for normal populations.

26.   Since the distribution is symmetric, use the Wilcoxon signed-rank test.

27.   Since we have a large sample and σ is unknown, use the t-test.

28.   (a)   The Wilcoxon signed-rank test is appropriate in Problem 26 since the distribution is symmetric. It is not appropriate in Problem 27 since the distribution is highly skewed to the left.

(b)   The sample size is large ($n=50$) and the distribution is symmetric in Problem 26, so either the z-test or the Wilcoxon signed-rank test could be used. Since the distribution appears to be more peaked with longer tails than a normal distribution would have, the Wilcoxon test is preferable.

29.   With the data in a column named POUNDS, we choose **Stat ▶ Basic Statistics ▶ 1-Sample z...**, select POUNDS in the **Variables** text box, click on the **Test mean** button, type 27.3 in the **Test mean** text box,

click on the arrow in the **Alternative** text box and select 'greater than,' type <u>6.9</u> in the Σ text box, and click **OK**. The result is

Test of μ = 27.30 vs μ > 27.30
The assumed σ = 6.90

Variable	N	Mean	StDev	SE Mean	Z	P-Value
POUNDS	35	27.80	6.48	1.17	0.43	0.33

Since the P-value of 0.33 is greater than 0.10, do not reject the null hypothesis.

30. (a) H_0: μ = 27.3 lb, H_a: μ > 27.3 lb

(b) THE ASSUMED SIGMA = 6.90

(c) STDEV = 6.48

(d) N = 35

(e) MEAN = 27.80 lb

(f) Z = 0.43

(g) P-VALUE = 0.33

(h) The smallest significance level at which the null hypothesis can be rejected is 0.33.

(i) Since 0.33 > 0.10, do not reject H_0. At the 10% significance level, the data do not provide sufficient evidence to conclude that last year's mean cheese consumption, μ, for all Americans has increased over the 1995 mean of 27.3 lb.

31. (a) With the data in a column named MILEAGES, choose **Calc ▶ Functions...**, select MILEAGES in the **Input Column** text box and type <u>NSCORE</u> in the **Result in** text box, click on the **Normal scores** button, and click **OK**. We now choose **Graph ▶ Plot...**, select NSCORE for the **Y** variable for **Graph 1** and MILEAGES for the **X** variable, and click **OK**.

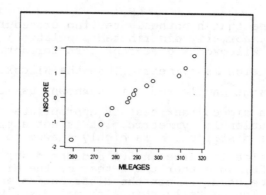

(b) To carry out the hypothesis test, we choose **Stat ▶ Basic Statistics ▶ 1-Sample t...**, select MILEAGES in the **Variables** text box, click on the **Test mean** button, type <u>29</u> in the **Test mean** text box, click on the arrow in the **Alternative** text box and select 'not equal to,' and click **OK**. The result is

Test of μ = 29.000 vs μ not = 29.000

Variable	N	Mean	StDev	SE Mean	T	P-Value
MILEAGES	15	28.753	1.595	0.412	-0.60	0.56

(c)   Based on the normal probability plot in (a), it seems reasonable to assume normality for the population of gas mileages.

32.   (a) H_0: μ = 29 mpg, H_a: $\mu \neq$ 29 mpg

(b)   STDEV = 1.595

(c)   n = 15

(d)   MEAN = 28.753

(e)   t = -0.60

(f)   P-VALUE = 0.56

(g)   The smallest significance level at which the null hypothesis can be rejected is 0.56.

(h)   Since 0.56 > 0.05, do not reject H_0. At the 5% significance level, the data do not provide sufficient evidence to conclude that the mean gas mileage μ for all cars of the model under consideration differs from the figure of 29 mpg reported by the company.

33.   With the data in a column named MILEAGES, we choose **Stat** ▶

Nonparametrics ▶ **1-Sample Wilcoxon...**, select MILEAGES in the **Variables** text box, click on the **Test median** button and type 29 in its text box, select 'not equal to' in the **Alternative** text box, and click **OK**. The result is

Test of Median = 29.00 Versus Median not = 29.00

	N	N for Test	Wilcoxon Statistic	P-value	Estimated Median
Mileages	15	15	48.5	0.532	28.65

34.   (a) H_0: η = 29 mpg, H_a: $\eta \neq$ 29 mpg

(b)   15   (c)   0   (d)   48.5

(e)   0.532 (f)   0.532

(g)   At the 5% significance level, the data do not provide sufficient evidence to conclude that the company's report was incorrect.

(h)   28.65 mpg

CHAPTER 10 ANSWERS

Exercises 10.1

10.1 (a) The variable is the age of buyers of new cars.

(b) The two populations are the buyers of new domestic cars and the buyers of new imported cars.

(c) H_0: $\mu_1 = \mu_2$, H_a: $\mu_1 > \mu_2$ **where** μ_1 = is the mean age of buyers of new domestic cars and μ_2 = the mean age of buyers of new imported cars.

10.3 Answers will vary.

10.5 (a) μ_1, σ_1, μ_2, and σ_2 are parameters; $\bar{x}_1$, s_1, $\bar{x}_2$, and s_2 are statistics.

(b) μ_1, σ_1, μ_2, and σ_2 are fixed numbers; $\bar{x}_1$, s_1, $\bar{x}_2$, and s_2 are random variables.

10.7 It is the sampling distribution of the difference of the two sample means that allows us to determine whether the difference of the sample means can reasonably be attributed to sampling error or whether it is large enough for us to conclude that the population means are different.

10.9 (a) The mean of $\bar{x}_1 - \bar{x}_2$ is $\mu_1 - \mu_2 = 40 - 40 = 0$

The standard deviation of $\bar{x}_1 - \bar{x}_2$ is

$$\sqrt{\sigma_1^2/n_1 + \sigma_2^2/n_2} = \sqrt{12^2/9 + 6^2/4} = 5$$

(b) No. The determination of the mean of $\bar{x}_1 - \bar{x}_2$ is the same for all populations. The formula for the standard deviation of $\bar{x}_1 - \bar{x}_2$ is dependent on the samples being independent, but not on the populations from which they came.

(c) No. It is not known that the two populations are normally distributed (which would lead to $\bar{x}_1 - \bar{x}_2$ being normally distributed), and the sample sizes of 9 and 4 are too small to claim that $\bar{x}_1$ and $\bar{x}_2$ are normally distributed (which would also lead to $\bar{x}_1 - \bar{x}_2$ being normally distributed).

10.11 (a) Population 1: Accounting, $n_1 = 32$, $\bar{x}_1 = 1111.6/32 = 34.74$, $\sigma_1 = 1.73$

Population 2: Liberal arts, $n_2 = 35$, $\bar{x}_2 = 1137.9/35 = 32.51$, $\sigma_2 = 1.82$

Step 1: H_0: $\mu_1 = \mu_2$, H_a: $\mu_1 > \mu_2$

Step 2: $\alpha = 0.05$

Step 3: Critical value = 1.645

Step 4: $z = (34.74 - 32.51)/\sqrt{(1.73^2/32) + (1.82^2/35)} = 5.14$

Step 5: Since $5.14 > 1.645$, reject H_0.

Step 6: At the 5% significance level, the data provide sufficient evidence to conclude that accounting graduates have a higher mean starting salary than liberal arts graduates.

For the P-value approach, $P(z > 5.14) = 0.0000$. Therefore, because the P-value is smaller than the significance level, reject H_0.

(b)   The 95% confidence interval for $\mu_1 - \mu_2$ is

$$(34.74 - 32.51) \pm 1.96 \sqrt{(1.73^2/32) + (1.82^2/35)}$$

$$= 2.23 \pm 0.85 = (1.38, 3.08)$$

10.13 A hypothesis test of H_0: $\mu_1 = \mu_2$ versus H_a: $\mu_1 \neq \mu_2$ at the significance level α will lead to rejection of the null hypothesis if and only if the number <u>zero</u> does not lie in the $(1 - \alpha)$-level confidence interval for $\mu_1 - \mu_2$.

10.15 (a)   We have Minitab take 1000 samples of size 12 observations from a normally distributed population having mean 640 and standard

deviation 70 by choosing **Calc ▶ Random Data ▶ Normal...**, typing <u>1000</u> in the **Generate rows of data** text box, typing <u>C1-C12</u> in the **Store in columns** text box, typing <u>640</u> in the **Mean** text box and <u>70</u> in the **Standard deviation** text box, and clicking **OK**. To get the

mean of each row (sample), select **Calc ▶ Row Statistics..**, select The **Mean** button, enter <u>C1-C12</u> in the **Input Variables:** text box, and type X1BAR in the **Store Result in:** text box. The sample means will be found in C13.

(b)   We repeat part (a) for a normally distributed population having

mean 715 and standard deviation 150 by choosing **Calc ▶ Random**

Data ▶ Normal..., typing <u>1000</u> in the **Generate rows of data** text box, typing <u>C14-C28</u> in the **Store in columns** text box, typing <u>715</u> in the **Mean** text box and <u>150</u> in the **Standard deviation** text box, and clicking **OK**. To get the mean of each row (sample), select

Calc ▶ Row Statistics.., select The **Mean** button, enter <u>C14-C28</u> in the **Input Variables:** text box, and type X2BAR in the **Store Result in:** text box. The sample means will be found in C29.

(c)   To find the 1000 differences between X1BAR and X2BAR, select **Calc**

▶ Calculator..., type <u>DIFF</u> in the **Store results in variable** text box, type 'X1BAR' - 'X2BAR' in the **Expression** text box, and click **OK**. The DIFF values will appear in C30.

(d)   To get the mean and standard deviation of the DIFF values, select

Stat ▶ Basic Statistics ▶ Display Descriptive statistics..., and select <u>DIFF</u> in the **Variables** text box. Click **OK**. The results for our simulation were

Variable	N	Mean	Median	TrMean	StDev	SE Mean
DIFF	1000	-75.52	-74.49	-75.47	42.46	1.34

Variable	Minimum	Maximum	Q1	Q3
DIFF	-206.85	55.47	-105.52	-46.99

We see that the simulated mean of $\overline{x}_1 - \overline{x}_2$ is -75.47 and the standard deviation is

42.46. Now select **Graph ▶**

Histogram and enter <u>DIFF</u> in the **X** column for **Graph 1** and click OK. The result is

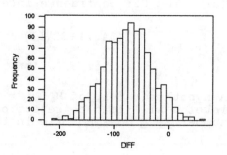

(e)   Theoretically, $\overline{x}_1 - \overline{x}_2$ is normally distributed with mean 740 - 715 = -75 and standard deviation $\sqrt{70^2/12 + 150^2/15}$ = 43.68   .

(f)   The histogram looks normal, the simulated mean of -75.47 is very close to -75, and the simulated standard deviation of 42.46 is very close to 43.68.

Section 10.2

10.17 s_p is called the pooled standard deviation because it combines information about the variability from both samples into one estimate of the common value of the population standard deviations.

10.19 Population 1: Fraud, $n_1 = 10$, $\overline{x}_1 = 10.12$, $s_1 = 4.90$

Population 2: Firearms, $n_2 = 10$, $\overline{x}_2 = 18.78$, $s_2 = 4.64$

Step 1: H_0: $\mu_1 = \mu_2$, H_a: $\mu_1 < \mu_2$

Step 2: $\alpha = 0.05$

Step 3:   df = 18, Critical value = -1.734

Step 4:

$$s_p^2 = [(9)4.90^2 + (9)4.64^2]/18 = 22.7698$$

$$t = (10.12 - 18.78)/\sqrt{(22.7698)(\frac{1}{10} + \frac{1}{10})} = -4.06$$

Step 5:   Since -4.06 < -1.734, reject H_0.

Step 6:   At the 5% significance level, the data do provide sufficient evidence to conclude that, on the average, the mean time served for fraud is less than that served for firearms offenses.

10.21 Population 1: 25-34 years, $n_1 = 10$, $\overline{x}_1 = 70.19$, $s_1 = 2.951$

Population 2: 45-54 years, $n_2 = 15$, $\overline{x}_2 = 68.58$, $s_2 = 3.543$

$$s_p = \sqrt{\frac{(10-1)\cdot 2.951^2 + (15-1)\cdot 3.543^2}{10 + 15 - 2}} = 3.324, \quad df = 10 + 15 - 2 = 23$$

Step 1: H_0: $\mu_1 = \mu_2$, H_a: $\mu_1 > \mu_2$

Step 2: $\alpha = 0.05$

Step 3:   Critical value = 1.714

Step 4: $t = (70.19 - 68.58)/(3.324 \cdot \sqrt{(1/10) + (1/15)}\,) = 1.188$

Step 5:   Since $1.188 < 1.714$, do not reject H_0.

Step 6:   At the 5% significance level, the data do not provide sufficient evidence to conclude that males in the age group 25-34 years are, on the average, taller than those who were in the same age group 20 years ago.

For the P-value approach, $P(t > 1.186) > 0.10$. Therefore, because the P-value is larger than the significance level, do not reject H_0.

10.23 Population 1: Natural gas, $n_1 = 30$, $\overline{x}_1 = 1497.60$, $s_1 = 160.35$

Population 2: Electricity, $n_2 = 36$, $\overline{x}_2 = 1243.64$, $s_2 = 165.13$

Step 1: H_0: $\mu_1 = \mu_2$, H_a: $\mu_1 \neq \mu_2$

$$s_p = \sqrt{\frac{(30-1)\cdot 160.35^2 + (36-1)\cdot 165.13^2}{30 + 36 - 2}} = 162.98$$

$$df = 30 + 36 - 2 = 64$$

Step 2: $\alpha = 0.05$

Step 3:   Critical values = ± 2.000

Step 4: $t = (1497.6 - 1243.6)/(162.98 \cdot \sqrt{(1/30) + (1/36)}\,) = 6.304$

Step 5:   Since $6.304 > 2.000$, reject H_0.

Step 6:   At the 5% significance level, the data provide sufficient evidence to conclude that last year's mean annual fuel expenditure for households using natural gas is different from that for households using only electricity.

For the P-value approach, $2\{P(t > 6.304)\} = 0.0000$ (to four decimal places). Therefore, because the P-value is smaller than the significance level, reject H_0.

10.25 From Exercise 10.19, $s_p = 4.772$, df = 18

(a)
$$(10.12 - 18.78) \pm 2.552 \cdot 4.772\sqrt{(1/10) + (1/10)}$$

$$-8.66 \pm 5.45$$

$$-14.11 \text{ to } -3.21 \text{ months}$$

(b)   We can be 98% confident that the difference, $\mu_1 - \mu_2$, between the mean times served by prisoners in the fraud and firearms offense categories is somewhere between -14.11 and -3.21 months.

10.27 From Exercise 10.21, $s_p = 3.324$, df = 23

(a)
$$(70.19 - 68.58) \pm 1.714 \cdot 3.324\sqrt{(1/10) + (1/15)}$$

$$1.61 \pm 2.33$$

$$-0.72 \text{ to } 3.94 \text{ inches}$$

(b)   We can be 90% confident that the difference, $\mu_1 - \mu_2$, between the mean height of males in the age group 25-34 years and the mean height of males in the age group 45-54 years is somewhere between -0.71 and 3.93 inches.

10.29 From Exercise 10.23, $s_p = 162.98$, df = 64

(a)
$$(1,497.6 - 1,243.6) \pm 2.000 \cdot 162.98\sqrt{(1/30) + (1/36)}$$

$$254.0 \pm 80.58$$

$$\$173.42 \text{ to } \$334.58$$

(b)   We can be 95% confident that the difference, $\mu_1 - \mu_2$, between last year's mean fuel expenditure for households using natural gas and those households using only electricity is somewhere between $173.42 and $334.58.

10.31 The parameter σ represents the common value of the standard deviations of the two populations. The standard deviation of most populations is unknown, and thus the common standard deviation of two populations is also unknown. One argument often given for the standard deviations being unknown in this situation is that in order to compute a population standard deviation, the mean μ is needed. In the comparison of means problem, then, both population means would be needed in order to find the common σ. But if both population means are known, the real difference $\mu_1 - \mu_2$ is already known and there is no longer any reason to compute a confidence interval for the difference.

10.33 (a)   With the data in columns named FRAUD and FIREARMS, we choose **Calc**

▶ **Functions...**, select FRAUD in the **Input Column** text box and type 'NSCOREFR' in the **Result in** text box, click on the **Normal**

scores button, and click **OK**. We now choose **Graph** ▶ **Plot...**, select NSCOREFR for the **Y** variable for **Graph 1** and FRAUD for the **X** variable, and click **OK**.

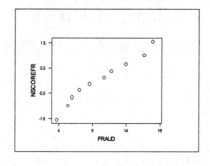

Now choose **Graph ▶ Boxplot...**, select FRAUD, and click **OK**. The result is

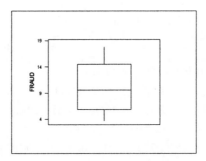

Finally, choose **Calc ▶ Column statistics...**, click on the **Standard deviation** button, select FRAUD in the **Input Variables** text box, and click **OK**. The result is

Standard deviation of FRAUD   =   4.8976

Repeating these processes for FIREARMS, we obtain the following results:

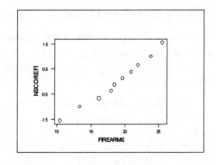   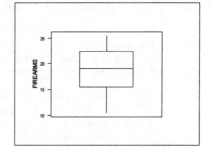

Standard deviation of FIREARMS=   4.6382

(b)   To carry out the hypothesis test and obtain a confidence interval

for the difference in means, we choose **Stat ▶ Basic Statistics ▶ 2-Sample t...**, click on the **Samples in different columns** option button, select FRAUD in the **First** text box and FIREARMS in the **Second** text box, click on **Assume equal variances** so that an X appears in the box, click on the arrow in the **Alternative** text box and select 'less than', choose 98 for the **Confidence level**, and click **OK**. The results are

Twosample T for FRAUD vs FIREARMS

	N	Mean	StDev	SE Mean
FRAUD	10	10.12	4.90	1.5
FIREARMS	10	18.78	4.64	1.5

98% C.I. for μ FRAUD - μ FIREARMS: (-14.1, -3.2)

T-Test μ FRAUD = μ FIREARMS (vs <): T= -4.06 P=0.0004 DF= 18

Both use Pooled StDev = 4.77

(c)　Based on the graphic displays and sample standard deviations in part (a), it appears reasonable to consider the assumptions for using pooled-t procedures satisfied. Those assumptions are : (1) Independent samples, (2) Normal populations, and (3) Equal population standard deviations.

10.35 (a)　$H_0: \mu_1 = \mu_2$, $H_a: \mu_1 < \mu_2$　　[T-TEST μ MINE = μ CONSTRUC (vs <)]

(b)　StDev for MINE = 2.25;　StDev for CONSTRUC = 2.36

(c)　n for MINE = 14;　n for CONSTRUC = 17

(d)　Mean for MINE = $15.93;　Mean for CONSTRUC = $16.42

(e)　t = -0.59

(f)　P = 0.28

(g)　The smallest significance level at which H_0 can be rejected is 0.28.

(h)　Since 0.28 > 0.05, do not reject H_0. At the 5% significance level, the data do not provide sufficient evidence to conclude that, on the average, the hourly earnings of mine workers are less than those of construction workers.

(i)　90% CI for μ MINE - μ CONSTRUC: (-1.91, 0.93)

(j)　Pooled StDev = 2.31

10.37 (a)　To simulate 1000 samples of size 4 from a normal distribution with mean 100 and standard deviation 16, choose **Calc ▶ Random data ▶ Normal...**, type <u>1000</u> in the **Generate rows of data** text box, type <u>C1-C4</u> in the **Store in Column(s)** text box, type <u>100</u> in the **Mean** text box and <u>16</u> in the **Standard deviation** text box, and Click **OK**.

Now choose **Calc ▶ Row Statistics...**, click on **Mean**, select C1, C2, C3, C4 in the **Input Variable(s)** text box, type **MEAN1** in the **Store result in** text box, and Click **OK**. Repeat this last step, but choose **Standard Deviation** and store it in SD1. The means and standard deviations will be stored in C5 and C6, respectively.

(b)　Repeat the steps of (a) using a normal distribution with mean 110 and standard deviation 16, generating the samples in columns C7-C9, and storing the sample means and standard deviations in MEAN2 and SD2 (C10 and C11, respectively).

(c)　To compute the pooled-t statistic for each pair of samples, we will first find the pooled standard deviation. Choose **Calc ▶ Calculator...**, type <u>SP</u> in the **Store result in Variable:** text box, type <u>SQRT((3*'SD1'**2+2*'SD2'**2)/5)</u> in the **Expression:** text box, and click **OK**. Then to compute t, choose **Calc ▶ Calculator...**, type T in the **Store result in Variable:** text box, type

<u>(('MEAN1'-'MEAN2')-(-10))/('SP'*SQRT(1/4+1/3))</u>

in the **Expression:** text box, and click **OK**. SP and T will be stored in C12 and C13, respectively.

(d)　Choose **Graph ▶ Histogram...**, select T for the **X** variable for **Graph 1,** and click **OK**. Our result is shown below. Yours may differ slightly from this one.

(e) The shape of the distribution should be that of a t-distribution with 5 degrees of freedom, that is, symmetric about a mean of zero and with a larger standard deviation than that of a standard normal distribution.

(f) The histogram shows a distribution that is symmetric about zero and is more widely spread out than a standard normal distribution. Thus the results agree with the theoretical.

Exercises 10.3

10.39 Use of the pooled procedure is based on the assumption that the two population standard deviations are equal, whereas the nonpooled procedure does require equal population standard deviations.

10.41 Population 1: Sophomores, $n_1 = 17$, $\overline{x}_1 = 2.84$, $s_1 = 0.52$

Population 2: Juniors, $n_2 = 13$, $\overline{x}_2 = 2.98$, $s_2 = 0.31$

$$
\Delta = \frac{[(s_1^2/n_1) + (s_2^2/n_2)]^2}{\dfrac{(s_1^2/n_1)^2}{n_1 - 1} + \dfrac{(s_2^2/n_2)^2}{n_2 - 1}}
$$

$$
= \frac{[(0.52^2/17) + (0.31^2/13)]^2}{\dfrac{(0.52^2/17)^2}{16} + \dfrac{(0.31^2/13)^2}{12}}
$$

$$
= 26.65 \approx 26 = DF.
$$

Step 1: $H_0: \mu_1 = \mu_2$, $H_a: \mu_1 \neq \mu_2$
Step 2: $\alpha = 0.05$

Step 3:   Critical value = ± 2.056

Step 4: $t = (2.84 - 2.98)/\sqrt{(0.52^2/17) + (0.31^2/13)} = -0.917$

Step 5:   Since $-2.056 < -0.917 < 2.056$, do not reject H_0.

Step 6:   At the 5% significance level, the data do not provide sufficient evidence to conclude that there is a difference between the mean GPA of sophomores and the mean GPA of juniors at the university.

For the P-value approach, $2\{P(t < -0.917)\} > 0.20$. Therefore, because the P-value is larger than the significance level, do not reject H_0.

10.43 Population 1: Dynamic, $n_1 = 14$, $\overline{x}_1 = 7.36$, $s_1 = 1.22$

Population 2: Static, $n_2 = 6$, $\overline{x}_2 = 10.50$, $s_2 = 4.59$

$$\Delta = \frac{[(s_1^2/n_1) + (s_2^2/n_2)]^2}{\dfrac{(s_1^2/n_1)^2}{n_1-1} + \dfrac{(s_2^2/n_2)^2}{n_2-1}}$$

$$= \frac{[(1.22^2/14) + (4.59^2/6)]^2}{\dfrac{(1.22^2/14)^2}{13} + \dfrac{(4.59^2/6)^2}{5}}$$

$$= 5.31 \approx 5 = DF.$$

Step 1: H_0: $\mu_1 = \mu_2$, H_a: $\mu_1 < \mu_2$
Step 2: $\alpha = 0.05$

Step 3:   Critical value = -2.015

Step 4: $t = (7.36-10.50)/\sqrt{(1.22^2/14) + (4.59^2/6)} = -1.651$

Step 5:   Since -1.651 > -2.015, do not reject H_0.

Step 6:   At the 5% significance level, the data do not provide sufficient evidence to conclude that the mean number of acute postoperative days in the hospital is smaller with the dynamic system than with the static system.

For the P-value approach, $0.05 < P(t<-1.651) < 0.10$. Therefore, since the P-value is larger than the significance level, do not reject H_0.

10.45 Population 1: Nevada, $n_1 = 32$, $\overline{x}_1 = 313.4$, $s_1 = 37.2$

Population 2: Idaho, $n_2 = 40$, $\overline{x}_2 = 279.6$, $s_2 = 34.6$

$$\Delta = \frac{[(s_1^2/n_1) + (s_2^2/n_2)]^2}{\dfrac{(s_1^2/n_1)^2}{n_1-1} + \dfrac{(s_2^2/n_2)^2}{n_2-1}}$$

$$= \frac{[(37.2^2/32) + (34.6^2/40)]^2}{\dfrac{(37.2^2/32)^2}{31} + \dfrac{(34.6^2/40)^2}{39}}$$

$$= 64.28 \approx 64 = DF.$$

Step 1: H_0: $\mu_1 = \mu_2$, H_a: $\mu_1 > \mu_2$
Step 2: $\alpha = 0.05$
Step 3:   Critical value = 1.671 (approximately)

Step 4: $t = (313.4-279.6)/\sqrt{(37.2^2/32) + (34.6^2/40)} = 3.951$

Step 5:   Since 3.951 > 1.671, reject H_0.

Step 6:   At the 5% significance level, the data provide sufficient evidence to conclude that Nevada has a larger mean potato yield than Idaho.

For the P-value approach, $P(t > 3.951) < 0.005$. Therefore, since the P-value is smaller than the significance level, reject H_0.

10.47 From Exercise 10.41, $\alpha = 0.05$, df = 26, $t_{0.025,26} = 2.056$

(a)
$$(2.84-2.98) \pm 2.056\sqrt{(0.52)^2/17 + (0.31)^2/13}$$

$$-0.14 \pm 0.31$$

$$-0.45 \text{ to } 0.17$$

(b) We can be 95% confident that the difference, $\mu_1 - \mu_2$, between the mean GPAs of sophomores and juniors is between -0.45 and 0.17.

10.49 From Exercise 10.43, $\alpha = 0.10$, df = 5, $t_{0.05,5} = 2.015$

(a)
$$(7.36-10.50) \pm 2.015\sqrt{(1.22)^2/14 + (4.59)^2/6}$$

$$-3.14 \pm 3.83$$

$$-6.97 \text{ to } 0.69$$

(b) We can be 90% confident that the difference, $\mu_1 - \mu_2$, between the mean numbers of acute postoperative days in the hospital with dynamic and static systems is somewhere between -6.97 and 0.69.

10.51 From Exercise 10.45, $\alpha = 0.10$, df = 64, $t_{0.05,64} = 1.671$

(a)
$$(313.4-279.6) \pm 1.671\sqrt{(37.2)^2/32 + (34.6)^2/40}$$

$$33.80 \pm 14.29$$

$$19.51 \text{ cwt to } 48.09 \text{ cwt}$$

(b) We can be 90% confident that the difference, $\mu_1 - \mu_2$, between the mean yields per acre of potatoes for Idaho and Nevada is somewhere between 19.51 and 48.09 cwt.

10.53 Population 1: Dynamic, $n_1 = 14$, $\bar{x}_1 = 7.36$, $s_1 = 1.22$

Population 2: Static, $n_2 = 6$, $\bar{x}_2 = 10.50$, $s_2 = 4.59$

Step 1: $H_0: \mu_1 = \mu_2$, $H_a: \mu_1 < \mu_2$

$$s_p = \sqrt{\frac{(14-1) \cdot 1.22^2 + (6-1) \cdot 4.59^2}{14 + 6 - 2}} = 2.6320$$

$$df = 14 + 6 - 2 = 18$$

Step 2: $\alpha = 0.05$

Step 3:   Critical value = -1.734

Step 4: $t = (7.36-10.50)/2.632\sqrt{(1/14)+(1/6)} = -2.445$

Step 5:   Since -2.445 < -1.734, reject H_0.

Step 6:   At the 5% significance level, the data do provide sufficient evidence to conclude that the mean number of acute postoperative days in the hospital is smaller with the dynamic system than with the static system.

For the P-value approach, $0.01 < P(t<-2.445) < 0.025$. Therefore, since the P-value is smaller than the significance level, reject H_0.

(b)   The pooled t-test resulted in rejecting the null hypothesis while the nonpooled t-test resulted in not rejecting the null hypothesis.

(c)   The nonpooled t-test is more appropriate. One sample standard deviation is almost four times as large as the other, making it highly unlikely that the two population standard deviations are equal. The fact that the two sample sizes are also quite different makes it essential that the pooled t-test not be used.

10.55 (a)   Pooled t-test. Both populations are normally distributed and the population standard deviations are equal.

(b)   Nonpooled t-test. Both populations are normally distributed, but the population standard deviations are unequal.

(c)   Neither. Both populations are skewed and it is given that both sample sizes are small.

(d)   Neither. Only one population is normally distributed; the other is skewed. Since the sample sizes are small, the non-normality of one sample rules out the use of either t-test.

10.57 (a)   With the data in columns named SOPH and JUNIOR, we choose **Calc ▶ Calculator...**, type 'NSCO S' in the **Store results in variable** text box (Note that the single quote marks must be typed if the column name has a space in it.), select **Normal scores** from the function list, select SOPH to replace **numbers** in the **Expression** text box, and click **OK**. Repeat this process for JUNIOR, using NSCO J. We

now choose **Graph ▶ Plot...**, select NSCO S for the **Y** variable for **Graph 1** and SOPH for the **X** variable, then select NSCO J for the **Y** variable for **Graph 2** and JUNIOR for the **X** variable, and click **OK**. Both graphs will be produced as shown below.

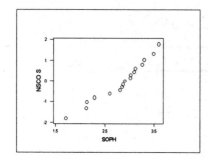 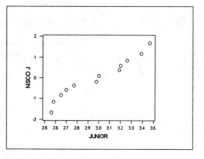

(b)   To carry out the hypothesis test and obtain a confidence interval for the difference in means, we choose **Stat ▶ Basic Statistics ▶ 2-Sample t...**, click on the **Samples in different columns** option button, select SOPH in the **First** text box and JUNIOR in the **Second** text box, make certain that no X appears in the **Assume equal variances** box, click on the arrow in the **Alternative** text box and select 'not equal to,' choose 95 for the **Confidence level**, and click **OK**. The results are

```
        Twosample T for SOPH vs JUNIOR
                 N      Mean     StDev    SE Mean
        SOPH    17     2.840     0.520      0.13
        JUNIOR  13     2.981     0.309      0.086
```

95% C.I. for μ SOPH - μ JUNIOR: (-0.45, 0.173)

T-Test μ SOPH = μ JUNIOR (vs not =): T= -0.92 P=0.36 DF= 26

- (c) Based on the normal plots in (a) for the two independent samples, it appears reasonable to consider the assumption of normality satisfied; thus, the use of the nonpooled-t procedure.

10.59 (a) $H_0: \mu_1 = \mu_2$, $H_a: \mu_1 > \mu_2$

- (b) $s_1 = 4.62$ yr, $s_2 = 7.25$ yr

- (c) $n_1 = 11$, $n_2 = 12$

- (d) $\overline{x_1} = 35.69$ yr, $\overline{x_2} = 32.81$ yr

- (e) $t = 1.15$

- (f) 0.13

- (g) 0.13

- (h) At the 5% significance level, the data do not provide sufficient evidence to conclude that the mean age at the time of first divorce for males is greater than that for females.

- (i) -1.5 to 7.2 yr

- (j) The age distributions under considerations here are probably not normal. But, since the nonpooled-t procedures are robust to moderate violations of the normality assumption, whether it is appropriate to apply the nonpooled-t here depends on how much the distributions deviate from normality.

Exercises 10.4

10.61 All normal distributions are symmetric about the population mean, and the standard deviation determines the spread (and hence the shape). Thus two normal distributions with the same standard deviation have the same shape.

10.63 (a) Use the pooled t-test. It is slightly more powerful than the Mann-Whitney when the conditions for its use (normal distributions with equal variances) are met.

- (b) Use the Mann-Whitney test since the distributions have the same shape, but are not normal.

10.65 Population 1: Students with fewer than two years of high-school algebra;

Population 2: Students with two or more years of high-school algebra.

Also, the subscript 1 is associated with that population having the smaller sample size.

Step 1: H_0: $\mu_1 = \mu_2$, H_a: $\mu_1 < \mu_2$

Step 2: $\alpha = 0.05$

Step 3:   We have $n_1 = 6$ and $n_2 = 9$. Since the hypothesis test is left-tailed with $\alpha = 0.05$, we use Table VIII to obtain the critical value, which is $M_1 = 33$. Thus, we reject H_0 if $M \leq 33$.

Step 4:   Construct a work table based upon the following: First, rank all the data from both samples combined. Adjacent to each column of data as it is presented in the Exercise, record the overall rank. Assign tied rankings the average of the ranks they would have had if there were no ties. For example, the two 81s in the table below are tied for eleventh smallest. Thus, each is assigned the rank $(11 + 12)/2 = 11.5$.

Fewer Than Two Years of High-School Algebra	Overall Rank	Two or More Years of High-School Algebra	Overall Rank
58	3	84	14
81	11.5	67	7
74	8.5	65	6
61	4	75	10
64	5	74	8.5
43	1	92	15
		83	13
		52	2
		81	11.5

Step 5:   The value of the test statistic is the sum of the ranks for the sample data from Population 1:

$M = 3 + 11.5 + 8.5 + 4 + 5 + 1 = 33.$

Step 6:   Since $M = 33$ equals the critical value, reject H_0.

Step 7:   At the 5% significance level, the data provide sufficient evidence to conclude that students with fewer than two years of high-school algebra have a lower mean semester average in this teacher's chemistry courses than do students with two or more years of high-school algebra.

10.67 Population 1: Volumes held by private colleges and universities;

Population 2: Volumes held by public colleges and universities.

Also, the subscript 1 is associated with that population having the smaller sample size.

Step 1: H_0: $\eta_1 = \eta_2$, H_a: $\eta_1 > \eta_2$

Step 2: $\alpha = 0.05$

Step 3:   We have $n_1 = 6$ and $n_2 = 7$. Since the hypothesis test is right-tailed with $\alpha = 0.05$, we use Table VIII to obtain the critical value, which is $M_r = 54$. Thus, we reject H_0 if $M \geq 54$.

Step 4:   Construct a work table based upon the following: First, rank all the data from both samples combined. Adjacent to each column of data as it is presented in the Exercise, record the overall rank.

Private	Overall Rank	Public	Overall Rank
27	3	15	1
67	5	24	2
113	7	41	4
139	8	79	6
500	11	265	9
603	13	411	10
		516	12

Step 5: The value of the test statistic is the sum of the ranks for the sample data from Population 1:

$$M = 3 + 5 + 7 + 8 + 11 + 13 = 47.$$

Step 6: Since $M < 54$, do not reject H_0.

Step 7: At the 5% significance level, the data do not provide sufficient evidence to conclude that the median number of volumes held by public colleges and universities is less than that held by private colleges and universities.

10.69 Population 1: Release times for prisoners with fraud offenses;

Population 2: Release times for prisoners with firearms offenses.

Also, the subscript 1 is associated with that population having the smaller sample size.

Step 1: H_0: $\mu_1 = \mu_2$, H_a: $\mu_1 < \mu_2$

Step 2: $\alpha = 0.05$

Step 3: We have $n_1 = 10$ and $n_2 = 10$. Since the hypothesis test is left-tailed with $\alpha = 0.05$, we use Table VIII to obtain the critical value, which is $M_1 = 83$. Thus, we reject H_0 if $M \le 83$.

Step 4: Construct a work table based upon the following: First, rank all the data from both samples combined. Adjacent to each column of data as it is presented in the Exercise, record the overall rank. Assign tied rankings the average of the ranks they would have had if there were no ties. For example, the two 17.9s in the table below are tied for thirteenth smallest. Thus, each is assigned the rank $(13 + 14)/2 = 13.5$.

Fraud	Overall Rank	Firearms	Overall Rank
3.6	1	10.4	6
5.3	2	13.3	9
5.9	3	16.1	11
7.0	4	17.9	13.5
8.5	5	18.4	15
10.7	7	19.6	16
11.8	8	20.9	17
13.9	10	21.9	18
16.6	12	23.8	19
17.9	13.5	25.5	20

Step 5:   The value of the test statistic is the sum of the ranks for the sample data from Population 1:

$$M = 1 + 2 + 3 + 4 + 5 + 7 + 8 + 10 + 12 + 13.5 = 65.5.$$

Step 6:   Since $M < 83$, reject H_0.

Step 7:   At the 5% significance level, the data provide sufficient evidence to conclude that the mean time served for fraud offenses is less than that for firearms offenses.

(b)   Normal distributions with the same standard deviations have the same shape, thus meeting the requirement for using the Mann-Whitney test. If the distributions are, in fact, normal with equal standard deviations, it is better to use the pooled t-test since it is slightly more powerful than the Mann-Whitney test in this situation.

10.71 (a)   Since the populations do not have the same shape, use the nonpooled t-test. This is appropriate because the populations are normally distributed.

(b)   Since the populations have the same shape, use the Mann-Whitney test.

(c)   Since both samples are large, use the two sample t-test.

10.73 Since the populations have the same shape and there is some question about the normality of the distributions, use the Mann-Whitney test. The presence of outliers in one of the samples strengthens this decision since the outliers could have a significant effect on the standard deviation and hence, on either the pooled or nonpooled t-test, whereas the presence of outliers will not have a great effect on the rank sum.

10.75 (a)   Step 1: H_0: $\mu_1 = \mu_2$, H_a: $\mu_1 < \mu_2$

Step 2: $\alpha = 0.05$

Step 3:   Critical value = -1.645

Step 4:   See Step 4 in the solution to Exercise 10.69.

Step 5:   From Step 5 in the solution to Exercise 10.69, $M = 65.5$. Now,

$$z = \frac{M - n_1(n_1 + n_2 + 1)/2}{\sqrt{n_1 n_2 (n_1 + n_2 + 1)/12}} = \frac{65.5 - 10(10 + 10 + 1)/2}{\sqrt{10 \cdot 10(10 + 10 + 1)/12}} = -2.99.$$

Step 6:   Since $-2.99 < -1.645$, reject H_0.

Step 7:   It appears that the new machine packs faster on the average.

(b)   The result in part (a) is the same as the one obtained in Exercise 10.69.

10.77 (a)

RANK						
1	2	3	4	5	6	M
A	A	A	B	B	B	6
A	A	B	A	B	B	7
A	A	B	B	A	B	8
A	A	B	B	B	A	9
A	B	A	A	B	B	8
A	B	A	B	A	B	9
A	B	A	B	B	A	10
A	B	B	A	A	B	10
A	B	B	A	B	A	11
A	B	B	B	A	A	12
B	A	A	A	B	B	9
B	A	A	B	A	B	10
B	A	A	B	B	A	11
B	A	B	A	A	B	11
B	A	B	A	B	A	12
B	A	B	B	A	A	13
B	B	A	A	A	B	12
B	B	A	A	B	A	13
B	B	A	B	A	A	14
B	B	B	A	A	A	15

(b)   5%

(c)   Given that $n_1 = 3$ and $n_2 = 3$, each row in part (a) has a 1/20

= 0.05 chance of occurring.

The probability distribution of the random variable M if $n_1 = 3$ and $n_2 = 3$ is presented below.

M	P(M)
6	0.05
7	0.05
8	0.10
9	0.15
10	0.15
11	0.15
12	0.15
13	0.10
14	0.05
15	0.05

(d)   A histogram for the probability distribution of M when $n_1 = 3$ and $n_2 = 3$ is:

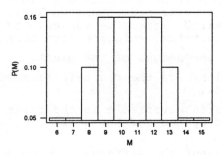

(e)   From part (c), we see that $P(M \leq 6) = 0.05$ and $P(M \geq 15) = 0.05$. These results correspond with the entries in Table VIII for $n_1 = 3$ and $n_2 = 3$. That is, $M_1 = 6$ at $\alpha = 0.05$, and $M_r = 15$ at $\alpha = 0.05$.

10.79 With the data in two columns named PUBLIC and PRIVATE, we choose **Stat ▶**

Nonparametrics ▶ Mann-Whitney..., select PUBLIC for **First sample** and PRIVATE for **Second sample**, select 95 for the **Confidence level**, select 'less than' for the **Alternative**, and click **OK**. The results are

```
PUBLIC      N =    7    Median =        79.0
PRIVATE     N =    6    Median =       136.0
Point estimate for ETA1-ETA2 is        -53.0
96.2 Percent C.I. for ETA1-ETA2 is (-459.1,278.1)
W = 44.0
Test of ETA1 = ETA2  vs.  ETA1 < ETA2 is significant at 0.2602
Cannot reject at α = 0.05
```

10.81 (a) H_0: $\eta_1 = \eta_2$, H_a: $\eta_1 < \eta_2$

(b) $16.615, $16.880

(c) $n_1 = 14$, $n_2 = 17$

(d)   W = 208 (Recall that Minitab uses W instead of M)

(e)   0.2692

(f)   0.2692

(g)   Do not reject H_0. At the 5% significance level, the data do not provide sufficient evidence to conclude that the median hourly earnings of the mine workers is less than that of the construction workers.

(h)   -$1.790 to $1.080

Exercises 10.5

10.83 Each pair in a paired sample consists of a member of one population and the member's corresponding member in the other population.

10.85 (a)   Ages of married people.

(b)   Married men and married women.

(c)   The pairs are married couples.

(d)   The paired difference variable is the difference between the age of the man and of the woman in each couple.

10.87 The samples must be paired (this is essential), and the population of all paired differences must be normally distributed or the sample must be large (the procedure works reasonably well even for small or moderate size samples if the paired-difference variable is not normally distributed provided that the deviation from normality is small).

10.89 (a)   The variable is the height of the plant.

(b)   The two populations are the cross-fertilized plants and the self-fertilized plants.

(c)   The paired-difference variable is the difference in heights of the cross-fertilized plants and the self-fertilized plants grown in the same pot.

(d)   Yes. They represent the difference in heights of two plants grown under the same conditions (same pot), one from each category of plants.

(e)   Step 1: H_0: $\mu_1 = \mu_2$, H_a: $\mu_1 \neq \mu_2$

Step 2:   α = 0.05

Step 3:   df = 14; critical values = ±2.145

Step 4:   The paired differences are given.

Step 5:

$$t = \frac{\overline{d}}{s_d/\sqrt{n}} = \frac{20.93}{37.74/\sqrt{15}} = 2.148$$

Step 6:   Since 2.148 > 2.145, reject H_0. The data do provide sufficient evidence at the 5% significance level that there is a difference between the mean heights of cross-fertilized and self-fertilized plants.

10.91 Population 1:   Measurements of tread wear using the Weight method

Population 2:   Measurements of tread wear using the Groove method

Step 1: H_0: $\mu_1 = \mu_2$, H_a: $\mu_1 \neq \mu_2$

Step 2: $\alpha = 0.01$

Step 3:   df = 10; critical values = ± 3.169

Step 4:   The paired differences 'Weight method – Groove method' are 1.8, 5.0, 8.6, 7.3, 3.6, -0.5, 8.4, 1.0, 3.7, 2.2, and 0.2. For these differences,

$\Sigma d = 41.3$ and $\Sigma d^2 = 258.83$.

Step 5:

$$\overline{d} = 41.3/11 = 3.75$$

$$s_d = \sqrt{\frac{\sum d^2 - (\sum d)^2/n}{n-1}} = \sqrt{\frac{258.83 - (41.3)^2/11}{10}} = 3.2213$$

$$t = \frac{\overline{d}}{s_d/\sqrt{n}} = \frac{3.75}{3.2213/\sqrt{11}} = 3.866$$

Step 6:   Since 3.866 > 3.169, reject H_0. The data do provide sufficient evidence at the 1% significance level that the two measurement methods give different results.

For the P-value approach, $2P(t > 3.866) < 0.01$. Since the P-value is smaller than the significance level, reject H_0.

10.93 Population 1: Husband; Population 2: Wife; df = 9

Step 1: H_0: $\mu_1 = \mu_2$, H_a: $\mu_1 > \mu_2$

Step 2: $\alpha = 0.05$

Step 3:   Critical value = 1.833

Step 4:   Paired differences, d = $x_1 - x_2$:

1	-1	6	1	13
-1	4	-2	4	8

Step 5: $\overline{d} = \dfrac{33}{10} = 3.3$; $s_d = \sqrt{\dfrac{(309) - (33)^2/10}{(9)}} = 4.715$

$$t = 3.3/(4.715/\sqrt{10}) = 2.213$$

Step 6:   Since 2.213 > 1.833, reject H_0.

Step 7:   At the 5% significance level, the data provide sufficient evidence to conclude that the mean age of married men is greater than the mean age of married women.

For the P-value approach, $0.025 < P(t > 2.213) < 0.05$. Since the P-value is smaller than the significance level, reject H_0.

10.95 From Exercise 10.89, df = 14.

(a)

$$20.93 \pm 2.145 \cdot (37.74/\sqrt{15})$$

$$20.93 \pm 20.90$$

0.03 to 41.83 eighths of an inch

(b)   We can be 95% confident that the difference, $\mu_1 - \mu_2$, between the mean heights of cross-fertilized and self-fertilized Zea mays is somewhere between 0.03 and 41.83 eighths of an inch.

10.97 From Exercise 10.91, df = 10.

(a)

$$3.75 \pm 3.169 \cdot (3.221/\sqrt{11})$$

$$3.75 \pm 3.08$$

0.67 to 6.83 thousands of miles

(b)   We can be 99% confident that the difference, $\mu_1 - \mu_2$, between the mean treadwear as measured by the Weight method and by the Groove method is somewhere between 0.67 and 6.83 thousands of miles.

10.99 From Exercise 10.93, df = 9.

(a)

$$3.3 \pm 1.833 \cdot (4.715/\sqrt{10})$$

$$3.3 \pm 2.73$$

0.6 to 6.0 years

(b)   We can be 90% confident that the difference, $\mu_1 - \mu_2$, between the mean age of married men and the mean age of married women is somewhere between 0.6 and 6.0 years.

10.101   Data is obtained from two populations whose members can be naturally paired. By letting d represent the difference between the values in each pair, we can reduce the data set from pairs of numbers to a single number representing each pair. Since the mean of the paired differences is the same as the difference of the two population means, we can test the equality of the two population means by testing to see if the mean of the paired differences is zero (or some other value if appropriate). If the paired differences are normally distributed, then the one-sample t-test can be used to carry out the test.

10.103   A nonparametric test should be used because the outlier will not have as much effect on the results. In general, nonparametric procedures are more resistant to outliers.

10.105   (a)   With the differences in a column named DIFF, choose **Calc ▶ Functions...**, select DIFF in the **Input Column** text box and type 'NSCO D' in the **Result in** text box (Note that the single quote marks must be typed if the column name has a space in it.), click on the **Normal scores** button, and click **OK**.

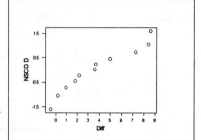

Now choose **Graph ▶ Plot...**, select NSCO D for the **Y** variable for **Graph 1** and DIFF for the **X** variable, and click **OK**.

(b)   Now choose **Stat ▶ Basic statistics ▶ Paired t...**, select WEIGHT in the **First Sample** text box, select GROOVE in the **Second Sample** text box , click on the **Options...** button, click in the **Confidence level** text box and type 99, click on the **Test Mean** button, click in the **Test mean** text box and type 0, select 'not equal to' in the **Alternative** box, click **OK**, and click **OK**. The results are

Paired T for WEIGHT - GROOVE

	N	Mean	StDev	SE Mean
WEIGHT	11	23.71	7.19	2.17
GROOVE	11	19.95	5.77	1.74
Difference	11	3.755	3.221	0.971

99% CI for mean difference: (0.675, 6.834)
T-Test of mean difference = 0 (vs not = 0): T-Value = 3.87 P-Value = 0.003

(c)   The normal probability plot indicates that the assumption of a normal distribution for the difference data is reasonable; therefore the 1-sample t-test is appropriate.

10.107   (a) $H_0: \mu_1 = \mu_2$ $H_a: \mu_1 \neq \mu_2$

(b)   The mean final grades were 69.73 for the lecture group and 67.82 for the PSI group.

(c)   The mean of the paired differences is 1.91.

(d) $s_d = 3.73$

(e) $t = 1.70$

(f)   P-value = 0.120

(g)   The smallest significance level at which the null hypothesis can be rejected is 0.120.

(h)   Since 0.120 > 0.05, do not reject the null hypothesis.

(i)   (-0.59, 4.41)

Exercises 10.6

10.109   (a)   Yes. The two requirements for a paired-t test, that the variable is a paired-difference variable and that it is normally distributed are met.

(b)   Yes. The Wilcoxon signed-rank test requires that the distribution of the differences be symmetrical. The normal distribution is symmetric.

(c)   The paired-t test is preferred when the underlying distribution of the differences is normal because it is the more powerful test.

10.111   (a)   Population 1: Cross-fertilized;

Population 2: Self-fertilized

Step 1: H_0: $\mu_1 = \mu_2$, H_a: $\mu_1 \neq \mu_2$

Step 2: $\alpha = 0.05$

Step 3:

d	|d|	Rank	Signed-rank
49	49	12	12
-67	67	14	-14
8	8	2	2
16	16	4	4
6	6	1	1
23	23	5	5
28	28	7	7
41	41	9	9
14	14	3	3
29	29	8	8
56	56	11	11
24	24	6	6
75	75	15	15
60	60	13	13
-48	48	10	-10

Step 4:   n = 15

Step 5:   The critical values are $W_l = 25$, $W_r = 95$

Step 6:   Work table prepared in Step 3.

Step 7:   The value of the test statistic is the sum of the positive ranks.

W = 12 + 2 + 4 + 1 + 5 + 7 + 9 + 3 + 8 + 11 + 6 + 15 + 13 = 96

Step 8:   Since 96 > 95, reject H_0.

Step 9:   At the 5% significance level, the data provide enough evidence to conclude that there is a difference in the mean heights of cross-fertilized Zea mays and self-fertilized Zea mays.

10.113 (a) Population 1: Weight Method; Population 2: Groove Method

Step 1: H_0: $\mu_1 = \mu_2$, H_a: $\mu_1 \neq \mu_2$

Step 2: $\alpha = 0.01$

Step 3:

d	|d|	Rank	Signed-rank
1.8	1.8	4	4
5	5.0	8	8
8.6	8.6	11	11
7.3	7.3	9	9
3.6	3.6	6	6
-0.5	0.5	2	-2
8.4	8.4	10	10
1.0	1.0	3	3
3.7	3.7	7	7
2.2	2.2	5	5
0.2	0.2	1	1

Step 4: n = 11

Step 5: The critical values: $W_1 = 5$, $W_r = 61$

Step 6: Work table prepared in Step 3.

Step 7: The value of the test statistic is the sum of the positive ranks.

W = 4 + 8 + 11 + 9 + 6 + 10 + 3 + 7 + 5 + 1 = 64

Step 8: Since 64 > 61, reject H_0.

Step 9: At the 1% significance level, the data provide enough evidence to conclude that the two measurement methods give different results.

10.115 (a) Population 1: Husbands; Population 2: Wives

Step 1: H_0: $\mu_1 = \mu_2$, H_a: $\mu_1 > \mu_2$

Step 2: $\alpha = 0.05$

Step 3:

d	|d|	Rank	Signed-rank
1	1	2.5	2.5
-1	1	2.5	-2.5
-1	1	2.5	-2.5
4	4	6.5	6.5
6	6	8	8
-2	2	5	-5
1	1	2.5	2.5
4	4	6.5	6.5
13	13	10	10
8	8	9	9

Step 4:   n = 10;

Step 5:   The critical value, $W_r = 44$

Step 6:   Work table prepared in Step 3.

Step 7:   The value of the test statistic is the sum of the positive ranks.

W = 2.5 + 6.5 + 8 + 2.5 + 6.5 + 10 + 9 = 45

Step 8:   Since 45 > 44, reject H_0.

Step 9:   At the 5% significance level, the data do provide enough evidence to conclude that the mean age of married men is greater than the mean age of married women.

10.117   The Wilcoxon paired-sample signed-rank test should be used because the outlier will not have as much effect on the results. In general, nonparametric procedures are more resistant to outliers.

10.119   (a)   Because the paired differences have a uniform distribution which is symmetric and nonnormal, use the Wilcoxon paired-sample signed-rank test.
        (b)   Because the population of paired differences is neither normal nor symmetric, but the sample size is large, use the paired t-test.
        (c)   Because the paired differences are roughly normally distributed, use the paired t-test.

10.121   (a)   With the differences in a column named DIFF, choose **Graph ▶ Boxplot...**, select DIFF for the **Y** variable for **Graph 1**, and click **OK**. Then choose **Graph ▶ Stem and Leaf...**, select DIFF in the **Variables:** text box and click **OK**. The results are

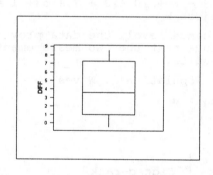

```
Stem-and-leaf of DIFF      N  = 11
Leaf Unit = 0.10

   1      -0 5
   2       0 2
   4       1 08
   5       2 2
  (2)      3 67
   4       4
   4       5 0
   3       6
   3       7 3
   2       8 46
```

        (b)   Choose **Stat ▶ Nonparametrics... ▶ 1-Sample Wilcoxon**, and specify DIFF in the **Variables** text box, click on the **Test Median** option button, click on the **Test Median** text box and type 0, select **less than** in the **Alternative** drop-down list box, and click **OK**. The results are

Test of median = 0.000000 versus median not = 0.000000

	N	N for Test	Wilcoxon Statistic	P	Estimated Median
DIFF	11	11	64.0	0.007	3.675

Since the P-value of 0.007 is less than 0.01, reject the null hypothesis.

(c) Since the boxplot and stem-and-leaf plot indicate that the distribution of the difference data is nearly symmetric, the paired Wilcoxon signed rank test is justified.

10.123

(a) H_0: $\mu_1 = \mu_2$, H_a: $\mu_1 \neq \mu_2$

(b) n = 11 students were selected

(c) Since only 10 students were included in the analysis, there was 1 pair in which both students had the same final grade.

(d) W = 42.5

(e) P-value = 0.139

(f) The smallest significance level at which the null hypothesis can be rejected is 0.139.

(g) Since 0.139 > 0.05, do not reject the null hypothesis.

(h) A point estimate for the difference between mean final grades of the two instructional methods is the median of the differences, 1.500.

10.125

(a) With the differences in a column named D, select **Calc ▶ Calculator...**, then type 'NSCORE D' in the **Store results in variable** text box, select **Normal scores** from the function list, select D to replace **numbers** in the **Expression** text

box, and click **OK**. Choose **Graph ▶ Plot...**, select NSCORE D for the **Y** variable for **Graph 1** and D for the **X** variable and

click **OK**. Choose **Graph ▶ Stem and Leaf...**, select D in the

Variables: text box and click **OK**. Choose **Graph ▶ Histogram...**, select D for the **X** variable for **Graph 1**, and

click **OK**. Finally, choose **Graph ▶ Boxplot...**, select D for the **Y** variable for **Graph 1**, and click **OK**. The results are

```
Stem-and-leaf of D        N = 10
Leaf Unit = 1.0

  1    -0 2
  3    -0 11
  5     0 11
  5     0
  5     0 44
  3     0 6
  2     0 8
  1     1
  1     1 3
```

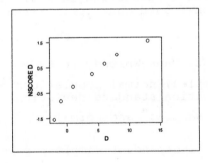

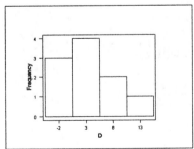

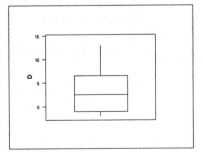

(b)   Choose **Stat ▶ Basic Statistics ▶ 1-Sample t...**, select D in the
       Variables text box, click on the **Test mean** button and type <u>0</u> in
       the **Test mean** text box, select **greater than** in the **Alternative**
       box, and click **OK**. The results are

Test of μ = 0.00 vs μ > 0.00

Variable	N	Mean	StDev	SE Mean	T	P-Value
D	10	3.30	4.72	1.49	2.21	0.027

(c)   Choose **Stat ▶ Nonparametrics ▶ 1-Sample Wilcoxon...**, select D in
       the **Variables** text box, click on the **Test median** button and type <u>0</u>
       in the **Test median** text box, select **greater than** in the
       Alternative box, and click **OK**. The results are

Test of median = 0.000000 versus median > 0.000000

	N	N for Test	Wilcoxon Statistic	P	Estimated Median
D	10	10	45.0	0.042	2.500

(d)   The P-values are 0.027 for the paired t-test and 0.042 for the
       paired Wilcoxon signed-rank test. Both are less than the
       significance level of 0.05, and therefore the null hypothesis is
       rejected in each case.

(e)   With the exception of one data value the distribution appears to
       be quite symmetrical, but not quite normal. No potential outliers
       are indicated. Given that non-parametric tests are less affected
       by non-normality, the paired Wilcoxon signed rank test is the more
       appropriate test for these data.

Exercises 10.7

10.127   (a)   Pooled t-test, nonpooled t-test, Mann-Whitney test

(b)   **Pooled t-test:** Independent samples, normal populations or
       large samples, and equal population standard deviations

Nonpooled t-test: Independent samples, and normal
populations or large samples

Mann-Whitney signed-rank test: Independent samples, same
shape populations, and $n_1 \leq n_2$

(c)   **Pooled t-test**

$$t = (\overline{x}_1 - \overline{x}_2) / s_p \sqrt{1/n_1 + 1/n_2} \text{ where}$$

$$s_p^2 = \frac{(n_1 - 1)s_1^2 + (n_2 - 1)s_2^2}{n_1 + n_2 - 2}$$

Nonpooled t-test

$$t = (\overline{x}_1 - \overline{x}_2)/\sqrt{s_1^2/n_1 + s_2^2/n_2}$$

where the degrees of freedom are given by

$$\Delta = \frac{[(s_1^2/n_1) + (s_2^2/n_2)]^2}{\dfrac{(s_1^2/n_1)^2}{n_1-1} + \dfrac{(s_2^2/n_2)^2}{n_2-1}}$$

Mann-Whitney test

M = the sum of the ranks for sample data from Population 1

10.129   (a)   One could use the pooled t-test, nonpooled t-test, or Mann-Whitney test.

         (b)   The pooled t-test is the most appropriate since its conditions are satisfied and it will be the most powerful under these circumstances.

10.131   (a)   Because the sample sizes are large, one could use the pooled t-test, nonpooled t-test, or Mann-Whitney test.

         (b)   Since the populations have the same shape but are not normally distributed, the Mann-Whitney test is the most appropriate since its conditions are satisfied and it will be the most powerful under these circumstances.

10.133   (a)   One could use the paired t-test or the paired Wilcoxon signed-rank test (paired W-test).

         (b)   The paired W-test is the more appropriate test when the distribution is symmetric, but non-normal.

10.135   To determine which procedure should be used to perform the hypothesis test, ask the following questions in sequence according to the flowchart in Figure 10.14:

	Answer to question	
Question to ask	Yes	No
Are the samples paired?		x
Are the populations normal?	x	
Are the populations the same shape?		x
Is the sample large?		x

Since the two independent populations are normal, use the nonpooled t-test.

10.137   To determine which procedure should be used to find the required confidence interval, ask the following questions in sequence according to the flowchart in Figure 10.16:

Question to ask	Answer to question	
	Yes	No
Are the samples paired?		x
Are the populations normal?		x
Are the samples large?	x	

It appears that one of the populations is not normal nor symmetric, but the sample size is large. Thus, use the non-pooled t-test.

10.139   To determine which procedure should be used to perform the hypothesis test, ask the following questions in sequence according to the flowchart in Figure 10.18:

Question to ask	Answer to question	
	Yes	No
Are the samples paired?	x	
Are the differences normal?		x
Are the differences symmetric?		x
Is the sample large?		x

It appears the paired differences are neither normal or symmetric. Thus, a statistician must be consulted.

REVIEW TEST FOR CHAPTER 10

1.   Randomly sample independently from both populations, compute the means of both samples and reject the null hypothesis if the sample means differ by too much. Otherwise, do not reject the null hypothesis.

2.   Sample pairs of observations from the two populations, compute the difference of the two observations in each pair, compute the mean of the differences, and reject the null hypothesis if the sample mean of the differences differs from zero by too much. Otherwise, do not reject the null hypothesis.

3.   (a)   The pooled t-test requires that the population standard deviations be equal whereas the nonpooled t-test does not.

     (b)   It is absolutely essential that the assumption of independence be satisfied.

     (c)   The normality assumption is especially important for both t-tests for small samples. With large samples, the Central Limit Theorem applies and the normality assumption is less important.

     (d)   Unless we are quite sure that the population standard deviations are equal, the nonpooled t-procedures should be used instead of the pooled t-procedures.

4.   (a)   No. If the two distributions are normal and have the same shape, then they have the same population standard deviations, and the

pooled t-test should be used. If the two distributions are not normal, but have the same shape, the Mann-Whitney test is preferred.

(b)   The pooled t-test is preferred to the Mann-Whitney test if both populations are normally distributed with equal standard deviations.

5.   A paired sample may reduce the estimate of the standard error of the mean of the differences, making it more likely that a difference of a given size will be judged significant.

6.   The paired t-test is preferred to the paired Wilcoxon signed-rank test if the distribution of differences is normal, or if the sample size is large and the distribution of differences is not symmetric. If the distribution of differences is symmetric but not normal, then the paired Wilcoxon signed-rank test is preferred.

7.   Population 1: Male; $n_1 = 13$, $\overline{x}_1 = 2127$, $s_1 = 513$

Population 2: Female; $n_2 = 14$, $\overline{x}_2 = 1843$, $s_2 = 446$

Step 1: H_0: $\mu_1 = \mu_2$, H_a: $\mu_1 > \mu_2$

Step 2: $\alpha = 0.05$,

Step 3:   df = 25, Critical value = 1.708

Step 4:

$$s_p = \sqrt{\frac{12 \cdot 513^2 + 13 \cdot 446^2}{25}} = 479.33$$

$$t = (2127-1843)/479.33 \cdot \sqrt{(1/13) + (1/14)} = 1.538$$

Step 5:   Since 1.538 < 1.708, do not reject H_0.

Step 6:   At the 5% significance level, the data do not provide enough evidence to conclude that the mean right-leg strength of males exceeds that of females.

(b)   For the P-value approach, the $0.05 < P(t > 1.538) < 0.10$. Therefore, because the P-value is larger than the significance level, do not reject H_0.

(c)   The evidence against the null hypothesis is moderate.

8.

$$(2127-1843) \pm 1.708 \cdot 479.33\sqrt{(1/13) + (1/14)}$$

$$284 \pm 315.33$$

$$-31.3 \text{ to } 599.3$$

We can be 90% confident that the difference, $\mu_1 - \mu_2$, between the mean right-leg strengths of males and females is between -31.3 and 599.3 newtons.

9.   Population 1: Germans, $n_1 = 10$, $\overline{x}_1 = 11.4$, $s_1 = 2.84$

Population 2: Russians, $n_2 = 15$, $\overline{x}_2 = 16.07$, $s_2 = 5.61$

$$\Delta = \frac{[\,(s_1{}^2/n_1) + (s_2{}^2/n_2)\,]^2}{\dfrac{(s_1{}^2/n_1)^2}{n_1-1} + \dfrac{(s_2{}^2/n_2)^2}{n_2-1}}$$

$$= \frac{[\,(2.84^2/10) + (5.61^2/15)\,]^2}{\dfrac{(2.84^2/10)^2}{9} + \dfrac{(5.61^2/15)^2}{14}}$$

$$= 21.82 \approx 21 = DF.$$

Step 1: H_0: $\mu_1 = \mu_2$, H_a: $\mu_1 < \mu_2$

Step 2: $\alpha = 0.05$

Step 3:   Critical value = -1.721

Step 4:

$$t = (11.40-16.07)/\sqrt{(2.84^2/10)+(5.61^2/15)} = -2.740$$

Step 5:   Since $-2.74 < -1.721$, reject H_0.

Step 6:   At the 5% significance level, the data provide sufficient evidence to conclude that last year the average German consumed less fish than the average Russian.

For the P-value approach, $0.005 < P(t < -2.74) < 0.01$. Therefore, because the P-value is smaller than the significance level, reject H_0.

10.

$$(11.40 - 16.07) \pm 1.721 \sqrt{\frac{2.84^2}{10} + \frac{5.61^2}{15}}$$

$$-4.67 \pm 2.933$$

$$-7.60 \text{ to } -1.74$$

We can be 90% confident that the difference, $\mu_1 - \mu_2$, between last year's mean fish consumption by Germans and Russians is somewhere between -7.60 and -1.74 kilograms.

11.   Population 1: Home prices in New York City;

Population 2: Home prices in Los Angeles.

Step 1: H_0: $\mu_1 = \mu_2$, H_a: $\mu_1 \neq \mu_2$

Step 2: $\alpha = 0.05$

Step 3:   We have $n_1 = 10$ and $n_2 = 10$. Since the hypothesis test is two-tailed with $\alpha = 0.05$, we use Table VIII to obtain the critical values, which are $M_l = 79$ and $M_r = 131$. Thus, we reject H_0 if $M \geq 131$ or $M \leq 79$.

Step 4:   Construct a work table based upon the following: First, rank all the data from both samples combined. Adjacent to each column of data as it is presented in the Exercise, record the overall rank.

NYC	Overall Rank	LA	Overall Rank
131.5	8	137.5	10
379.8	20	290.8	18
132.8	9	215.9	16
145.8	13	127.7	6
235.8	17	335.9	19
118.1	5	130.0	7
168.5	14	101.7	1
105.2	3	140.9	11
141.8	12	104.5	2
174.8	15	107.0	4

Step 5: The value of the test statistic is the sum of the ranks for the sample data from Population 1:

$M = 8 + 20 + 9 + 13 + 17 + 5 + 14 + 3 + 12 + 15 = 116.$

Step 6: Since $79 < M < 131$, do not reject H_0.

Step 7: At the 5% significance level, the data do not provide sufficient evidence to conclude that the mean costs for existing single-family homes differ in New York City and Los Angeles.

12. **Step 1:** $H_0: \mu_1 = \mu_2$, $H_a: \mu_1 \neq \mu_2$

Step 2: $\alpha = 0.10$

Step 3: Critical values $= \pm 1.833$

Step 4: Paired differences, $d = x_1 - x_2$:

82	-95	-49	0	-36
-152	49	-38	-43	-118

Step 5: $\overline{d} = \dfrac{-400}{10} = -40; \quad s_d = \sqrt{\dfrac{(62,168) - (-400)^2/10}{(9)}} = 71.622$

$t = -40/(71.622/\sqrt{10}) = -1.766$

Step 6: Since $-1.833 < -1.766 < 1.833$, do not reject H_0.

Step 7: At the 10% significance level, the data do not provide sufficient evidence to conclude that there is a difference in mean results for the two speed reading programs.

For the P-value approach, $0.10 < 2\{P(t < -1.766)\} < 0.20$. Therefore, because the P-value is larger than the significance level, do not reject H_0.

13.

$$-40 \pm 1.833 \cdot (71.622/\sqrt{10})$$

$$-40 \pm 41.52$$

$$-81.52 \text{ to } 1.52 \text{ words per minute}$$

We can be 90% confident that the difference, $\mu_1 - \mu_2$, between the mean reading speed of people using Program 1 and the mean reading speed of people using Program 2 is somewhere between -81.5 and 1.5 words per minute.

14.   Population 1: South; Population 2: Midwest;

Step 1: $H_0: \mu_1 = \mu_2$, $H_a: \mu_1 > \mu_2$

Step 2: $\alpha = 0.01$

Step 3:

d	|d|	Rank	Signed-rank
24	24	8	8
27	27	10	10
4	4	1.5	1.5
19	19	7	7
46	46	12	12
-16	16	6	-6
26	26	9	9
66	66	14	14
12	12	5	5
36	36	11	11
48	48	13	13
-4	4	1.5	-1.5
-11	11	4	-4
78	78	15	15
8	8	3	3

Step 4: $n = 15$

Step 5:   The critical value, $W_r = 100$

Step 6:   Work table shown in Step 3.

Step 7:   The value of the test statistic is the sum of the positive ranks.

$$W = 8 + 10 + 1.5 + 7 + 12 + 9 + 14 + 5 + 1 + 13 + 15 + 3 = 108.5$$

Step 8:   Since $108.5 > 100$, reject H_0.

Step 9:   At the 5% significance level, the data do provide sufficient evidence to conclude that the mean monthly rent for renter-occupied housing units in the South exceeds that for those in the Midwest.

15.   (a)   Choose **Calc ▶ Calculator...**, type <u>NSCORE M</u> in the **Store results in variable** text box, select **Normal scores** from the function list, select MALE to replace **numbers** in the **Expression** text box, and click **OK**. We now choose **Graph ▶ Plot...**, select NSCORE M for the **Y** variable for **Graph 1** and MALE for the **X** variable, and click

OK. Next choose **Graph ▶ Stem-and-Leaf...**, select MALE in the

Variables text box, and click **OK**. Now choose **Graph ▶ Boxplot...**, select MALE for the **Y** Variable for **Graph 1**, and click **OK**. Now

choose **Stat ▶ Column statistics...**, click on the **Standard deviation** button, select MALE in the **Input Variables** text box, and click **OK**. Then repeat this entire process for the FEMALE data.

The results for MALES are

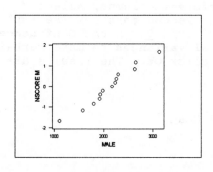

```
Stem-and-leaf of MALE      N  = 13
Leaf Unit = 100

    1       1  1
    1       1
    2       1  5
    3       1  7
    6       1  999
   (1)      2  1
    6       2  222
    3       2
    3       2  66
    1       2
    1       3  1
```

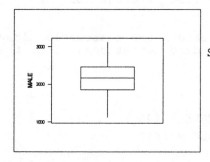

Standard deviation of MALE   =   512.99

For the FEMALE data, the results are

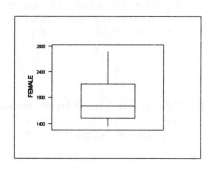

```
Stem-and-leaf of FEMALE    N  = 14
Leaf Unit = 100

    3       1  333
    5       1  55
   (3)      1  667
    6       1  88
    4       2  0
    3       2  3
    2       2  4
    1       2  7
```

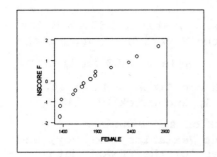

Standard deviation of FEMALE =   445.96

(b)   To carry out the pooled t-test, choose **Stat ▶ Basic Statistics ▶ 2-Sample t**, click on **Samples in different columns**, select MALE in the **First** text box and FEMALE in the **Second** text box, select **greater than** in the **Alternative** box, type <u>90</u> in the **Confidence level** text box, click on **Assume equal variances** to make certain that there is an **X** in the box, and click **OK**.   The results are

Twosample T for MALE vs FEMALE

	N	Mean	StDev	SE Mean
MALE	13	2127	513	142
FEMALE	14	1843	446	119

90% C.I. for μ MALE - μ FEMALE: (-31,   600)

T-Test μ MALE = μ FEMALE (vs >): T= 1.54   P=0.068   DF= 25

Both use Pooled StDev = 479

(c)   The plots indicate that the distributions are reasonable close to normal with approximately equal standard deviations.   Thus the pooled t-test is appropriate.

16.   1 = MALE; 2 = FEMALE

(a) H_0: $\mu_1 = \mu_2$, H_a: $\mu_1 > \mu_2$

(b)   StDev for MALE = 513; StDev for FEMALE = 446

(c)   Mean for MALE = 2127; Mean for FEMALE = 1843

(d)   t = 1.54

(e)   P = 0.068

(f)   The smallest significance level at which H_0 can be rejected is 0.068.

(g)   Since 0.068 > 0.05, do not reject H_0.   At the 5% significance level, the data do not provide sufficient evidence to conclude that the mean right-leg strengths of males exceeds that of females.

(h)   -31 to 599

(i)   POOLED STDEV = 479

17.   (a)   Choose   **Calc ▶ Calculator...**, type <u>NSCORE R</u> in the **Store results in variable** text box, select **Normal scores** from the function list, select RUSSIANS to replace **numbers** in the **Expression** text box and click **OK**.   We now choose **Graph ▶ Plot...**, select NSCORE R for the **Y** variable for **Graph 1** and RUSSIANS for the **X** variable, and click **OK**.   The result is

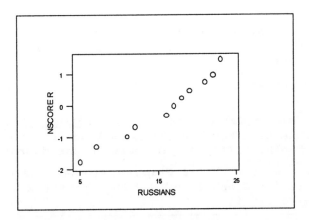

Now choose **Graph ▶ Stem-and-Leaf...**, select RUSSIANS in the **Variables** text box, and click **OK**. The result is

```
Stem-and-leaf of RUSSIANS   N  = 16
Leaf Unit = 1.0

  1      0 5
  2      0 7
  2      0
  3      1 1
  5      1 22
  5      1
 (4)     1 6677
  7      1 899
  4      2 1
  3      2 233
```

Now choose **Graph ▶ Boxplot...**, select RUSSIANS for the **Y** Variable for **Graph 1**, and click **OK**.

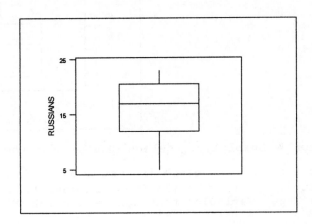

Now choose **Stat ▶ Column statistics...**, click on the **Standard deviation** button, select RUSSIANS in the **Input Variables** text box, and click **OK**.

Standard deviation of RUSSIANS=   5.4268

Choose   **Calc ▶ Calculator...**, type <u>NSCORE G</u> in the **Store results in variable** text box, select **Normal scores** from the function list, select GERMANS to replace **numbers** in the **Expression** text box, and click **OK**. We now choose **Graph ▶ Plot...**, select NSCORE G for the **Y** variable for **Graph 1** and GERMANS for the **X** variable, and click **OK**. The result is

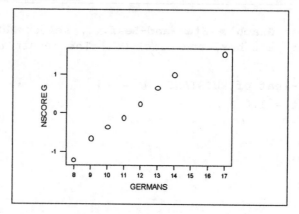

Now choose **Graph ▶ Stem-and-Leaf...**, select GERMANS in the **Variables** text box, and click **OK**. The result is

```
Stem-and-leaf of GERMANS    N  = 10
Leaf Unit = 0.10

    2      8 00
    3      9 0
    4     10 0
    5     11 0
    5     12 00
    3     13 0
    2     14 0
    1     15
    1     16
    1     17 0
```

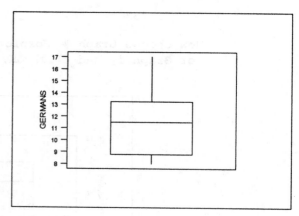

Now choose **Graph ▶ Boxplot...**, select GERMANS for the **Y** Variable for **Graph 1**, and click **OK**. The graph is just above. Now choose **Stat ▶ Column statistics...**, click on the **Standard deviation** button, select GERMANS in the **Input Variables** text box, and click **OK**.

Standard deviation of GERMANS =   2.8363

(b)   Now choose **Stat ▶ Basic Statistics ▶ 2-Sample t...**, click on the **Samples in different columns** button, select GERMANS in the **First** text box and RUSSIANS in the **Second** text box, select 'less than' in the **Alternative** text box, type <u>90</u> in the **Confidence level** text box, and click **OK**. The result is

```
Twosample T for GERMANS vs RUSSIANS
            N      Mean     StDev    SE Mean
GERMANS    10     11.40      2.84      0.90
RUSSIANS   16     16.12      5.43      1.4
```

90% C.I. for μ GERMANS - μ RUSSIANS: (-7.51, -1.9)

T-Test μ GERMANS = μ RUSSIANS (vs <): T= -2.91 P=0.0040 DF= 23

(c)   Based on the graphic displays and sample standard deviations in part (a) for two independent samples, it appears reasonable to consider only the assumption of normality satisfied; thus, the use of the nonpooled-t procedure.

18.   1 = German; 2 = Russian

(a) $H_0: \mu_1 = \mu_2$, $H_a: \mu_1 < \mu_2$

(b) $s_1 = 2.84$; $s_2 = 5.61$

(c) $n_1 = 10$; $n_2 = 15$

(d) $\overline{x_1} = 11.40$; $\overline{x_2} = 16.07$

(e) $t = -2.74$

(f) $P = 0.0062$

(g)   The smallest significance level at which H_0 can be rejected is 0.0062.

(h)   Since $0.0062 < 0.05$, reject H_0. At the 5% significance level, the data provide sufficient evidence to conclude that last year the Germans averaged less fish consumption than the Russians.

(i)   -7.60 to -1.7

19.   (a)   With the data in columns named NY and LA, choose **Calc ▶ Calculator...**, type <u>'NSCO NY'</u> in the **Store results in variable** text box, select **Normal scores** from the function list, select NY to replace **numbers** in the **Expression** text box, and click **OK**. We now choose **Graph ▶ Plot...**, select NSCO NY for the **Y** variable for **Graph 1** and NY for the **X** variable, and click **OK**. Next, choose **Graph ▶ Stem-and-Leaf...**, select NY in the **Variables** text box, and click **OK**. Then, choose **Graph ▶ Boxplot...**, select NY for the **Y** Variable for **Graph 1**, and click **OK**. Finally, choose **Stat ▶ Column statistics...**, click on the **Standard deviation** button, select NY in the **Input Variables** text box, and click **OK**. Repeat the above process for the variable LA in place of NY and NSCO LA in place of NSCO NY. The results for New York are

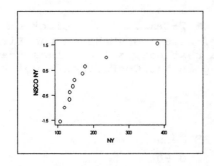

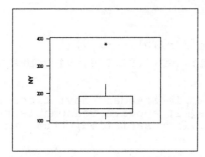

```
Stem-and-leaf of NY          N  = 10
Leaf Unit = 10

   (6)    1  013344
    4     1  67
    2     2  3
    1     2
    1     3
    1     3  7
```

Standard deviation of NY   =   81.236

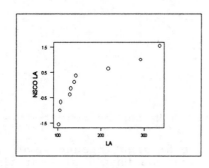

The results for Los Angeles are

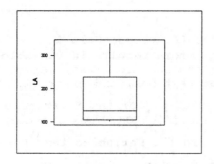

```
Stem-and-leaf of LA          N  = 10
Leaf Unit = 10

   (7)    1  0002334
    3     1
    3     2  1
    2     2  9
    1     3  3
```

Standard deviation of LA = 83.288

(b)   Now choose **Stat ▶ Nonparametrics ▶ Mann-Whitney...**, select NY in
       the **First Sample** text box and LA in the **Second sample** text box,
       select 'not equal to' in the **Alternative** text box, type 95 in the
       Confidence level text box, and click **OK**. The result is

```
NY          N =  10     Median =          143.8
LA          N =  10     Median =          133.8
Point estimate for ETA1-ETA2 is           11.4
95.5 Percent C.I. for ETA1-ETA2 is (-55.0,44.8)
W = 116.0
Test of ETA1 = ETA2  vs  ETA1 not = ETA2 is significant at 0.4274

Cannot reject at α = 0.05
```

(c)   Based on the graphic displays and sample standard deviations in part (a), it appears that the two populations are nonnormal; however, they do have the same shape. Thus, the use of the Mann-Whitney test is appropriate.

20. 1 = NEW YORK; 2 = LOS ANGELES

(a) H_0: $\eta_1 = \eta_2$, H_a: $\eta_1 \neq \eta_2$

(b)   Median for New York is 143.8; Median for Los Angeles is 133.8.

(c) $n_1 = 10$; $n_2 = 10$

(d)   W = 116.0

(e)   P = 0.4274

(f)   The smallest significance level at which H_0 can be rejected is 0.4274.

(h)   Since 0.4274 > 0.05, do not reject H_0. At the 5% significance level, the data do not provide sufficient evidence to conclude that there is a difference in the median costs of existing single-family homes in New York City and Los Angeles.

(i)   -55.0 to 44.8

21. (a)   With the differences in a column named DIFF, choose **Calc ▶ Calculator...**, type <u>NSCODIFF</u> in the **Store results in variable** text box, select **Normal scores** from the function list, select NY to replace **numbers** in the **Expression** text box, and click **OK**. We now

choose **Graph ▶ Plot...**, select NSCO NY for the **Y** variable for **Graph 1** and NY for the **X** variable, and click **OK**. Next, choose

Graph ▶ Stem-and-Leaf..., select NY in the **Variables** text box, and click **OK**. Then, choose **Graph ▶ Boxplot...**, select NY for the **Y** Variable for **Graph 1**, and click **OK**. The results are

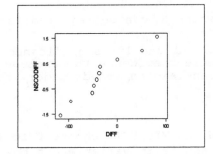

```
Stem-and-leaf of DIFF      N  = 10
Leaf Unit = 10

  1     -1 1
  3     -0 95
 (4)    -0 4433
  3      0 0
  2      0 58
```

(b)   Now choose **Stat ▶ Basic statistics ▶ Paired t...**, select PROGRAM1 in the **First Sample** text box, select PROGRAM2 in the **Second Sample** text box , click on the **Options...** button, click in the **Confidence level** text box and type <u>90</u>, click on the **Test Mean** button, click in the **Test mean** text box and type <u>0</u>, select 'not equal to' in the **Alternative** box, click **OK**, and click **OK**. The results are

Paired T for PROGRAM1 - PROGRAM2

	N	Mean	StDev	SE Mean
PROGRAM1	10	1021.4	80.3	25.4
PROGRAM2	10	1061.4	53.4	16.9
Difference	10	-40.0	71.6	22.6

90% CI for mean difference: (-81.5, 1.5)

T-Test of mean difference = 0 (vs not = 0): T-Value = -1.77   P-Value = 0.111

(c)   The graphs in part (a) indicate that the normality assumption is reasonable for the differences, leading to the use of the paired t-test and paired difference confidence interval.

22.   1 = PROGRAM1; 2 = PROGRAM2

(a) H_0: $\mu_1 = \mu_2$, H_a: $\mu_1 \neq \mu_2$

(b)   Mean for Program 1 is 1021.4; Mean for Program 2 is 1061.4.

(c)   Mean of the paired differences = -40.0.

(d)   Standard deviation of the paired differences = 71.6

(e)   t = -1.77

(f)   P = 0.111

(g)   The smallest significance level at which H_0 can be rejected is 0.111.

(h)   Since 0.111 > 0.10, do not reject H_0. At the 10% significance level, the data do not provide sufficient evidence to conclude that there is a difference in the mean reading speeds under the two programs.

(i)   -81.5 to 1.5

23.   (a)   With the data in columns named SOUTH and MIDWEST, choose **Calc ▶ Calculator...**, type <u>D</u> in the Store results in variable ext box, type <u>'SOUTH' - 'MIDWEST'</u> in the **Expression** text box, and click **OK**.

Then choose **Stat ▶ Nonparametrics ▶ 1-Sample Wilcoxon...**, select D in the **Variables** text box, click on the **Test median** button and type <u>0</u> in the **Test median** text box, select 'greater than' in the **Alternative** box, and click **OK**. The result is

Test of median = 0.000000 versus median > 0.000000

	N	N For Test	Wilcoxon Statistic	P	Estimated Median
D	15	15	108.5	0.003	23.50

(b)   Choose **Calc ▶ Calculator...**, type <u>NSCORE D</u> in the **Store results in variable** text box, select **Normal scores** from the function list, select D to replace **numbers** in the **Expression** text box, and click

OK. We now choose **Graph ▶ Plot...**, select NSCORE D for the **Y** variable for **Graph 1** and D for the **X** variable, and click **OK**.

Next, choose **Graph ▶ Stem-and-Leaf...**, select D in the **Variables**

text box, and click **OK**. Then, choose **Graph ▶ Boxplot...**, select D for the **Y** Variable for **Graph 1**, and click **OK**. The results are

Stem-and-leaf of D   N = 15
Leaf Unit = 1.0

```
   2    -1 61
   3    -0 4
   5     0 48
   7     1 29
  (3)    2 467
   5     3 6
   4     4 68
   2     5
   2     6 6
   1     7 8
```

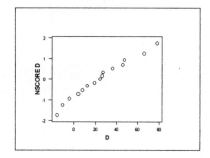

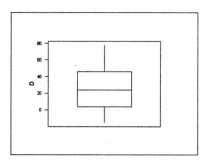

(b)   Since the graphs in part (b) indicate that the distribution of differences is symmetric, the conditions for using a paired Wilcoxon signed-rank test are satisfied.

(c)   The graphs indicate that the assumption of a normal distribution for the differences is reasonable. Thus the conditions for a paired t-test are satisfied.

(e)   Since the conditions for a paired t-test are satisfied, the paired t-test is the preferred test since it is more powerful than the Wilcoxon test when the distribution of differences is normal.

24.   1 = SOUTH; 2 = MIDWEST

(a) H_0: $\mu_1 = \mu_2$, H_a: $\mu_1 > \mu_2$

(b)   There were no pairs in which both houses had the same monthly rent.

(c)   W = 108.5

(d)   P = 0.003

(e)   The smallest significance level at which H_0 can be rejected is 0.003

(f)   Since 0.003 < 0.01, reject H_0. At the 1% significance level, the data do provide sufficient evidence to conclude that the mean monthly rents are greater in the South than in the Midwest.

(g)   23.50

CHAPTER 11 ANSWERS

Exercises 11.1

11.1 A variable has a chi-square distribution if its distribution has the shape of a right-skewed curve called a chi-square curve.

11.3 The curve with 20 degrees of freedom more closely resembles a normal distribution. By Property 4 of Key Fact 11.1, as the degrees of freedom increases, the distributions look increasing like normal distribution curves.

11.5 (a) $\chi^2_{0.025} = 32.852$　　　(b) $\chi^2_{0.95} = 10.117$

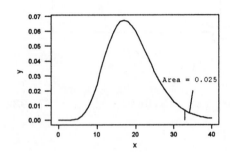

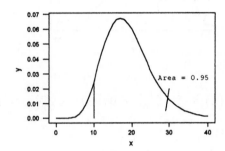

11.7 (a) $\chi^2_{0.05} = 18.307$　　　(b) $\chi^2_{0.975} = 3.247$

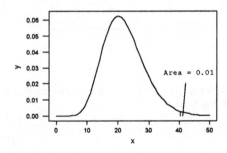

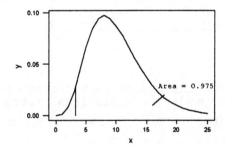

11.9 (a)   A left area of 0.01 is equivalent to a right area of 0.99: $\chi^2_{0.99} = 1.646$

(b)   A left area of 0.95 is equivalent to a right area of 0.05: $\chi^2_{0.05} = 15.507$

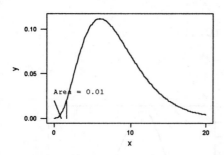
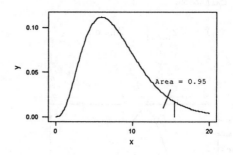

11.11 (a) $\chi^2_{0.975} = 0.831$ $\chi^2_{0.025} = 12.832$ (b) $\chi^2_{0.975} = 13.844$ $\chi^2_{0.025} = 41.923$

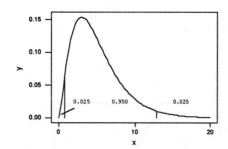
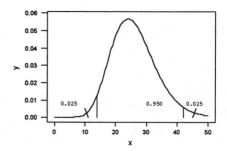

11.13 The chi-square test for one population standard deviation is not robust to moderate violations of the normality assumption.

11.15 Step 1: H_0: $\sigma = 100$ H_a: $\sigma \neq 100$

Step 2: $\alpha = 0.05$

Step 3:   The critical values with $n-1 = 24$ degrees of freedom are 12.401 and 39.364.

Step 4: $\chi^2 = \dfrac{n-1}{\sigma_0^2} s^2 = \dfrac{24}{100^2} 85.5^2 = 17.545$

Step 5: Since $12.401 < 17.545 < 39.364$, we do not reject H_0.

Step 6: There is insufficient evidence at the 0.05 level to claim that σ has changed from the 1941 standard of 100.

11.17 Step 1: H_0: $\sigma = 0.27$ H_a: $\sigma > 0.27$

Step 2: $\alpha = 0.01$

Step 3:   The critical value with $n-1 = 9$ degrees of freedom is 21.666.

Step 4: $\chi^2 = \dfrac{n-1}{\sigma_0^2}s^2 = \dfrac{9}{.27^2}0.756^2 = 70.560$

Step 5: Since $70.560 > 21.666$, we reject H_0.

Step 6: There is sufficient evidence at the 0.01 level to claim that the process variation for this piece of equipment exceeds the analytical capability of 0.27.

11.19 Step 1: H_0: $\sigma = 0.2$ H_a: $\sigma < 0.2$

Step 2: $\alpha = 0.05$

Step 3:   The critical value with $n-1 = 14$ degrees of freedom is 6.571.

Step 4: $\chi^2 = \dfrac{n-1}{\sigma_0^2}s^2 = \dfrac{14}{0.2^2}0.154^2 = 8.301$

Step 5: Since $8.301 > 6.571$, we do not reject H_0.

Step 6: There is insufficient evidence at the 0.05 level to claim that the standard deviation of the amounts dispensed is less than 0.2 fluid ounces.

11.21 The 95% confidence interval for σ of last year's verbal SAT scores is

$$\left(\sqrt{\frac{n-1}{\chi_{.025}^2}}s,\sqrt{\frac{n-1}{\chi_{.975}^2}}s\right) = \left(\sqrt{\frac{24}{39.364}}85.5,\sqrt{\frac{24}{12.401}}85.5\right) = (66.761,118.944)$$

11.23 The 98% confidence interval for the process variation of the piece of equipment under consideration is

$$\left(\sqrt{\frac{n-1}{\chi_{.01}^2}}s,\sqrt{\frac{n-1}{\chi_{.99}^2}}s\right) = \left(\sqrt{\frac{9}{21.666}}0.756,\sqrt{\frac{9}{2.088}}0.756\right) = (0.487,1.570)$$

11.25 The 90% confidence interval for the standard deviation of the amounts of coffee being dispensed is

$$\left(\sqrt{\frac{n-1}{\chi_{.05}^2}}s,\sqrt{\frac{n-1}{\chi_{.95}^2}}s\right) = \left(\sqrt{\frac{14}{23.685}}0.154,\sqrt{\frac{14}{6.571}}0.154\right) = (0.118,0.225)$$

11.27 If the standard deviation is too large, some cups will be filled with too little coffee (making for dissatisfied customers), and some cups may overflow or be in danger of being spilled by the customer. Customers don't appreciate hot coffee spilled on them; in addition, the company loses money (or makes less) if too much coffee is dispensed.

11.29 If the alternative hypothesis is $\sigma < 0.09$, the advantage is that the manufacturer can be quite sure, when the null hypothesis is rejected, that the standard deviation of the bolt diameters is acceptable, that the manufacturing process is "in control." The disadvantage is that σ could actually be smaller than 0.09 (acceptable), but not enough smaller to trigger rejection of the null hypothesis. Without strong evidence that $\sigma < 0.09$, the manufacturer may unnecessarily shut down the manufacturing process to fix a problem that doesn't exist.

If the alternative hypothesis is $\sigma > 0.09$, the advantage is that the manufacturer can be quite sure, when the null hypothesis is rejected, that the standard deviation of the bolt diameters is unacceptable, that the process is "out of control," and that it is worthwhile to shut down the process to fix a problem. The disadvantage is that σ could actually be larger than 0.09 (unacceptable), but not enough larger to trigger rejection of the null hypothesis. Thus the manufacturer may allow the manufacturing process to continue even though the variation is great enough to cause more unacceptable bolts than desired. See the next exercise for information on how to reduce these problems.

11.31 (a)   With the data in a column named SATVERB, we choose **Calc ▶ Calculator...**, type <u>NSCORE</u> in the **Store result in variable** text box, select the function NSCOR from the **Function** list, select SATVERB to replace **number** in the **Expression** text box, and click

OK. Then choose **Graph ▶ Plot...**, select NSCORE in the **Y** column for **Graph1** and SATVERB in the **X** column, and click **OK**. Then choose

Graph ▶ Boxplot..., select SATVERB in the **Y** column for **Graph1**, and click **OK**. The two graphs are

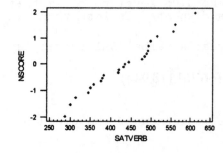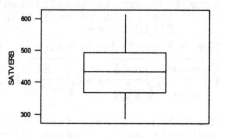

(b)   In Minitab, click in the Sessions Window, and if there is no MTB prompt showing, choose **Editor ▶ Enable command language**. Then with the *DataDisk* in drive A, return to the MTB prompt and type <u>%A:\Macro\1stdev.mac 'SATVERB'</u> and press the ⌷ENTER⌷ key. Then proceed as follows.

1 In response to Do you want to perform a hypothesis test (Y/N)?, type <u>Y</u> and press the ⌷ENTER⌷ key.

2 In response to Enter the null hypothesis population standard deviation, type <u>100</u> and press the ⌷ENTER⌷ key.

3 In response to Enter 0, 1, or -1, respectively, for a two-tailed, right-tailed, or left-tailed test., type <u>0</u> and press the ⌷ENTER⌷ key. The output is

Test of σ =   100 vs σ not =   100

Row	Variable	n	StDev	Chi-Sq	P
1	SATVERB	25	85.492	17.541	0.351

The macro continues with the confidence-interval procedure.

1 In response to Do you want a confidence interval (Y/N)?, type Y and press the ENTER key.

2 Since we want a 95% confidence interval, type 95 in response to Enter the confidence level, as a percentage and press the ENTER key. The result is

Row	Variable	n	StDev	Level	CI for σ
1	SATVERB	25	85.492	95.0%	(66.755, 118.933)

(c) The procedure in (b) depends on the data coming from a normal distribution. The normal probability plot and the box plot indicate that this is a reasonable assumption for the SAT verbal data.

11.33 (a) We will generate the 1000 samples as 1000 rows in Minitab. Choose **Calc ▶ Random data ▶ Normal...**, type 1000 in the **Generate rows of data** text box, type IQ1 IQ2 IQ3 IQ4 in the **Store in columns** text box, type 100 in the **Mean** text box, type 16 in the **Standard deviation** text box, and click **OK**. Now compute the standard deviation of each of the 1000 rows.

(b) Choose **Calc ▶ Calculator**, type SD in the **Store result in variable** text box, click in the **Expression** text box and select **Std. Dev. (Rows)** from the function list, select IQ1, IQ2, IQ3, and IQ4 to replace Number in the function, making sure that the final expression is RSTDEV(IQ1,IQ2,IQ3,IQ4), and click **OK**.

(c) Choose **Calc ▶ Calculator**, type CHI2 in the **Store result in variable** text box, click in the **Expression** text box, type 3*'SD'**2/256, and click **OK**.

(d) Choose **Graph ▶ Histogram...**, select CHI2 for **Graph1** of **X** and click **OK**. The result is

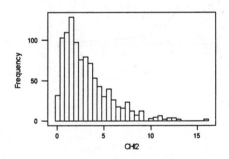 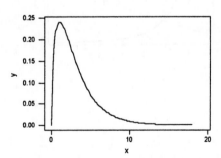

(e) The distribution of the variable in part (c) is a χ^2 distribution with n − 1 = 4 − 1 = 3 degrees of freedom (as shown above right).

(f) A Chi-square distribution with 3 degrees of freedom is very right-skewed. The histogram in part (d) corresponds to such a distribution.

Exercises 11.2

11.35 We identify an F-distribution and its corresponding F-curve by stating its number of degrees of freedom.

11.37 $F_{0.05}$, $F_{0.025}$, F_α

11.39 (a)   12   (b)   7

11.41 (a) $F_{0.05} = 1.89$ (b) $F_{0.01} = 2.47$ (c) $F_{0.025} = 2.14$

11.43 (a) $F_{0.01} = 2.88$ (b) $F_{0.05} = 2.10$ (c) $F_{0.10} = 1.78$

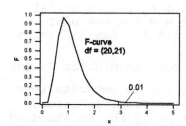 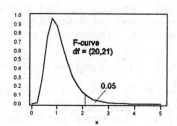 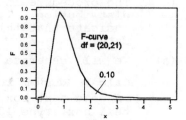

11.45 (a)   The F value for a 0.01 left   (b)   The F Value for a left
area for df = (6,8) equals   area of 0.95 equals F
1/F for df = (8,6) for a   for a right area of
0.01 right area. F = 1/8.10   0.05 = 3.58
= 0.12

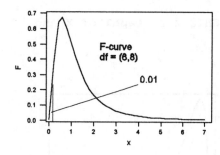   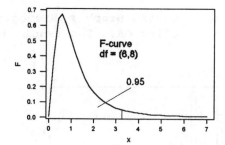

11.47 (a) $F_{0.025} = 9.07; F_{0.975} = 1/F_{0.025}$ (4,7)   (b) $F_{0.025} = 2.68; F_{0.975} =$

$=1/F_{0.025}$ (4,7) $1/F_{0.025}$ (20,12) =

$=1/5.52 = 0.18$ $1/3.07 = 0.33$

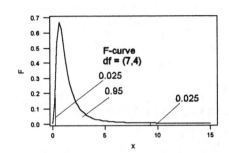
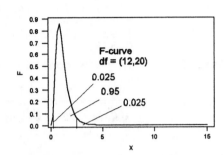

11.49 The F procedures are extremely nonrobust to even moderate violations of the normality assumption for both populations.

11.51 1=Control, 2=Experimental

Step 1: H_0: $\sigma_1 = \sigma_2$ H_a: $\sigma_1 > \sigma_2$

Step 2: $\alpha = 0.05$

Step 3: The critical value is $F_{0.05}$ with df = (40,19) or 2.03

Step 4: $F = 7.813^2/5.286^2 = 2.18$

Step 5: Since 2.18 > 2.03, we reject the null hypothesis.

Step 6:   There is sufficient evidence at the 0.05 significance level to claim that the variation in the control group is greater than that in the experimental group.

11.53 1=Relaxation tapes, 2=Neutral tapes

Step 1: H_0: $\sigma_1 = \sigma_2$ H_a: $\sigma_1 \neq \sigma_2$

Step 2: $\alpha = 0.10$

Step 3: The critical values are $F_{0.05}$ with df = (30,24) or 1.94 and $F_{0.95}$ = $1/F_{0.05}$ with df =(24,30) = 1/1.89 = 0.53

Step 4: $F = 10.154^2/9.197^2 = 1.22$

Step 5: Since 0.53 < 1.22 < 1.94, we do not reject the null hypothesis.

Step 6:   There is not sufficient evidence at the 0.10 significance level to claim that the variation in anxiety test scores for patients seeing videotapes showing progressive relaxation exercises is different from that in patients seeing neutral videotapes.

11.55 Step 1:   df = (30,24). $F_{0.05}$ is 1.94; $F_{0.95} = 1/F_{0.05}$ for df = (24,30) = 1/1.89 = 0.529.

Step 2:   The confidence interval for σ_1/σ_2 is

$$\frac{1}{\sqrt{F_{0.05}}} \cdot \frac{s_1}{s_2} \, to \, \frac{1}{\sqrt{F_{0.95}}} \cdot \frac{s_1}{s_2} = \frac{1}{\sqrt{1.94}}\frac{10.154}{9.197} \, to \, \frac{1}{\sqrt{0.529}}\frac{10.154}{9.197} = (0.79, 1.52)$$

11.57 I would use linear interpolation using the two sets of degrees of freedom closest to (25,20) - namely (24,20) and (30,20).

11.59 Step 1:   df = (40,19). $F_{0.05}$ = 2.03. $F_{0.95}$ = $1/F_{0.05}$ for df = (19,40). $F_{0.05}$ for df = (19,40) will require double linear interpolation since 19 is not in the list for dfn and 40 is not in the list for dfd. First we interpolate between df = (15,30) and (20,30) to estimate $F_{0.05}$ with (19,30) as 2.01 + (4/5)(1.93-2.01) = 1.95; then we interpolate between df = (15,60) and (20,60) to estimate $F_{0.05}$ with (19,60) as 1.84 + (4/5)(1.75-1.84) = 1.77. Finally, we interpolate between df = (19,30) and (19,60) to estimate $F_{0.05}$ with df = (19,40) as 1.95 + (10/30)(1.77-1.95) = 1.89. Thus $F_{0.95}$ = 1/1.89 = 0.529.

Step 2:   The 90% confidence interval for σ_1/σ_2 is

$$\frac{1}{\sqrt{F_{0.05}}} \cdot \frac{s_1}{s_2} \, to \, \frac{1}{\sqrt{F_{0.95}}} \cdot \frac{s_1}{s_2} = \frac{1}{\sqrt{2.03}} \frac{7.813}{5.286} \, to \, \frac{1}{\sqrt{0.529}} \frac{7.813}{5.286} = (1.037, 2.032)$$

11.61 (a) With the data in columns named CONTROL and EXPER, we choose **Calc ▶ Calculator...**, type <u>NCONTROL</u> in the **Store result in variable** text box, select the function NSCOR from the **Function** list, select CONTROL to replace **number** in the **Expression** text box, and click **OK**. Then choose

Graph ▶ Plot..., select NCONTROL in the **Y** column for **Graph1** and CONTROL

in the **X** column, and click **OK**. Then choose **Graph ▶ Boxplot...**, select CONTROL in the **Y** column for **Graph1**, and click **OK**. Repeat the entire process with EXPER and NEXPER. The four graphs are

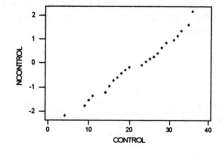 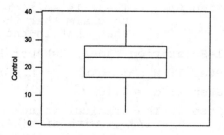

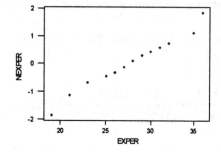 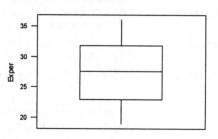

(b)   To run the hypothesis test and obtain the confidence interval

using *DataDisk* in floppy drive A, we type in the Sessions Window at the MTB> prompt,

%A:\Macro\2stdev.mac 'CONTROL' 'EXPER'

and press the ⌈ ENTER ⌉ key. Proceed as follows

1 In response to do you want to perform a hypothesis test (Y/N)?, type Y and press the ⌈ ENTER ⌉ key.

2 In response to Enter 0, 1, or -1, respectively, for a two-tailed, right-tailed, or left-tailed test., type 1 and press the ⌈ ENTER ⌉ key. The output is

F-Test of sigma1 = sigma2 (vs >)

Row	Variable	n	StDev	F	P
1	CONTROL	41	7.813	2.19	0.035
2	EXPER	20	5.286		

Since $0.035 < 0.05$, we reject the null hypothesis and claim that $\sigma_1 > \sigma_2$. At the 5% significance level, the data do provide sufficient evidence to conclude that the population standard deviation of final exam scores is greater for the control group than for the experimental group. Continuing with the confidence-interval aspect,

1 In response to Do you want a confidence interval (Y/N)?, type Y and press the ⌈ ENTER ⌉ key.

2 Since we want a 90% confidence interval, type 90 in response to Enter the confidence level, as a percentage, and then press the ⌈ ENTER ⌉ key. The resulting output is

Row	Variable	n	StDev	Level	CI for ratio
1	CONTROL	41	7.813	90.0%	(1.038, 2.012)
2	EXPER	20	5.286		

(c) The graphs in part (a) indicate that the normality assumption is reasonable for both variables. Thus performing the F-procedures is valid.

11.63 (a) With the data in columns named RELAX and NEUTRAL, we choose **Calc** ▶ **Calculator...**, type NRELAX in the **Store result in variable** text box, select the function NSCOR from the **Function** list, select RELAX to replace **number** in the **Expression** text box, and click **OK**. Then choose

Graph ▶ **Plot...**, select NRELAX in the **Y** column for **Graph1** and RELAX in

the **X** column, and click **OK**. Then choose **Graph** ▶ **Boxplot...**, select RELAX in the **Y** column for **Graph1**, and click **OK**. Repeat the entire process with NEUTRAL and NNEUTRAL. The four graphs are

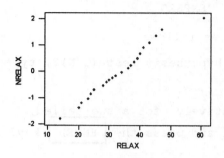

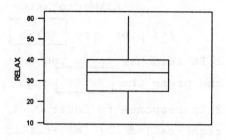

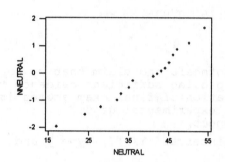

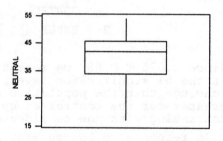

(b)   To run the hypothesis test and obtain the confidence interval using *DataDisk* in floppy drive A, we type in the Sessions Window at the MTB> prompt,

%A:\Macro\2stdev.mac 'RELAX' 'NEUTRAL'

and press the ☐ ENTER ☐ key.   Proceed as follows

1 In response to do you want to perform a hypothesis test (Y/N)?, type Y and press the ☐ ENTER ☐ key.

2 In response to Enter 0, 1, or -1, respectively, for a two-tailed, right-tailed, or left-tailed test., type 0 and press the ☐ ENTER ☐ key. The output is

F-Test of sigma1 = sigma2 (vs not =)

Row	Variable	n	StDev	F	P
1	RELAX	31	10.154	1.22	0.624
2	NEUTRAL	25	9.197		

Since $0.624 > 0.10$, we do not reject the null hypothesis that $\sigma_1 = \sigma_2$.

At the 10% significance level, the data do not provide sufficient evidence to conclude that the population standard deviation of anxiety-test scores of patients seeing videotapes showing progressive relaxation exercises is different from that for those seeing neutral videotapes. Continuing with the confidence-interval aspect,

1 In response to Do you want a confidence interval (Y/N)?, type Y and press the ⌞ENTER⌟ key.

2 Since we want a 90% confidence interval, type 90 in response to Enter the confidence level, as a percentage, and then press the ⌞ENTER⌟ key. The resulting output is

Row	Variable	n	StDev	Level	CI for ratio
1	RELAX	31	10.154	90.0%	(0.793, 1.517)
2	NEUTRAL	25	9.197		

(c) The graphs in part (a) indicate that the normality assumption is reasonable for both variables. Thus performing the F-procedures is valid.

11.65 Using Minitab, we choose **Calc ▶ Probability distributions... ▶ F**, click on the **Cumulative probability** button, type 9 in the **Numerator degrees of freedom** text box and 9 in the **Denominator degrees of freedom** text box, click on the **Input constant** button and type .41 in the text box, and click **OK**. The result in the Sessions window is

F distribution with 9 DF in numerator and 9 DF in denominator

x	P(X <= x)
0.4100	0.1001

Since the test is two-tailed, the P-value is twice the probability shown in the output, or 0.2002.

11.67 (a) With the data in columns named PUBLIC and PRIVATE, using *DataDisk* in floppy drive A, we type in the Sessions Window at the MTB> prompt,

%A:\Macro\2stdev.mac 'PUBLIC' 'PRIVATE'

and press the ⌞ENTER⌟ key. Proceed as follows

1 In response to do you want to perform a hypothesis test (Y/N)?, type Y and press the ⌞ENTER⌟ key.

2 In response to Enter 0, 1, or -1, respectively, for a two-tailed, right-tailed, or left-tailed test., type 0 and press the ⌞ENTER⌟ key. The output is

F-Test of sigma1 = sigma2 (vs not =)

Row	Variable	n	StDev	F	P
1	DYNAMIC	14	84.750	4.93	0.089
2	STATIC	6	38.166		

Since $0.089 > 0.01$, we do not reject the null hypothesis. There is insufficient evidence to conclude that the variation in operative time differs for the dynamic system and the static system.

(b)   We choose **Calc ▶ Calculator...**, type <u>NDYNAMIC</u> in the **Store result in variable** text box, select the function NSCOR from the **Function** list, select DYNAMIC to replace **number** in the **Expression** text box,

and click **OK**. Then choose **Graph ▶ Plot...**, select NDYNAMIC in the **Y** column for **Graph1** and DYNAMIC in the **X** column, and click **OK**. Repeat this process for STATIC and NSTATIC. The resulting graphs are

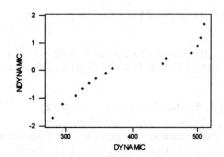
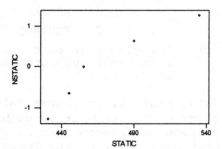

(c)   The normal probability plot for DYNAMIC exhibits a distinct non-linear pattern. It is doubtful that DYNAMIC is normally distributed and this makes the test performed in part (a) very questionable.

REVIEW TEST FOR CHAPTER 11

1.   Chi-squared distribution (χ^2)

2.   (a)   A χ^2-curve is <u>right</u> skewed.

    (b)   As the number of degrees of freedom becomes larger, a χ^2-curve looks increasingly like a <u>normal</u> curve.

3.   The variable must be normally distributed. That assumption is very important because the χ^2 procedures are not robust to moderate violations of the normality assumption.

4.   (a)   6.408   (b)   33.409   (c)   27.587   (d)   8.672

    (e)   7.564 and 30.191

5.   The F distribution is used when making inferences comparing two population standard deviations.

6.   (a)   An F-curve is <u>right</u>-skewed

    (b)   <u>reciprocal</u>, (<u>5,14</u>)

    (c)   0

7.   Both variables must be normally distributed. This is very important since the F procedures are not robust to violations of the normality assumption.

8.   (a)   7.01

    (b) $F_{0.99} = 1/F_{0.01}$ where $F_{0.01}$ has df $= (8,4)$.

        Thus $F_{0.99} = 1/14.80 = 0.068$

    (c)   3.84

(d)   The F value with 0.05 to its left = $F_{0.95}$ = $1/F_{0.05}$ where $F_{0.05}$ has df = (8,4).

Thus $F_{0.95}$ = 1/6.04 = 0.166

(e)   The F value with 0.025 to its left = $F_{0.975}$ = $1/F_{0.025}$ where $F_{0.025}$ has df = (8,4).

Thus $F_{0.975}$ = 1/8.98 = <u>0.111</u>; $F_{0.025}$ = <u>5.05</u>

9.   (a)   Step 1: H_0: σ = 16 H_a: $\sigma \neq$ 16

Step 2: α = 0.10

Step 3:   The critical values with n-1 = 24 degrees of freedom are 13.848 and 36.415.

Step 4: $\chi^2 = \dfrac{n-1}{\sigma_0^2} s^2 = \dfrac{24}{16^2} 15.006^2 = 21.111$

Step 5: Since 13.848 < 21.111 < 36.415, we do not reject H_0.

Step 6: There is insufficient evidence at the 0.10 level to claim that σ for IQs measured on the Stanford revision of the Binet-Simon Intelligence Scale is different from 16.

(b)   Normality is crucial for the hypothesis test in (a) since the procedure is not robust to moderate deviations from the normality assumption.

10.   The 90% confidence interval for σ of IQs measured on the Stanford Revision of the Binet-Simon Intelligence Scale is

$(\sqrt{\dfrac{n-1}{\chi_{.05}^2}} s, \sqrt{\dfrac{n-1}{\chi_{.95}^2}} s) = (\sqrt{\dfrac{24}{36.415}} 15.006, \sqrt{\dfrac{24}{13.848}} 15.006) = (12.182, 19.755)$

11.   (a)   F distribution with df = (14,19)

(b)   To find the value of $F_{0.99}$ with df = (14,19), we would need the reciprocal of $F_{0.01}$ with df = (19,14). Table X does not contain a value for that pair of degrees of freedom, so we cannot determine the critical value for the test using Table X.

(c)   1=Runners, 2=Others

Step 1: H_0: $\sigma_1 = \sigma_2$ H_a: $\sigma_1 < \sigma_2$

Step 2: α = 0.01

Step 3: For df = (14,19), the critical value is $F_{0.99}$ = 0.28

Step 4: F = $1.798^2/6.606^2$ = 0.074

Step 5: Since 0.074 < 0.28, we reject the null hypothesis.

Step 6:   There is sufficient evidence at the 0.01 significance level to claim that the variation in skinfold thickness among runners is less than that among others.

(d)   We are assuming that skinfold thickness is a normally distributed variable. This assumption can be checked by looking at a normal probability plot for each set of data. If both plots are linear, the normality assumption is reasonable.

(e)   We also assume that the samples are independent.

12.   (a)   Step 1:   df = (14,19). $F_{0.01}$ is 3.19; $F_{0.99}$ = 0.28 from Review Exercise 11.

Step 2:   The confidence interval for σ_1/σ_2 is

$\dfrac{1}{\sqrt{F_{0.01}}} \cdot \dfrac{s_1}{s_2} \; to \; \dfrac{1}{\sqrt{F_{0.99}}} \cdot \dfrac{s_1}{s_2} = \dfrac{1}{\sqrt{3.19}} \dfrac{1.798}{6.606} \; to \; \dfrac{1}{\sqrt{0.28}} \dfrac{1.798}{6.060} = (0.152, 0.514)$

(b)   We can be 98% confident that the ratio of the population standard deviation of skinfold thickness for runners to the standard deviation for others lies between 0.15 and 0.51.

13.   With the data in a column named IQ and with the *DataDisk* in drive A, at the MTB> prompt in the Sessions Window, type

%A:\1stdev.mac 'IQ'

and press the ⎿ ENTER ⏌ key.   Then proceed as follows.

1 In response to Do you want to perform a hypothesis test (Y/N)?, type Y and press the ⎿ ENTER ⏌ key.

2 In response to Enter the null hypothesis population standard deviation, type 16 and press the ⎿ ENTER ⏌ key.

3 In response to Enter 0, 1, or -1, respectively, for a two-tailed, right-tailed, or left-tailed test., type 0 and press the ⎿ ENTER ⏌ key.   The output is

Test of $\sigma = $ 16 vs σ not = 16

Row	Variable	n	StDev	Chi-Sq	P
1	IQ	25	15.006	21.110	0.736

Since 0.736 > 0.10, we do not reject the null hypothesis.

The macro continues with the confidence-interval procedure.
1 In response to Do you want a confidence interval (Y/N)?, type Y and press the ⎿ ENTER ⏌ key.

2 Since we want a 90% confidence interval, type 90 in response to Enter the confidence level, as a percentage and press the ⎿ ENTER ⏌ key. The result is

Row	Variable	n	StDev	Level	CI for σ
1	IQ	25	15.006	90.0%	(12.182, 19.755)

14.   To run the hypothesis test and obtain the confidence interval using *DataDisk* in floppy drive A, we type in the Sessions Window at the MTB> prompt,

%A:\2STDEV.MAC 'RUNNERS' 'OTHERS'

and press the ⎿ ENTER ⏌ key.   Proceed as follows

1 In response to do you want to perform a hypothesis test (Y/N)?, type Y and press the ⎿ ENTER ⏌ key.

2 In response to Enter 0, 1, or -1, respectively, for a two-tailed, right-tailed, or left-tailed test., type -1 and press the ENTER key. The output is

F-Test of sigma1 = sigma2 (vs <)

Row	Variable	n	StDev	F	P
1	RUNNERS	15	1.798	0.07	0.000
2	OTHERS	20	6.606		

Since the P-value of 0.000 is less than 0.01, we reject the null hypothesis.

Continuing with the confidence-interval aspect,

1 In response to Do you want a confidence interval (Y/N)?, type Y and press the ENTER key.

2 Since we want a 90% confidence interval, type 90 in response to Enter the confidence level, as a percentage, and then press the ENTER key. The resulting output is

Row	Variable	n	StDev	Level	CI for ratio
1	RUNNERS	15	1.798	98.0%	(0.152, 0.511)
2	OTHERS	20	6.606		

Exercises 12.1

12.1 Answers will vary.

12.3 A population proportion p is a parameter since it is a descriptive measure for a population. A sample proportion $\hat{p}$ is a statistic since it is a descriptive measure for a sample.

12.5 (a) $p = 2/5 = 0.4$

(b)

Sample	Number of females x	Sample proportion $\hat{p}$
J,G	1	0.5
J,P	0	0.0
J,C	0	0.0
J,F	1	0.5
G,P	1	0.5
G,C	1	0.5
G,F	2	1.0
P,C	0	0.0
P,F	1	0.5
C,F	1	0.5

(c)   The population proportion is marked by the vertical line below.

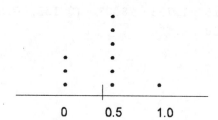

(d) $\mu_{\hat{p}} = (\Sigma\hat{p})/10 = 4.0/10 = 0.4$

(e)   The answers to (a) and (d) are the same. $\hat{p}$ is a sample mean. The mean of the sampling distribution of $\hat{p}$ is the same as the mean of the population which is p.

12.7 (b)

Sample	Number of females x	Sample proportion $\hat{p}$
J,P,C	0	0
J,P,G	1	1/3
J,P,F	1	1/3
J,C,G	1	1/3
J,C,F	1	1/3
J,G,F	2	2/3
P,C,G	1	1/3
P,C,F	1	1/3
P,G,F	2	2/3
C,G,F	2	2/3

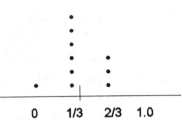

(c)   Dot plot at the right.

(d) $\mu_{\hat{p}} = (\Sigma\hat{p})/10 = (12/3)/10 = 0.4$

(e)   The answers to (a) and (d) are the same. $\hat{p}$ is a sample mean. The mean of the sampling distribution of $\hat{p}$ is the same as the mean of the population which is p.

12.9 (b)

Sample	Number of females x	Sample proportion $\hat{p}$
J,P,C,G,F	2	0.4

(c)   The population proportion is marked by the vertical line below.

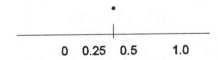

(d) $\mu_{\hat{p}} = (\Sigma\hat{p})/5 = 0.4/1 = 0.4$

(d)   The answers to (a) and (d) are the same. $\hat{p}$ is a sample mean. The mean of the sampling distribution of $\hat{p}$ is the same as the mean of the population which is p.

12.11 (a)   The population consists of all 1997 U.S. governors.

(b)   The specified attribute is being Republican.

(c)   The 0.64 is a population proportion since all of the governors and their political affiliations are known. There is no need to sample this group.

12.13 (a) $\hat{p} = 452/779 = 0.58$; $z_{0.025} = 1.96$; the margin of error is

$$E = z_{0.025} \cdot \sqrt{\hat{p}(1-\hat{p})/n} = 1.96\sqrt{0.58(0.42)/779} = 0.0347$$

(b)   The margin of error will be smaller for a 90% confidence interval. Specifically, 1.96 will be replaced by 1.645 in the formula for E and everything else stays the same. More generally speaking, in order to have a higher level of confidence in an interval, one needs to have a wider interval.

12.15 (a)   0.4   (b)   0.5   (c)   0.7   (d)   0.2   (e)   0.5   (f)   0.5

(g)   For (a), $0.4 < \hat{p} < 0.6$

For (b), none

For (c), $0.3 < \hat{p} < 0.7$

For (d), $0.2 < \hat{p} < 0.8$

For (e), none

For (d), none

12.17 n = 1,516, x = 985, and n - x = 531. Both x and n - x are at least 5.

$\hat{p} = x/n = 985/1,516 = 0.65$; $z_{\alpha/2} = z_{0.025} = 1.96$

(a)

$$0.65-1.96\sqrt{0.65(1-0.65)/1,516}, \ 0.65+1.96\sqrt{0.65(1-0.65)/1,516}$$

$$0.626 \text{ to } 0.674$$

(b)   We can be 95% confident that the percentage of Americans who drink beer, wine, or hard liquor, at least occasionally, is somewhere between 62.6% and 67.4%.

12.19 n = 1,528, x = 1,238, and n - x = 290. Both x and n - x are at least 5.

$\hat{p} = x/n = 1,238/1,528 = 0.81$; $z_{\alpha/2} = z_{0.005} = 2.575$

(a)

$$0.81-2.575\sqrt{0.81(1-0.81)/1,528}, \ 0.81+2.575\sqrt{0.81(1-0.81)/1,528}$$

$$78.4\% \text{ to } 83.6\%$$

(b)   We can be 99% confident that the percentage of adult Americans who are in favor of "right to die" laws is somewhere between 78.4% and 83.6%.

12.21 Here, n = 500, $\hat{p} = 0.8\% = 0.008$, x = $\hat{p} \cdot$ n = 4, and n - x = 496. To use Procedure 12.1, both n and n - x must be at least 5. Since x < 5, I should not have used Procedure 12.1.

12.23 $\hat{p} = 0.34$, E = 0.024

$\hat{p}$ - E   to $\hat{p}$ + E

0.34 -0.024 to 0.34 + 0.024   In percentage terms, this is

31.6%   to   36.4%

12.25 (a)   The confidence interval from Exercise 12.17 was (0.626, 0.674). To find the error, take the width of the confidence interval and divide by 2. E = (0.674 - 0.626)/2 = 0.024.

(b) $E = 0.02$; $z_{\alpha/2} = z_{0.025} = 1.96$;

$$n = (0.5)^2 \frac{z_{\frac{\alpha}{2}}^2}{E^2} = (0.5)^2 \frac{1.96^2}{0.02^2} = 2401$$

(c) $n = 2401$; $\hat{p} = 0.63$; $z_{\alpha/2} = z_{0.025} = 1.96$

$0.63 \pm 1.96\sqrt{0.63(1-0.63)/2401}$

0.611 to 0.649

(d)   The margin of error for the estimate is 0.019, which is less than what is required in part (b).

(e) $\hat{p}_g = 0.60$; $z_{\alpha/2} = z_{0.025} = 1.96$

part (b) $E = 0.02$, $z_{0.025} = 1.96$; sample size is:

$$n = \hat{p}(1 - \hat{p}) \frac{z_{\frac{\alpha}{2}}^2}{E^2} = (0.60)(1 - 0.60) \frac{1.96^2}{0.02^2} = 2304.96$$

Thus the required sample size is n = 2305.
part (c)

$0.63 \pm 1.96\sqrt{0.63(1-0.63)/2305}$

0.610 to 0.650

part (d)
The margin of error is 0.020, which is the same as what is specified in part (b).

(f)   By employing the guess for $\hat{p}$ in part (d) we can reduce the required sample size (from 2401 to 2305), saving considerable time and money. Moreover, the margin of error only rises from 0.019 to 0.020. The risk of using the guess 0.60 for $\hat{p}$ is that if the actual value of $\hat{p}$ turns out to be less than 0.60 (but not less than 0.40) then the achieved margin of error will exceed the specified 0.02.

12.27 (a)   The confidence interval from Exercise 12.19 was (78.4%, 83.6%).
To find the error, take the width of the confidence interval and divide by 2.   E = (83.6% - 78.4%)/2 = 2.6%.

(b) $E = 0.01$; $z_{\alpha/2} = z_{0.005} = 2.575$;

$$n = (0.5)^2 \frac{z_{\frac{\alpha}{2}}^2}{E^2} = (0.5)^2 \frac{2.575^2}{0.01^2} = 16,576.56 \approx 16,577$$

(c) $n = 16,577$; $\hat{p} = 0.825$; $z_{\alpha/2} = z_{0.005} = 2.575$

$$0.825 - 2.575\sqrt{0.825(1-0.825)/16577}, \; 0.825 + 2.575\sqrt{0.825(1-0.825)/16577}$$

$$0.817 \text{ to } 0.833$$

(d)   The margin of error for the estimate is 0.008, which is less than what is required in part (b).

(e) $\hat{p}_g$ is between 75% and 90%; $z_{\alpha/2} = z_{0.005} = 2.575$

part (b) $E = 0.01$, $z_{0.005} = 2.575$; sample size is:

$$n = \hat{p}(1 - \hat{p})\,\frac{z_{\frac{\alpha}{2}}^2}{E^2} = (0.75)(1 - 0.75)\,\frac{2.575^2}{0.01^2} = 12,432.4$$

Thus the required sample size is $n = 12,433$.

part (c)

$$0.825 - 2.575\sqrt{0.825(1-0.825)/12433}, \; 0.825 + 2.575\sqrt{0.825(1-0.825)/12433}$$

$$0.816 \text{ to } 0.834$$

part (d)   The margin of error is 0.009, which is less than what is specified in part (b).

(f)   By employing the guess for $\hat{p}$ in part (d) we can reduce the required sample size (from 16,577 to 12,433), saving considerable time and money. Moreover, the margin of error only rises from 0.008 to 0.009. The risk of using the guess 0.75 for $\hat{p}$ is that if the actual value of $\hat{p}$ turns out to be less than 0.75 (but not less than 0.25) then the achieved margin of error will exceed the specified 0.01.

12.29 We will assume that both polls used a 95% confidence level. Then at the 95% confidence level, both polls are giving a range of believable values for the true population (Arizonians) proportion that felt Symington should resign. The Research Resources poll gave its range to be 0.531 to 0.639. The Behavior Research Center poll gave its range to be 0.496 to 0.584. We note that both ranges of values have common believable values from 0.531 to 0.584. Thus, it is possible that both of these polls were correct in their conclusions.

12.31 The sample size is directly proportional to the quantity $p(1 - p)$, which takes on its maximum value of 0.25 when $p = .5$. Sample sizes increase as p approaches 0.5 from either above or below. Thus to achieve a sample size adequate for any p in a given range, we should choose the largest sample that could result for any p in the range. This will always happen when we choose the value of p in the range that is closest to 0.5.

12.33 (a) Np

(b) $\mu_y = \sum y P(Y=y) = 0 \cdot (1-p) + 1 \cdot p = p$

(c)
$$\sigma_y^2 = \sum y^2 P(Y=y) - \mu^2 = [0^2 \cdot (1-p) + 1^2 \cdot p] - p^2 = p - p^2 = p \cdot (1-p)$$

Therefore

$$\sigma_y = \sqrt{p(1-p)}$$

(d) In a sample of size n, there are x ones and (n-x) zeros. The ones and zeros are the y values. Therefore the mean of the y values is
$\bar{y} = [x \cdot 1 + (n-x) \cdot 0]/n = x/n = \hat{p}$

(e) From Key Fact 7.4, when n is large, the sampling distribution of the mean is approximately normal with mean μ and standard deviation $\sigma / \sqrt{n}$, i.e., $\bar{y}$ has a sampling distribution with mean μ and standard deviation $\sigma / \sqrt{n}$. But $\bar{y}$ is $\hat{p}$, μ is p, and $\sigma / \sqrt{n}$ is

$$\frac{\sqrt{p(1-p)}}{\sqrt{n}} = \sqrt{\frac{p(1-p)}{n}}$$. Thus when n is large, the sampling distribution of $\hat{p}$ is approximately normal with mean p and standard deviation $\sqrt{\frac{p(1-p)}{n}}$.

12.35 (a) Choose **Calc ▶ Basic statistics ▶ 1 Proportion...**, select the **Summarized data** option button, click in the **Number of trials** text box and type 1516, click in the **Number of successes** text box and type 985, click the **Options...** button, click in the **Confidence level** text box and type 95, click **OK**, and click **OK**. The resulting output is

Test of p = 0.5 vs p not = 0.5

```
                                    Exact
Sample    X     N  Sample p     95.0 % CI        P-Value
1       985  1516  0.649736  (0.625120, 0.673770)  0.000
```

(b) The confidence interval in Exercise 12.17 was (0.626, 0.674). The slight discrepancy between that result and this one is due to this one being an exact result based on the binomial distribution while the previous result was based on a normal approximation to the binomial distribution.

12.37 (a) n = 1004 (b) 670 (c) 66.7331% (d) (63.7230%,69.6447%)

12.39 (a) Choose **Calc ▶ Basic statistics ▶ 1 Proportion...**, select the **Summarized data** option button, click in the **Number of trials** text box and type 500, click in the **Number of successes** text box and type 4, click the **Options...** button, click in the **Confidence level** text box and type 90, click **OK**, and click **OK**. The resulting output is

Test of p = 0.5 vs p not = 0.5

```
                                    Exact
Sample    X     N  Sample p     90.0 % CI        P-Value
1         4   500  0.008000  (0.002737, 0.018213)  0.000
```

(b)   The confidence interval in Exercise 12.21 was (0.001, 0.015). The large discrepancy between that result and this one is due to this one being an exact result based on the binomial distribution while the previous result was based on an improper use (because x = 4, which is not larger than the recommended guideline of 5) of the normal approximation to the binomial distribution.

Exercises 12.2

12.41 (a)   The sample proportion is $\hat{p}$ = x/n = 394/758 = 0.520.

(b) α = 0.05, p_0 = 0.50

np_0 = 758(0.50) = 379; n(1 - p_0) = 758(1-0.50) = 379

Since both are at least 5, we can employ Procedure 12.2.

Step 1: H_0: p = 0.50, H_a: p > 0.50

Step 2: α = 0.05

Step 3:   Since α = 0.05, the critical value is z_α = 1.645

Step 4:   z = (0.520 - 0.50)/ $\sqrt{0.50(1 - 0.50)/758}$ = 1.10

Step 5:   Since 1.10 < 1.645, do not reject H_0. Note, for the p-value approach, P(z > 1.10) = 0.1357; so p-value > α. Thus we do not reject H_0.

Step 6:   The test results are not statistically significant at the 5% level; that is, at the 5% significance level, the data do not provide sufficient evidence to conclude that a majority of Arizona families who celebrate Christmas wait until Christmas Day to open their presents.

12.43 The sample proportion is $\hat{p}$ = x/n = 146/1283 = 0.1138.

α = 0.10, p_0 = 0.12

np_0 = 1283(0.12) = 154.0; n(1 - p_0) = 1283(1-0.12) = 1129.0

Since both are at least 5, we can employ Procedure 12.2.

Step 1: H_0: p = 0.12, H_a: p ≠ 0.12

Step 2: α = 0.10

Step 3:   Since α = 0.10, the critical values are z_α = ±1.28

Step 4:   z = (0.1138 - 0.1200)/ $\sqrt{0.12(1 - 0.12)/1283}$ = -0.68

Step 5:   Since -0.68 > -1.28, do not reject H_0. Note, for the p-value approach, 2P(Z < -0.68) = 2(0.2483) = 0.4966; so p-value > α. Thus we do not reject H_0.

Step 6:   The test results are not statistically significant at the 10% level; that is, at the 10% significance level, the data do not provide sufficient evidence to conclude that the percentage of 18-25 year-olds who currently use marijuana or hashish has changed from the 1995 percentage of 12.0%.

12.45 The sample proportion is $\hat{p}$ = x/n = 2081/2544 = 0.818.

α = 0.05, p_0 = 0.831

np_0 = 2544(0.831) = 2114; n(1 - p_0) = 2544(1-0.831) = 430.

Since both are at least 5, we can employ Procedure 12.2.

Step 1: H_0: p = 0.831, H_a: p < 0.831

Step 2: α = 0.05

Step 3: Since $\alpha = 0.05$, the critical value is $z_\alpha = -1.645$

Step 4: $z = (0.818 - 0.831)/\sqrt{0.831(1 - 0.831)/2544} = -1.75$

Step 5: Since $-1.75 < -1.645$, reject H_0. Note, for the p-value approach, $P(Z < -1.75) = 0.0401$; so p-value $< \alpha$. Thus we reject H_0.

Step 6: The test results are statistically significant at the 5% level; that is, at the 5% significance level, the data do provide sufficient evidence to conclude that this year's percentage of home buyers purchasing single-family houses has decreased from the 1995 figure of 83.1%.

12.47 Choose **Calc ▶ Basic statistics ▶ 1 Proportion...**, select the **Summarized data** option button, click in the **Number of trials** text box and type <u>1283</u>, click in the **Number of successes** text box and type <u>146</u>, click the **Options...** button, click in the **Test proportion** text box and type <u>0.12</u>, click the arrow button at the right of the **Alternative** drop-down list box and select **not equal**, click **OK**, and click **OK**. The resulting output is

Test of p = 0.12 vs p not = 0.12

Sample	X	N	Sample p	90.0 % CI	Exact P-Value
1	146	1283	0.113796	(0.099490, 0.129430)	0.493

Since **0.493 > 0.10, do not reject the null hypothesis.**

12.49 (a) H_0: p = 0.5, H_a: p > 0.5

(b) 779

(c) 452

(d) 58.0231%

(e) P-value = 0.000

(f) Since P-value $< \alpha$, we reject the null hypothesis and claim that more than 50% (a majority) of the population think that it makes sense to abolish the U.S. tax code and start all over.

12.51 In Example 12.6, we used the z-test based on the normal approximation to the binomial distribution. The Minitab output in Printout 12.5 is the result of an exact computation based directly on the binomial distribution and is the correct value.

Exercises 12.3

12.53 (a) Attending church at least once a week

(b) Children in the U.S. and children in Germany

(c) The two population proportions under consideration are the proportion of children in the U.S. who attend church at least once a week and the proportion of children in Germany who attend church at least once a week.

12.55 (a) Using sunscreen before going out in the sun

(b) Teen-age girls and teen-age boys

(c) The two proportions are sample proportions. The reference is specifically to those teen-age girls and boys who were surveyed.

12.57 (a)   The parameters are p_1 and p_2. The rest are statistics.

(b)   The fixed numbers are p_1 and p_2. The rest are variables.

12.59 (a)   Population 1:   Women who took multivitamins containing folic acid

$$\hat{p}_1 = 35/2,701 = 0.013$$

Population 2:   Women who received only trace elements

$$\hat{p}_2 = 47/2,052 = 0.023$$

$$\hat{p}_p = (35 + 47)/(2,701 + 2,052) = 0.017$$

Step 1: $H_0: p_1 = p_2,$ $H_a: p_1 < p_2$

Step 2: $\alpha = 0.01$

Step 3:   Since $\alpha = 0.01$, the critical value is $z_\alpha = -2.33$

Step 4: $z = \dfrac{0.013 - 0.023}{\sqrt{0.017(1-0.017)}\ \sqrt{1/2701 + (1/2052)}} = -2.64$

Step 5:   Since $-2.64 < -2.33$, reject H_0. Note: For the p-value approach, $p(z < -2.64) = 0.0041$; so p-value $< \alpha$. Thus, we reject H_0.

Step 6:   The test results are significant at the 1% level; that is, at the 1% significance level, the data do provide sufficient evidence to conclude that the women who take folic acid are at lesser risk of having children with major birth defects.

(b)   This is an designed experiment. The researchers decided which women would take daily multivitamins.

(c)   Yes. By using the basic principles of design (control, randomization, and replication) the doctors could conclude that a difference in the rates of major birth defects between the two groups not reasonably attributable to chance is likely caused by the folic acid.

12.61 Population 1: 1980 American men 20-34 years old, $\hat{p}_1 = 130/750 = 0.173$

Population 2: 1990 American men 20-34 years old, $\hat{p}_2 = 160/700 = 0.229$

$$\hat{p}_p = (130 + 160)/(750 + 700) = 0.200$$

Step 1: $H_0: p_1 = p_2,$ $H_a: p_1 < p_2$

Step 2: $\alpha = 0.05$

Step 3:   Since $\alpha = 0.05$, the critical value is $z_\alpha = -1.645$

Step 4: $z = \dfrac{0.173 - 0.229}{\sqrt{0.2(1-0.2)}\ \sqrt{1/750 + (1/700)}} = -2.66$

Step 5:   Since $-2.66 < -1.645$, reject H_0. Note: For the P-value approach, $P(z < -2.66) = 0.0039$; so P-value $< \alpha$. Thus, we reject H_0.

Step 6:   The test results are significant at the 5% level; that is, at the 5% significance level, the data do provide sufficient evidence to conclude that a higher percentage of men 20-34 years old were overweight in 1990 than in 1980.

12.63 Population 1:   Employed, $\hat{p}_1 = 262/400 = 0.655$

Population 2:   Unemployed, $\hat{p}_2 = 224/450 = 0.498$

$$\hat{p}_p = (262 + 224)/(400 + 450) = 0.572$$

Step 1: H_0: $p_1 = p_2$, H_a: $p_1 \neq p_2$

Step 2: $\alpha = 0.05$

Step 3:   Since $\alpha = 0.05$, the critical values are $z_{\alpha/2} = \pm 1.96$

Step 4: $z = \dfrac{0.655 - 0.498}{\sqrt{0.572(1-0.572)}\ \sqrt{(1/400)+(1/450)}} = 4.62$

Step 5:   Since 4.62 > 1.96, reject H_0. Note for the P-value approach, $P(z > 4.62) = 0.000$; so P-value < α. Thus we reject H_0.

Step 6:   The test is significant at the 5% level; that is the data do provide evidence to conclude that the percentage of employed workers who have registered to vote differs from the percentage of unemployed workers who have registered to vote.

12.65 (a) From Exercise 12.59,

$(0.013-0.023) \pm 2.33 \cdot \sqrt{0.013(1-0.013)/2{,}701 + 0.023(1-0.023)/2{,}052}$

-0.010 ± 0.009

-0.019 to -0.001

(b) We can be 98% confident that the difference $p_1 - p_2$ between the rates of major birth defects for babies born to women who have taken folic acid and those born to women who have not taken folic acid is somewhere between -0.019 and -0.001.

12.67 (a) From Exercise 12.61,

$(0.173-0.229) \pm 1.645 \cdot \sqrt{0.173(1-0.173)/750 + 0.229(1-0.229)/700}$

-0.056 ± 0.035

-0.091 to -0.021

(b) We can be 90% confident that the difference $p_1 - p_2$ between the proportions of men 20-34 years old who were overweight in 1980 and 1990 is somewhere between -0.091 and -0.021.

12.69 (a) From Exercise 12.63,

$(0.655-0.498) \pm 1.96 \cdot \sqrt{0.655(1-0.655)/400 + 0.498(1-0.498)/450}$

0.157 ± 0.066

0.091 to 0.223

(b) We can be 95% confident that the difference $p_1 - p_2$ between the proportions of employed and unemployed workers who have registered to vote is somewhere between 0.091 and 0.223.

12.71 The first formula in Key Fact 12.2: $\mu_{\hat{p}_1 - \hat{p}_2} = p_1 - p_2$. This formula says that the mean of $\hat{p}_1 - \hat{p}_2$ is $p_1 - p_2$.

12.73 Using Minitab, choose **Calc ▶ Basic statistics ▶ 2 Proportions...**, select the **Summarized data** option button, click in the **Trials** text box for **First sample** and type 750, click in the **Successes** text box for **First sample** and type 130, click in the **Trials** text box for **Second sample** and type <u>700</u>, click in the **Successes** text box for **Second sample** and type <u>160</u>, click the **Options...** button, click in the **Confidence level** text box and type <u>90</u>, click in the **Test difference** text box and type <u>0</u>, click the arrow button at the right of the **Alternative** drop-down list box and select **less than**, select the **Use pooled estimate of p for test** check box, click **OK**, and click **OK**. The resulting output is

```
Sample      X      N    Sample p
1          130    750   0.173333
2          160    700   0.228571
```

Estimate for p(1) - p(2): -0.0552381

90% CI for p(1) - p(2): (-0.0898562, -0.0206200)

Test for p(1) - p(2) = 0 (vs < 0): Z = -2.63 P-Value = 0.004

Note: The slight differences between the solutions here and in 12.61 and 12.67 are due to round-off error in the earlier exercises.

12.75 (a)   272 and 229

(b)   193 and 165

(c)   70.96% and 72.05%

(d)   -0.0903 to 0.0683

(e) $p_1 - p_2 = 0$ vs. $p_1 - p_2 \neq 0$

(f)   0.787

(g)   Do not reject H_0. There is not sufficient evidence to conclude that the percentage of people who were in favor of a military strike against Iraq changed during the two weeks that elapsed between the polls.

CHAPTER 12 REVIEW TEST

1.   (a)   Favorite recreation is golf

(b)   Chief financial officers

(c)   Proportion of all chief financial officers whose favorite recreation is golf

(d)   Proportion of CFOs in the sample whose favorite recreation is golf

2.   (a)   Washing their vehicle at least once a month

(b)   Car owners

(c)   Proportion of all car owners who wash their car at least once a month

(d)   Proportion of car owners in the sample who wash their car at least once a month

(e)   It is a sample proportion. USA Today does not have the resources to take a census of all car owners, and even if it did, it would not make sense to spend the amount of money required to get the perfect answer to this question.

3.   It is often impossible to take a census of an entire population. It is also expensive and time-consuming.

4.   (a)   "Number of successes" stands for the number of members of the sample that exhibit the specified attribute.

(b)   "Number of failures" stands for the number of members of the sample that do not exhibit the specified attribute.

5.   (a)   population proportion

      (b)   normal

      (c)   number of successes, number of failures, 5

6.   The margin of error for the estimate of a population proportion tells us what the maximum difference between the sample proportion and the population proportion is <u>likely</u> to be. It is not an absolute maximum, and how likely it is depends on the confidence level used.

7.   (a)   Getting the "holiday blues"

      (b)   Men and women

      (c)   The proportion of men in the population who get the "holiday blues" and the proportion of women in the population who get the "holiday blues"

      (d)   The proportion of men in the sample who get the "holiday blues" and the proportion of women in the sample who get the "holiday blues"

      (e)   They are sample proportions since the information came from a poll, not a census. Also, it could not be a population proportion because I was not asked.

8.   (a)   difference of the population proportions

      (b)   normal

9.   (a) $n = 0.5 \dfrac{(1.96)^2}{(0.01)^2} = 19208$

      (b) $n = (0.75(1-0.75) + 0.75(1-0.75))(\dfrac{1.96}{0.01})^2 = 14{,}406$

10.   n = 1006, $\hat{p}$ = 0.43, $z_{\alpha/2}$ = $z_{0.025}$ = 1.96

      (a)

$$0.43 - 1.96 \cdot \sqrt{0.43(1-0.43)/1006} \quad to \quad 0.43 + 1.96 \cdot \sqrt{0.43(1-0.43)/1006}$$

$$0.399 \quad to \quad 0.461$$

      (b)   We can be 95% confident that the proportion, p, of all Americans who thought news organizations were criticizing the Clinton administration unfairly is somewhere between 0.399 and 0.461.

11.   (a)   The error is found by taking the width of the confidence interval and dividing by 2. So E = (0.461 - 0.399)/2 = 0.031.

      (b)   E = 0.02; $z_{\alpha/2}$ = $z_{0.025}$ = 1.96; $\hat{p}_g$ = 0.50

$$n = (0.5)^2 \frac{z_{\frac{\alpha}{2}}^2}{E^2} = (0.5)^2 \frac{1.96^2}{0.02^2} = 2401$$

      (c)   n = 2401; $\hat{p}$ = 0.416; $z_{\alpha/2}$ = $z_{0.025}$ = 1.96

$$0.416 - 1.96 \cdot \sqrt{0.416(1-0.416)/2401} \quad to \quad 0.416 + 1.96 \cdot \sqrt{0.416(1-0.416)/2401}$$

$$0.396 \quad to \quad 0.436$$

(d)   The margin of error for the estimate is 0.02, the same as what is required in part (b).

(e) $\hat{p}_g = 0.45$; $z_{\alpha/2} = z_{0.025} = 1.96$

part (b)   E = 0.01, $z_{0.025} = 1.96$; sample size is:

$$n = \hat{p}(1 - \hat{p})\frac{z_{\frac{\alpha}{2}}^2}{E^2} = (0.45)(1 - 0.45)\frac{1.96^2}{0.02^2} = 2376.99$$

Thus the required sample size is n = 2377.

part (c)

$$0.416 - 1.96 \cdot \sqrt{0.416(1-0.416)/2377} \quad \text{to} \quad 0.416 + 1.96 \cdot \sqrt{0.416(1-0.416)/2377}$$

$$0.396 \quad \text{to} \quad 0.436$$

part (d)

The margin of error is 0.02, which is the same as what is specified in part (b).

(f)   By employing the guess for $\hat{p}$ in part (d) we can reduce the required sample size (from 2401 to 2377), saving a little time and money. Moreover, the margin of error stays the same. The risk of using the guess 0.45 for $\hat{p}$ is that if the actual value of $\hat{p}$ turns out to be larger than .45 (but not more than .55), then the achieved margin of error will exceed the specified 0.02.

12. $\hat{p} = 0.17$, E = 0.04, so the confidence interval is:

$$\hat{p} - E \text{ to } \hat{p} + E$$

0.17 - 0.04 to 0.17 + 0.04   In terms of percentages, this is
13% to 21%

13.   n = 2512, x = 578, $\alpha = 0.05$

$np_0 = 2512(0.25) = 628$; $n(1 - p_0) = 2512(1-0.25) = 1884$

Since both are at least 5, we can employ Procedure 12.2.

(a)   Step 1: H_0: p = 0.25, H_1: p < 0.25

Step 2: $\alpha = 0.05$

Step 3:   Since $\alpha = 0.05$, the critical value is $z_\alpha = -1.645$

Step 4: $z = (0.23 - 0.25)/\sqrt{0.25(1 - 0.25)/2512} = -2.31$

Step 5:   Since $-2.31 < -1.645$, reject H_0.

Step 6:   The test results are statistically significant at the 5% level; that is, at the 5% significance level, the data do provide sufficient evidence to conclude that less than one in four Americans believe that juries "almost always" convict the guilty and free the innocent.

(b)   P(z < -2.31) = 0.0104. Since P-value < α, reject H_0. Same conclusion as (a).

(c)   The strength of the evidence against the null hypothesis is strong.

14.   (a)   Observational study. The researchers had no control over any of the factors of the study.

(b)   Height may not be the only factor to be considered. Although there does appear to be an association, we don't know if it is a direct association.

15.   Population 1: first poll, $\hat{p}_1 = 0.48$,

      Population 2: second poll, $\hat{p}_2 = 0.60$

$$\hat{p}_p = (0.48 + 0.60)/2 = 0.54$$

(a)   Step 1: $H_0: p_1 = p_2, \quad H_a: p_1 < p_2$

     Step 2: $\alpha = 0.01$

     Step 3: Since $\alpha = 0.01$, the critical value $z_\alpha = -2.33$

     Step 4: $z = \dfrac{0.48 - 0.60}{\sqrt{0.54(1-0.54)}\sqrt{(1/600)+(1/600)}} = -4.17$

     Step 5: Since $-4.17 < -2.33$, reject H_0.

     Step 6: The test is significant at the 1% level; that is the data do provide evidence to conclude that the percentage of Maricopa County residents who thought that the state's economy would improve over the next 2 years was less during the time of the first poll than during the time of the second poll.

(b) $P(z < -4.17) = 0.0000$. So P-value $< \alpha$. Thus, we reject H_0. Same conclusion as (a).

(c)   The strength of the evidence against the null hypothesis is very strong.

16.   (a)   From Exercise 15

$$(0.48 - 0.60) \pm 2.33 \cdot \sqrt{0.48(1-0.48)/600 + 0.60(1-0.60)/600}$$

$$-0.12 \pm 0.067$$

$$-0.187 \text{ to } -0.053$$

(b)   We can be 95% confident that the difference $p_1 - p_2$ between the proportions of Maricopa County residents who thought that the state's economy would improve over the next 2 years during the time of the 1992 poll and during the time of the 1993 poll is somewhere between -0.187 and -0.053.

17.   (a) $E = (-0.053 - (-0.187))/2 = 0.067$

We can be 98% confident that the error in estimating the difference between the two population proportions, $p_1 - p_2$, by the difference between the two sample proportions, -0.12, is at most 0.067.

(b) $$E = z_{\frac{\alpha}{2}}\sqrt{\frac{\hat{p}_1(1-\hat{p}_1)}{n_1} + \frac{\hat{p}_2(1-\hat{p}_2)}{n_2}}$$

$$= 2.33\sqrt{\frac{(0.48)(1-0.48)}{600} + \frac{(0.60)(1-0.60)}{600}}$$

$$= 0.067$$

(c) $E = 0.03$, $\alpha = 0.02$

$$n = 0.50 \frac{z^2_{\frac{\alpha}{2}}}{E^2} = 0.50 \frac{2.33^2}{0.03^2} = 3016.05 \approx 3017$$

(d) $n = 3017$, $\hat{p}_1 = 0.475$, $\hat{p}_2 = 0.603$

$$(0.475 - 0.603) \pm 2.33 \cdot \sqrt{0.475(1-0.475)/3017 + 0.603(1-0.603)/3017}$$

$$-0.128 \pm 0.030$$

$$-0.158 \text{ to } -0.098$$

(e) $E = 0.030$ which is the same as what is required in part (c).

18.   Choose **Calc ▶ Basic statistics ▶ 1 Proportion...**, select the **Summarized data** option button, click in the **Number of trials** text box and type 1006, click in the **Number of successes** text box and type 433, click the **Options...** button, click in the **Confidence level** text box and type 95, click **OK**, and click **OK**. The resulting output is

Test of p = 0.5 vs p not = 0.5

Sample	X	N	Sample p	95.0 % CI	Exact P-Value
1	433	1006	0.430417	(0.399566, 0.461677)	0.000

19.   Choose **Calc ▶ Basic statistics ▶ 1 Proportion...**, select the **Summarized data** option button, click in the **Number of trials** text box and type 2512, click in the **Number of successes** text box and type 578, click the **Options...** button, click in the **Test proportion** text box and type 0.25, click the arrow button at the right of the **Alternative** drop-down list box and select **less than**, click **OK**, and click **OK**. The resulting output is

Test of p = 0.25 vs p < 0.25

Sample	X	N	Sample p	95.0 % CI	Exact P-Value
1	578	2512	0.230096	(0.213759, 0.247063)	0.011

20.   Using Minitab, choose **Calc ▶ Basic statistics ▶ 2 Proportions...**, select the **Summarized data** option button, click in the **Trials** text box for **First sample** and type 600, click in the **Successes** text box for **First sample** and type 288, click in the **Trials** text box for **Second sample** and type 600, click in the **Successes** text box for **Second sample** and type 360, click the **Options...** button, click in the **Confidence level** text box and type 98, click in the **Test difference** text box and type 0, click the arrow button at the right of the **Alternative** drop-down list box and select **less than**, select the **Use pooled estimate of p for test** check box, click **OK**, and click **OK**. The resulting output is

Sample	X	N	Sample p
1	288	600	0.480000
2	360	600	0.600000

Estimate for p(1) - p(2): -0.12
98% CI for p(1) - p(2): (-0.186454, -0.0535462)
Test for p(1) - p(2) = 0 (vs < 0): Z = -4.17 P-Value = 0.000

21. (a) 1006
     (b) 433
     (c) 43.0417%
     (d) 0.399566 to 0.461677

22. (a) 2512
     (b) 578
     (c) 23.0096%
     (d) p = 0.25 vs p < 0.25
     (e) P-value = 0.011
     (f) Since the P-value < α, reject the null hypothesis and claim that the percentage of people who believe that juries 'almost always' convict the guilty and free the innocent is less than 25%.

23. (a) 600 and 600
     (b) 288 and 360
     (c) 48% and 60%
     (d) -0.186454 to -0.0535462
     (e) H_0: $p_1 = p_2$, H_a: $p_1 < p_2$; or H_0: $p_1 - p_2 = 0$, H_a: $p_1 - p_2 < 0$
     (f) P-value = 0.000
     (g) Since P-value < α, reject the null hypothesis and claim that the percentage of Maricopa County residents who thought the state's economy would improve during the next two years was greater at the time of the second poll than at the time of the first one a year earlier.

CHAPTER 13 ANSWERS

Exercises 13.1

13.1 A variable has a chi-square distribution if its distribution has the shape of right-skewed curve, called a chi-square curve.

13.3 The χ^2-curve with 20 degree of freedom more closely resembles a normal curve. This follows from Property 4 of Key Fact 13.1, " As the number of degrees of freedom becomes larger, χ^2-curves look increasingly like normal curves.

13.5 (a) $\chi^2_{0.025} = 32.852$ (b) $\chi^2_{0.95} = 10.117$

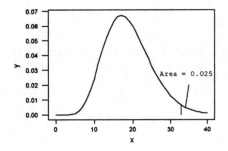

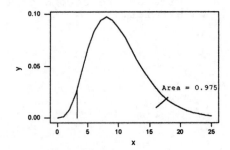

13.7 (a) $\chi^2_{0.05} = 18.307$ (b) $\chi^2_{0.975} = 3.247$

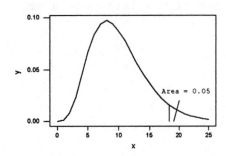

13.9 The χ^2-value having area 0.05 to its left has 0.95 to its right. Table V gives values with specified areas to their right. If df = 26,

$$\chi^2_{0.95} = 15.379$$

Exercises 13.2

13.11 The term "goodness-of-fit" is used to describe the type of hypothesis test considered in this section because the test is carried out by determining how well the observed frequencies match or fit the expected frequencies.

13.13 The observed frequencies are random variables since their values depend on chance; namely, on which sample is obtained. The expected frequencies are not random variables—they are fixed numbers, computed using the formula $E = np$.

13.15

Buyer Type	O	p	E = np	$(O - E)^2/E$
Consumer	1422	0.497	1491	3.193
Business	1521	0.485	1455	2.994
Government	57	0.018	54	0.167
	500		500	6.354

Step 1: H_0: Last year's type-of-buyer distribution for U.S. cars is the same as the 1995 distribution.

H_a: Last year's type-of-buyer distribution for U.S. cars is different from the 1995 distribution.

Step 2: Expected frequencies are presented in column 4 of the table.

Step 3: Assumptions 1 and 2 are satisfied since all expected frequencies are at least 5.

Step 4: $\alpha = 0.05$

Step 5: Critical value = 5.991

Step 6: $\chi^2 = 6.354$ (See column 5 of the table.)

Step 7: Since 6.354 > 5.991, reject H_0.

Step 8: The data do provide sufficient evidence to conclude that last year's type-of-buyer distribution for U.S. cars is different from the 1995 distribution.

For the P-value approach, $0.025 < P(\chi^2 > 6.354) < 0.05$. Since the P-value is smaller than the significance level, reject H_0.

13.17

Race	O	p	E = np	$(O - E)^2/E$
White	239	0.515	257.5	1.329
Black	183	0.331	165.5	1.850
Hispanic	72	0.144	72.0	0.000
Other	6	0.010	5.0	0.200
	500		500	3.379

Step 1: H_0: The distribution of AIDS deaths by race in 1995 is the same as the 1992 distribution.

H_a: The distribution of AIDS deaths by race in 1995 is different from the 1992 distribution.

Step 2: Expected frequencies are presented in column 4 of the table.

Step 3: Assumptions 1 and 2 are satisfied since all expected frequencies are at least 5.

Step 4: $\alpha = 0.10$

Step 5: Critical value = 6.251

Step 6: $\chi^2 = 3.379$ (See column 5 of the table.)

Step 7:   Since $3.379 < 6.251$, do not reject H_0.

Step 8:   The data do not provide sufficient evidence to conclude that the distribution of AIDS deaths by race in 1995 is different from the 1992 distribution.

For the P-value approach, the $P(\chi^2 > 3.379) > 0.10$. Since the P-value is larger than the significance level, do not reject H_0.

13.19

Number	O	p	E = np	$(O - E)^2/E$
1	23	1/6	25	0.16
2	26	1/6	25	0.04
3	23	1/6	25	0.16
4	21	1/6	25	0.64
5	31	1/6	25	1.44
6	26	1/6	25	0.04
	150		150	2.48

Step 1: H_0:   The die is not loaded.

             H_a: The die is loaded.

Step 2:   Expected frequencies are presented in column 4 of the table.

Step 3:   Assumptions 1 and 2 are satisfied since all expected frequencies are at least 5.

Step 4: $\alpha = 0.05$

Step 5:   Critical value = 11.070

Step 6: $\chi^2 = 2.48$ (See column 5 of the table.)

Step 7:   Since $2.48 < 11.070$, do not reject H_0.

Step 8:   The data do not provide sufficient evidence to conclude that the die is loaded.

For the P-value approach, $P(\chi^2 > 2.48) > 0.10$. Since the P-value is larger than the significance level, do not reject H_0.

13.21 (a)   The observed frequencies and the expected frequencies should both sum to n. By definition, n is the total number of observed frequencies. If we let p_i and E_i be the probability and expected frequency for category i, then $\sum E_i = \sum np_i = n \sum p_i = n \cdot 1 = n$.

(b)   zero

(c)   From part (a), the sums of the observed and expected frequencies <u>must</u> be the same. If they are not the same (except for possible round-off error), you can conclude that you have made an error either in finding the expected frequencies or in summing them.

(d)   No. It is possible (though admittedly not very likely) that you made two or more compensating errors in finding the expected frequencies and they still sum to the correct number.

13.23 Using Minitab, click in the Session window, choose **Editor ▶ Enable**

Command Language if it is not already enabled, and type in the Session window after the MTB> prompt, a percent sign (%) followed by the path for *fittest.mac*. If *Datadisk* is in drive a, we type

%a:\macro\fittest.mac and press the ENTER key. Then proceed as follows: Type <u>3</u> and press the ENTER key in response to Enter the number of possible values for the variable under consideration; type <u>0.497 0.485 0.018</u> and press the ENTER key in response to Enter the relative frequencies (or probabilities) for the null hypothesis; and type <u>1422 1521 57</u> and press the ENTER key in response to Enter the observed frequencies. The resulting output is

```
     X-square goodness-of-fit test

       Row        n     k    ChiSq      P-value

        1       3000     3   6.35364    0.0417181
```

13.25 (a) n = 500   (b) χ^2 = 4.82575 (c)   P-value = 0.185010

 (d) 0.185010   (e) Do not reject the null hypothesis. There is not sufficient evidence to conclude that the distribution has changed from the 1995 distribution.

Exercises 13.3

13.27 One example of univariate data arises if we record the religious affiliation of those in a sample. An example of bivariate data arises if we record the dominant eye and dominant hand of each of those in the sample.

13.29 Cells

13.31 To obtain the total number of observations of bivariate data in a continency table, one can sum the individual cell frequencies, sum the row subtotals, or sum the column subtotals.

13.33 Yes. If there were no association between the gender of the physician and specialty of the physician, then the same percentage of male and female physicians would choose internal medicine. Since different percentages of male and female physicians chose internal medicine, there is an association between the variables "sex" and "specialty".

13.35 (a)

	M	F	Total
BUS	2	7	9
ENG	10	2	12
LIB	3	1	4
Total	15	10	25

(b)

	BUS	ENG	LIB	Total
M	0.222	0.833	0.750	0.600
F	0.778	0.167	0.250	0.400
Total	1.000	1.000	1.000	1.000

(c)

	M	F	Total
BUS	0.133	0.700	0.360
ENG	0.667	0.200	0.480
LIB	0.200	0.100	0.160
Total	1.000	1.000	1.000

(d)   Yes. The conditional distributions of sex are different within each college.

13.37 (a)

	Fre	Sop	Jun	Sen	Total
Dem	2	6	8	4	20
Rep	3	9	12	6	30
Other	1	3	4	2	10
Total	6	18	24	12	60

(b)

	Fre	Sop	Jun	Sen
Dem	0.333	0.333	0.333	0.333
Rep	0.500	0.500	0.500	0.500
Other	0.167	0.167	0.167	0.167
Total	1.000	1.000	1.000	1.000

(c)   There is no association between party affiliation and class level. All of the conditional distributions of political party within class level are identical.

(a)   The marginal distribution of party affiliation will be identical to each of the conditional distributions, that is

Party	Frequency
Dem	0.333
Rep	0.500
Other	0.167
Total	1.000

(e)   True. Since party affiliation and class level are not associated, all of the conditional distributions of class level within political party will be identical and will be the same as the marginal distribution of class level.

13.39 (a)   8

(b)   First complete the first row, then the first column total, then the third row total, then the grand total.

	Public	Private	Total
Northeast	266	555	821
Midwest	359	504	863
South	533	502	1035
West	313	242	555
Total	1471	1803	3274

(c)   3274   (d)   863   (e)   1471   (f)   502

13.41 (a)   73,970   (b)   15,540   (c)   12,328

(d)   73970 + 15540 - 12328 = 77182

(e)   12,225   (f)   6,734 (g)   103,601 - 25,901 = 77,700

13.43 The table from Exercise 13.39 was first transposed so that the columns became the rows and vice versa. Then each cell entry was divided by the column total to produce the table below.

	Northeast	Midwest	South	West	Total
Public	0.324	0.416	0.515	0.564	0.449
Private	0.676	0.584	0.485	0.436	0.551
Total	1.000	1.000	1.000	1.000	1.000

(a)   The conditional distributions of type of institutional within each region are given in columns 2, 3, 4, and 5 of the above table.

(b)   The marginal distribution of type of institution is given by the last column of the above table.

(c)   Yes. The conditional distributions for type of institution are different for the four regions.

(d)   55.1% of the institutions of higher education are private.

(e)   67.6% of the institutions of higher education in the Northeast are private.

(f)   True. Since there is an association between type of institution and region, the conditional distributions of region within type of institution cannot be identical (If they were identical, there would be no association between the two variables.).

(g)   Directly from the table in Exercise 13.39, divide each cell entry by the column total below it to obtain the table below.

	Public	Private	Total
Northeast	0.181	0.308	0.251
Midwest	0.244	0.280	0.264
South	0.362	0.278	0.316
West	0.213	0.134	0.170
Total	1.000	1.000	1.000

The conditional distributions of region with types of educational institutions are given in columns 2 and 3 of the table. The marginal distribution of region is given in the last column of the table. The conditional distributions of region within types of educational institutions indicate that the South has the largest percentage of public institutions while the Northeast has the largest percentage of private institutions. The marginal distribution of region indicates that the South has the largest percentage of educational institutions of either kind while the West has the lowest percentage. Other interpretations are possible as well.

13.45 (a)

	Office	Hospital	Other	Total
General Surgery	0.326	0.472	0.445	0.367
Obstetrics/ gynecology	0.326	0.260	0.306	0.309
Orthopedics	0.181	0.164	0.111	0.174
Ophthal- mology	0.167	0.104	0.139	0.150
Total	1.000	1.000	1.000	1.000

(b)   Yes. The conditional distributions of specialty within the base-of-practice categories (columns 2, 3, and 4 of the table in part a) are not identical.

(c)   The marginal distribution of specialty is given in the last column of the table in part (a).

(d)

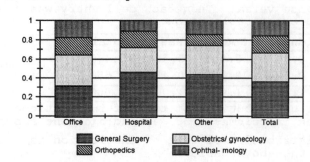

If the conditional distributions of specialty within the base-of-practice categories (first three bars) had been identical, each of the bars would have been segmented identically and would be identical to the fourth bar representing the marginal distribution of specialty. The fact that they are not segmented identically means that there is an association between specialty and base of practice.

(e)   False. Since specialty and base of practice are associated, the conditional distributions of base-of-practice within specialty will be different.

(f)   We interchanged the rows and columns of the table in Exercise 13.41 and then divided each cell entry by its associated column total to obtain the following table.

	General Surgery	Obstetrics/ gynecology	Orthopedics	Ophthal- mology	Total
Office	0.635	0.754	0.741	0.793	0.714
Hospital	0.322	0.210	0.236	0.173	0.250
Other	0.044	0.036	0.023	0.033	0.036
Total	1.000	1.000	1.000	1.000	1.000

The conditional distributions of base-of-practice within specialty are given in columns 2, 3, 4, and 5. The marginal distribution of base-of-practice is given by the last column.

(g)   From the table in part (f), 25.0% of the surgeons are hospital based.

(h)   From the table in part (f), 21.0% of the OB/GYNs are hospital based.

(i)   From the table in part (a), 26.0% of the hospital-based surgeons are OB/GYNs.

13.47 (a)   No. Every conditional distribution of y within x is identical.

(b)   The marginal distribution of y is the same as each of the conditional distributions of y within x. That is,

y	Frequency
0	0.316
1	0.422
2	0.211
3	0.047
4	0.004
Total	1.000

(c)   No. To determine the marginal distribution of x, the actual cell
counts are needed. These could be determined from the cell
percentages and the A, B, and C column total counts if those
counts were known. Since the total counts for the three columns
are not known, there is no way to obtain the marginal distribution
of x.

13.49 (a)   With the data from Table 13.11 in three columns of Minitab named

SEX, CLASS, and COLLEGE, choose **Stat ▶ Tables ▶ Cross
tabulation...**, specify SEX and COLLEGE in the **Classification
variables** text box, select the **Counts** check box in the **Display**
list, and click **OK**. The resulting printout is

```
Rows: SEX      Columns: COLLEGE


          BUS       ENG       LIB       All


F           7         2         1        10

M           2        10         3        15

All         9        12         4        25
    Cell Contents --

                    Count
```

(b)   To determine the conditional distribution of SEX within COLLEGE
and the marginal distribution of SEX, we proceed as in part (a)
except that we select the **Column Percents** check box instead of the
Counts check box (and unselect the **Counts** check box). The results
are shown in the following table.

```
          BUS       ENG       LIB       All


F        77.78     16.67     25.00     40.00

M        22.22     83.33     75.00     60.00

All     100.00    100.00    100.00    100.00
    Cell Contents --

                   % of Col
```

(c)   To determine the conditional distribution of COLLEGE within SEX
and the marginal distribution of COLLEGE, we proceed as in part
(b) except that we select the **Row Percents** check box instead of
the **Column Percents** check box (and unselect the **Column Percents**
check box). The results are shown in the following table.

	BUS	ENG	LIB	All
F	70.00	20.00	10.00	100.00
M	13.33	66.67	20.00	100.00
All	36.00	48.00	16.00	100.00

Cell Contents --

% of Row

13.51 (a)   With the data from Table 13.12 in two columns of Minitab named PARTY and CLASS, click anywhere in the CLASS column, then choose

Editor ▶ Set Column ▶ Value Order..., select the **User-specified order** option button; in the **Define an order (one value per line)** text box, edit the order of the possible classes so that it reads Fre, Sop, Jun, Sen, and click **OK**. Then repeat the process with the PARTY column so that the order is Dem, Rep, Other. Now choose

Stat ▶ Tables ▶ Cross tabulation..., specify CLASS and PARTY in the **Classification variables** text box, select the **Counts** check box in the **Display** list, and click **OK**. The resulting printout is

Rows: PARTY   Columns: CLASS

	Fre	Sop	Jun	Sen	All
Dem	2	6	8	4	20
Rep	3	9	12	6	30
Other	1	3	4	2	10
All	6	18	24	12	60

Cell Contents -

Count

(b)   To determine the conditional distribution of PARTY within CLASS and the marginal distribution of PARTY, we proceed as in part (a) except that we select the **Column Percents** check box instead of the **Counts** check box (and unselect the **Counts** check box). The results are shown in the following table.

	Fre	Sop	Jun	Sen	All
Dem	33.33	33.33	33.33	33.33	33.33
Rep	50.00	50.00	50.00	50.00	50.00
Other	16.67	16.67	16.67	16.67	16.67
All	100.00	100.00	100.00	100.00	100.00

Cell Contents -

% of Col

Exercises 13.4

13.53 In most cases, data for an entire population are not available. Thus we must apply inferential methods to decide whether an association exists between two variables.

13.55 In Section 13.3, we learned that if there is no association between two variables, all of the conditional distributions for the first variable within the values of the other variable will be identical, and those conditional distributions will be the same as the marginal distribution of the first variable. The same statement is true if the order of the variables in above statement is reversed.

13.57 The degrees of freedom equal $(r - 1)(c - 1) = (6 - 1)(4 - 1) = 15$.

13.59 If two variables are not associated, then their frequencies do not rise and fall together, whether by causation or as the result of some third variable. If the variables did have a causal relationship, then they would be associated.

13.61 Step 1:　　H_0: Family income and educational attainment of the householder are not associated.

　　　　　　　　H_a: Family income and educational attainment of the householder are associated.

　　Step 2:　　Calculate the expected frequencies using the formula

　　　　　　　　$E = RC/n$ where R = row total, C = column total, and n = sample size. The results are shown in the following table.

	Not HS grad	HS grad	College grad	Total
Under 25	32.9	61.3	47.8	142.0
25<70	36.4	67.7	52.9	157.0
50<75	22.7	42.3	33.0	98.0
75 or more	23.0	42.7	33.3	99.0
Total	115.0	214.0	167.0	496.0

　　Step 3:　　All of the expected frequencies are greater than 5, so the assumptions for the χ-square test are met.

　　Step 4:　　$\alpha = 0.01$

　　Step 5:　　The degrees of freedom are $(4 - 1)((3 - 1) = 6$, so the critical value from Table V is $\chi^2_{0.01} = 16.812$.

　　Step 6:　　Compute the value of the test statistic $\chi^2 = \sum \dfrac{(O - E)^2}{E}$,

　　　　　　　　Where O and E represent the observed and expected frequencies respectively. We show the contributions to this sum from each of the cells of the contingency table in the table below.

	Not HS grad	HS grad	College grad
Under 25	31.252	0.049	23.910
25<70	0.054	3.904	4.178
50<75	6.047	0.070	3.033
75 or more	15.696	9.098	44.856

The total of the 12 table entries above is the value of the χ-square statistic, that is $\chi^2 = 142.147$. Depending on rounding, your answer could differ slightly.

　　Step 7:　　Since 142.147 > 16.812, we reject the null hypothesis.

　　Step 8:　　We conclude that there is an association between family income and the educational attainment of householder.

13.63 The expected frequencies are shown below the observed frequencies in the following contingency table.

Political Affiliation	Good Idea	Poor Idea	No Opinion	Total
Republican	266 282.9	266 255.2	186 180.0	718
Democrat	308 289.2	250 260.8	176 184.0	734
Independent	28 29.9	27 27.0	21 19.0	76
Total	602	543	383	1528

The table below gives the contributions from each cell to the χ-square statistic.

Row, Column	O	E	$(O - E)^2/E$
1,1	266	282.9	1.010
1,2	266	255.2	0.457
1,3	186	180.0	0.200
2,1	308	289.2	1.222
2,2	250	260.8	0.447
2,3	176	184.0	0.348
3,1	28	29.9	0.121
3,2	27	27.0	0.000
3,3	21	19.0	0.211
	1528		4.016

Step 1: H_0: The feelings of adults on the issue of regional primaries are independent of political affiliation.

H_a: The feelings of adults on the issue of regional primaries are dependent on political affiliation.

Step 2: Observed and expected frequencies are presented in the frequency contingency table. Each expected frequency is placed below its corresponding observed frequency. Expected frequencies are calculated using the formula $E = (R \cdot C)/n$.

The same information about the Os and Es is presented in the table below the contingency table. Column 4 of this table is useful for Step 6.

Step 3: Assumptions 1 and 2 are satisfied since all expected frequencies are at least 5.

Step 4: $\alpha = 0.05$

Step 5: Critical value = 9.488

Step 6: $\chi^2 = 4.016$ (See column 4 of the table below the contingency table.)

Step 7: Since 4.016 < 9.488, do not reject H_0.

Step 8:   The data do not provide sufficient evidence to conclude that the feelings of adults on the issue of regional primaries are dependent on political affiliation.

For the P-value approach, $P(\chi^2 > 4.016) > 0.10$. Since the P-value is larger than the significance level, do not reject H_0.

13.65

Marital Status

Net Worth	Married	Single/ Divorced	Widowed	Total
$100,000– $249,999	227 223.9	54 53.0	63 67.1	344
$250,000– $499,999	60 63.1	15 14.9	22 18.9	97
$500,000– $999,999	20 20.2	4 4.8	7 6.0	31
$1,000,000 or more	10 9.8	2 2.3	3 2.9	15
Total	317	75	95	487

Step 1: H_0: Net worth and marital status for top wealthholders are statistically independent.

H_a: Net worth and marital status for top wealthholders are statistically dependent.

Step 2:   Observed and expected frequencies are presented in the frequency contingency table. Each expected frequency is placed below its corresponding observed frequency. Expected frequencies are calculated using the formula $E = (R \cdot C)/n$.

Step 3:   Assumption 2 is violated since 25% (3/12) of the expected frequencies are less than 5. Thus, we do not proceed with the test.

13.67 (a)   After combining the last two rows of the contingency table in Exercise 13.65, the table becomes

Marital Status

Net Worth	Married	Single/ Divorced	Widowed	Total
$100,000– $249,999	227 223.9	54 53.0	63 67.1	344
$250,000– $499,999	60 63.1	15 14.9	22 18.9	97
$500,000 or more	30 30.0	6 7.1	10 8.9	46
Total	317	75	95	487

(b)

Step 1: H_0: Net worth and marital status for top wealthholders are statistically independent.

H_a: Net worth and marital status for top wealthholders are statistically dependent.

Step 2:   Observed and expected frequencies are presented in the frequency contingency table. Each expected frequency is placed below its corresponding observed frequency. Expected frequencies are calculated using the formula $E = (R \cdot C)/n$.

Step 3:   All of the expected frequencies are greater than 5, so both assumptions are satisfied.

Step 4: $\alpha = 0.05$

Step 5:   Critical value = 11.143

Step 6:   The contributions to the χ-square statistic are shown in the table below.

Row, Column	O	E	$(O - E)^2/E$
1,1	227	223.9	0.043
1,2	54	53.0	0.019
1,3	63	67.1	0.251
2,1	60	63.1	0.152
2,2	15	14.9	0.001
2,3	22	18.9	0.508
3,1	30	30.0	0.000
3,2	6	7.1	1.120
3,3	10	8.9	0.170
	487		1.280

Step 7:   Since $1.280 < 11.143$, we do not reject the null hypothesis.

Step 8:   There is not sufficient evidence to conclude that for top wealthholders, net worth and marital status are statistically dependent.

(c)   After eliminating the last row of the contingency table in Exercise 13.65 and recalculating all of the expected frequencies, the table becomes

Marital Status

Net Worth	Married	Single/ Divorced	Widowed	Total
$100,000– $249,999	227 223.7	54 53.2	63 67.1	344
$250,000– $499,999	60 63.1	15 15.0	22 18.9	97
$500,000– $999,999	20 20.2	4 4.8	7 6.0	31
Total	307	73	92	472

(d)   Steps 1,2,4, and 5 are the same as in part (b).

Step 3:   All of the expected frequencies are greater than 1 and only 11.1% (1 out of 9) are less than 5, so the assumptions for the χ-square test are satisfied.

Step 6:

Row, Column	O	E	$(O - E)^2/E$
1,1	227	223.7	0.049
1,2	54	53.2	0.012
1,3	63	67.1	0.251
2,1	60	63.1	0.152
2,2	15	15.0	0.000
2,3	22	18.9	0.508
3,1	20	20.2	0.002
3,2	4	4.8	0.133
3,3	7	6.0	0.167
	487		1.274

Step 7:   Since 1.274 < 11.143, we do not reject the null hypothesis.

Step 8:   There is not sufficient evidence to conclude that for top wealthholders, net worth and marital status are statistically dependent.

13.69 With the data in columns named 'GOODIDEA', 'POORIDEA', and 'NOOPIN', choose **Stat ▶ Tables ▶ Chisquare test...**, select 'GOODIDEA', 'POORIDEA', and 'NOOPIN' as the variables in the **Columns containing the table** text box, and click **OK**. The result is

Expected counts are printed below observed counts

	GOODIDEA	POORIDEA	NOOPIN	Total
1	266	266	186	718
	282.88	255.15	179.97	
2	308	250	176	734
	289.18	260.84	183.98	
3	28	27	21	76
	29.94	27.01	19.05	
Total	602	543	383	1528

ChiSq = 1.007 + 0.461 + 0.202 +
        1.225 + 0.450 + 0.346 +
        0.126 + 0.000 + 0.200 =   4.017

df = 4, p = 0.404

13.71 (a)   The number of students in the sample who studied at least 10 hours per week is 34. This number occurs at the intersection of the column labeled "C4" and the row labeled "Total."

(b)   The number of students in the sample who received a grade of B in the class is 164. This number occurs at the intersection of the row labeled "2" and the column labeled "Total."

(c)   The observed number of students in the sample who studied at least 10 hours per week and received a grade of B in the class is 9. This number is the *top* number of the two occurring at the intersection of the column labeled "C4" and the row labeled "2."

(d)   The expected number of students in the sample who studied at least ten hours per week and received a grade of B in the class is 13.21. This number is the *bottom* number of the two occurring at the intersection of the column labeled "C4" and the row labeled "2."

(e)   The X-square subtotal 1.343 (located at the intersection of the fourth column and second row of the X-square statistic) is the subtotal for the "at least 10 hours per week *and* grade of B" cell (located at the intersection of the fourth column and the second row of the contingency table).

(f)   The number of degrees of freedom is 6. This is the last item presented in the printout; i.e., df = 6.

(g)   The null and alternative hypotheses for this exercise are:

H_0:   There is no association between grade and study time for intermediate algebra students at Arizona State University; i.e., grade and study time are statistically independent.

H_a: There is an association between grade and study time for intermediate algebra students at Arizona State University; i.e., grade and study time are statistically dependent.

(h) $\chi^2 = 10.574$

(i)   P-value = 0.104

(j)   We are given $\alpha = 0.05$. Also, df = 6 from part (f). The critical value of X-square is 12.592. Since 10.574 < 12.592, do not reject H_0. The data do not provide sufficient evidence to conclude that an association exists between grade and study time for intermediate algebra students at Arizona State University.

REVIEW TEST FOR CHAPTER 13

1.   The distributions and curves are distinguished by their degrees of freedom.

2.   (a) zero   (b) skewed right   (c)   normal curve

3.   (a)   No. The degrees of freedom for the χ-square goodness-of-fit test depends on the number of categories, not the number of observations.

(b)   No. The degrees of freedom for the χ-square independence test is $(r-1)(c-1)$ where r and c are the number of rows and columns, respectively, in the contingency table.

4.   Values of the test statistic near zero arise when the observed and expected frequencies are in close agreement. It is only when these frequencies differ enough to produce large values of the test statistic that the null hypothesis is rejected. These values are in the right tail of the χ-square distribution.

5.   Zero

6.   (a)   All expected frequencies are 1 or greater, and at most 20% of the expected frequencies are less than 5.

(b)   Very important. If either of these assumptions are not met, the test should not be carried out by these procedures.

7.   (a)   5.3%

(b)   5.3% of 58.5 million, or 3.1 million

(c)   Since the observed number of American living in the West who use public transportation is not equal to the number expected if there were no association between means of transportation and area of residence, we conclude that there *is* an association between means of transportation and area of residence.

8.   (a)   Compare the conditional distributions of one of the variables within categories of the other variable. If all of the conditional distributions are identical, there is no association between the variables; if not, there is an association.

      (b)   No. Since the data are for an entire population, we are not making an inference from a sample to the population. The association (or non-association) is a fact.

9.   (a)   Perform a χ-square test of independence. If the null hypothesis (of non-association) is rejected, we conclude that there is an association between the variables.

      (b)   Yes. It is possible (with probability α) that we could reject the null hypothesis when it is, in fact, true. It is also possible, that, even though there is actually an association between the variables, the evidence is not strong enough to draw that conclusion. Either of these types of errors are due to randomness in selecting a sample which does not exactly reflect the characteristics of the population.

10.   For df = 17:

      (a) $\chi^2_{0.99} = 6.408$ (b) $\chi^2_{0.01} = 33.409$ (c) $\chi^2_{0.05} = 27.587$

      (d) $\chi^2_{0.95} = 8.672$ (e) $\chi^2_{0.975} = 7.564$ and $\chi^2_{0.025} = 30.191$

11.

Highest level	O	p	E = np	$(O-E)^2/E$
Not HS graduate	92	0.248	124.0	8.258
HS graduate	168	0.300	150.0	2.160
Come college	86	0.187	93.5	0.602
Associate's degree	36	0.062	31.0	0.806
Bachelor's degree	79	0.131	65.5	2.782
Advanced degree	39	0.072	36.0	0.250
Total	500		500.0	14.859

Step 1: H_0:   The educational attainment distribution for adults 25 years old and over this year is the same as in 1990.

          H_a: The educational attainment distribution for adults 25 years old and over this year differs from that of 1990.

Step 2:   Expected frequencies are presented in column 4 of the table.

Step 3:   Assumptions 1 and 2 are satisfied since all of the expected frequencies are at least 1, and none of the expected frequencies are less than 5.

Step 4: $\alpha = 0.05$

Step 5:   Critical value = 11.070

Step 6: $\chi^2 = 14.859$ (See column 5 of the table.)

Step 7:   Since $14.859 > 11.070$, reject H_0.

Step 8:   It appears that the educational attainment distribution for adults 25 years old and over this year differs from that of 1990.

    (b)   For the P-value approach, $0.01 < P(\chi^2 > 14.859) < 0.025$. Since the P-value is smaller than the significance level, reject H_0. The evidence against the null hypothesis is strong.

    (c)   No. The critical value for a 1% significance level is 15.086. Since $14.859 < 15.086$, the null hypothesis cannot be rejected at the 1% significance level.

12. (a)

	DEM	REP	IND	TOTAL
MW	3	9	0	12
NE	2	6	1	9
SO	6	10	0	16
WE	6	7	0	13
TOTAL	17	32	1	50

(b)

	DEM	REP	IND	TOTAL
MW	0.176	0.281	0.000	0.240
NE	0.118	0.188	1.000	0.180
SO	0.353	0.313	0.000	0.320
WE	0.353	0.219	0.000	0.260
TOTAL	1.000	1.000	1.000	1.000

The conditional distributions of region within party are given in columns 2, 3, and 4 of the above table. The marginal distribution of region is given in the last column of the table.

(c)

	MW	NE	SO	WE	TOTAL
DEM	0.250	0.222	0.375	0.462	0.340
REP	0.750	0.667	0.625	0.538	0.640
IND	0.000	0.111	0.000	0.000	0.020
TOTAL	1.000	1.000	1.000	1.000	1.000

The conditional distributions of party within region are given in columns 2, 3, 4, and 5 of the above table. The marginal distribution of party is given in the last column of the table.

(d) There is an association between region and party of governor for the states of the U.S. since the conditional distributions of party within the several regions are not identical.

(e) From part (c), 64.0% of the states have Republican governors.

(f) If there were no association between region and party of governor, 64% of the Midwest states would have Republican governors.

(g) From the table in part (c), 75.0% of the Midwest states have Republican governors.

(h) From the table in part (b), 24.0% of the states are in the Midwest.

(i) If there were no association between region and party of governor, the percentage of states with Republican governors that would be in the Midwest would be the same as the percentage of states that are in the Midwest, i.e. 24%.

(j) In reality, from the table in part (b), 28.1% of the states with Republican governors are in the Midwest.

13. (a) 2046   (b) 737   (c) 266   (d) 3046   (e) 6580 - 1167 = 5413
    (f) 5403 + 1167 - 660 = 5910

14. (a)

	General	Psychiatric	Chronic	Tuberculosis	Other	Total
GOV	0.314	0.361	0.808	0.750	0.144	0.311
PROP	0.122	0.486	0.038	0.000	0.361	0.177
NP	0.564	0.153	0.154	0.250	0.495	0.512
Total	1.000	1.000	1.000	1.000	1.000	1.000

The conditional distributions of control type with facility type are given in columns 2, 3, 4, 5, and 6 of the table.

(b)   Yes. The conditional distributions of control type within facility type are not all identical.

(c)   The marginal distribution of control type is given by the last column of the table above.

(d)

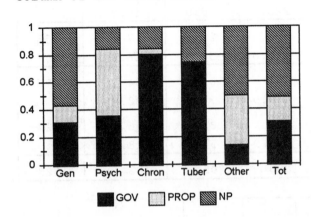

The conditional distributions of control type within facility type are shown by the first five bars of the graph. The marginal distribution of control type is given by the last bar. Since the bars are not identically shaded, control type and facility type are associated variables.

(e)   False. Since we have established that facility type and control type are associated, the conditional distributions of facility type within control types will not be identical.

(f)   After interchanging the rows and columns of the table given in Problem 12, we divided each cell entry by its associated column total to obtain the following table.

	GOV	PROP	NP	Total
General	0.829	0.566	0.905	0.821
Psychiatric	0.130	0.307	0.034	0.112
Chronic	0.010	0.001	0.001	0.004
Tuberculosis	0.001	0.000	0.000	0.001
Other	0.029	0.127	0.060	0.062
Total	1.000	1.000	1.000	1.000

The conditional distributions of facility type within control type are given in columns 2, 3, and 4 of the table. The marginal distribution of facility type is given by the last column of the table.

(g)   From the table in part (a), 17.7% of the hospitals are under proprietary control.

(h)   From the table in part (a), 48.6% of the psychiatric hospitals are under proprietary control.

(i)   From the table in part (f), 30.7% of the hospitals under proprietary control are psychiatric hospitals.

15.

Educational Level	Favor	Oppose	No Opinion	Total
College grad	264 229.1	17 38.5	6 19.3	287
Some college	205 190.0	26 31.9	7 16.0	238
High-school grad	461 459.9	81 77.3	34 38.8	576
Non-high-school grad	290 340.9	81 57.3	56 28.8	427
Total	1220	205	103	1528

Row, Column	O	E	$(O - E)^2/E$
1,1	264	229.1	5.317
1,2	17	38.5	12.007
1,3	6	19.3	9.165
2,1	205	190.0	1.184
2,2	26	31.9	1.091
2,3	7	16.0	5.063
3,1	461	459.9	0.003
3,2	81	77.3	0.177
3,3	34	38.8	0.594
4,1	290	340.9	7.600
4,2	81	57.3	9.803
4,3	56	28.8	25.689
	1528		77.693

Step 1: H_0:   Response and educational level are statistically independent.

 H_a: Response and educational level are statistically dependent.

Step 2:   Observed and expected frequencies are presented in the frequency contingency table. Each expected frequency is placed below its corresponding observed frequency. Expected frequencies are calculated using the formula $E = (R \cdot C)/n$.

The same information about the Os and Es is presented in the table below the contingency table. Column 4 of this table is useful for Step 6.

Step 3:   Assumptions 1 and 2 are satisfied since all expected frequencies are at least 5.

Step 4: $\alpha = 0.01$

Step 5:   Critical value = 16.812

Step 6: $\chi^2 = 77.693$ (See column 4 of the table below the contingency table.)

Note: Answers may vary slightly due to rounding.

Step 7:   Since 77.693 > 16.812, reject H_0.

Step 8:   The data provide sufficient evidence to conclude that response and educational level are associated.

For the P-value approach, $P(\chi^2 > 77.693) < 0.005$. Since the P-value is smaller than the significance level, reject H_0.

16. Using Minitab, click in the Session window, choose **Editor ▶ Enable Command Language** if it is not already enabled, and type in the Session window after the MTB> prompt, a percent sign (%) followed by the path for *fittest.mac*. If *Datadisk* is in drive a, we type

%a:\macro\fittest.mac and press the ⎣ENTER⎦ key. Then proceed as follows: type 6 and press the ⎣ENTER⎦ key in response to Enter the number of possible values for the variable under consideration; type

0.248 0.300 0.187 0.062 0.131 0.072 and press the ⎣ENTER⎦ key in response to Enter the relative frequencies (or probabilities) for the null hypothesis; and type 92 168 86 36 79 39 and press the ⎣ENTER⎦ key in response to Enter the observed frequencies. The resulting output is

    X-square goodness-of-fit test

    Row   n   k   ChiSq   P-value

     1   500   6   14.8586   0.0109841

17. (a) 500 (b) 14.8586 (c) 0.0109841 (d) 0.0109841

(e) Since the P-value is less than 0.05, reject the null hypothesis and conclude that the educational attainment distribution for adults 25 years old and over this year is different from the 1990 distribution.

18. (a) With the data from Problem 12 in three columns of Minitab named STATE, REGION, and PARTY, choose **Stat ▶ Tables ▶ Cross tabulation...**, specify REGION and PARTY in the **Classification variables** text box, select the **Counts** check box in the **Display** list, and click **OK**. The resulting printout is

	D	I	R	All
MW	3	0	9	12
NE	2	1	6	9
SO	6	0	10	16
WE	6	0	7	13
All	17	1	32	50

Cell Contents –

Count

(b) To determine the conditional distribution of REGION within PARTY and the marginal distribution of REGION, we proceed as in part (a) except that we select the **Column Percents** check box instead of the **Counts** check box (and unselect the **Counts** check box). The results are shown in the following table.

	D	I	R	All
MW	17.65	--	28.12	24.00
NE	11.76	100.00	18.75	18.00
SO	35.29	--	31.25	32.00
WE	35.29	--	21.87	26.00
All	100.00	100.00	100.00	100.00

Cell Contents -

% of Col

(c)   To determine the conditional distribution of PARTY within REGION and the marginal distribution of PARTY, we proceed as in part (b) except that we select the **Row Percents** check box instead of the **Column Percents** check box (and unselect the **Column Percents** check box). The results are shown in the following table.

	D	I	R	All
MW	25.00	--	75.00	100.00
NE	22.22	11.11	66.67	100.00
SO	37.50	--	62.50	100.00
WE	46.15	--	53.85	100.00
All	34.00	2.00	64.00	100.00

Cell Contents -

% of Row

19.   With the data in columns named 'FAVOR', 'OPPOSE', and 'NOOPIN', choose **Stat ▶ Tables ▶ Chisquare test...**, select 'FAVOR', 'OPPOSE', and 'NOOPIN' as the variables in the **Columns containing the table** text box, and click **OK**. The result is

Expected counts are printed below observed counts

	FAVOR	OPPOSE	NOOPIN	TOTAL
1	264	17	6	287
	229.15	38.50	19.35	
2	205	26	7	238
	190.03	31.93	16.04	
3	461	81	34	576
	459.90	77.28	38.83	
4	290	81	56	427
	340.93	57.29	28.78	
Total	1220	205	103	1528

```
Chisq = 5.300 + 12.010 +  9.207 +
        1.180 +  1.102 +  5.097 +
        0.003 +  0.179 +  0.600 +
        7.608 +  9.185 + 25.735 = 77.837
```

df = 6, p = 0.000

20. (a) The number of people sampled who would be opposed to the New Jersey Supreme Court ruling in their state is 205. This number occurs at the intersection of the column labeled "OPPOSE" and the row labeled "Total."

(b) The number of people sampled who had some college is 238. This number occurs at the intersection of the row labeled "2" and the column labeled "Total."

(c) The observed number of people sampled who had some college and would be opposed to such a ruling in their state is 26. This number is the *top* number of the two occurring at the intersection of the row labeled "2" and the column labeled "OPPOSE."

(d) The expected number of people sampled who had some college and would be opposed to such a ruling in their state is 31.93. This number is the *bottom* number of the two occurring at the intersection of the row labeled "2" and the column labeled "OPPOSE."

(e) The X-square subtotal 1.102 (located at the intersection of the second row and second column of the X-square statistic) is the subtotal for the "Some college *and* Oppose" cell (located at the intersection of the second row and the second column of the contingency table).

(f) The number of degrees of freedom is 6. This is the last item presented in the printout; i.e., df = 6.

(g) $\chi^2 = 77.837$

(h) P-value = 0.000

(i) We are given $\alpha = 0.01$. Also, df = 6 from part (f). The critical value of X-square is 16.812. Since 77.837 > 16.812, reject H_0. The data provide sufficient evidence to conclude that response and educational level are associated.

CHAPTER 14 ANSWERS

Exercises 14.1

14.1 (a) $y = b_0 + b_1x$ (b) Constants are b_0, b_1; variables are x, y

    (c) The independent variable is x and the dependent variable is y.

14.3 (a) b_0 is the y-intercept; it is the value of y where the line crosses the y-axis.

    (b) b_1 is the slope; it indicates the change in the value of y for every 1 unit increase in the value of x.

14.5 (a) $y = 45.90 + 0.25x$ (b) $b_0 = 45.90$, $b_1 = 0.25$

    (c)   (d)

Miles x	Cost (\$) y
50	$45.90 + 0.25(50) = 58.40$
100	$45.90 + 0.25(100) = 70.90$
250	$45.90 + 0.25(250) = 108.40$

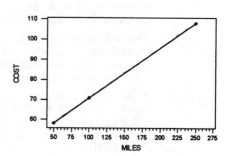

    (e) The visual cost estimate of driving the car 150 miles is about \$85. The exact cost is

$$y = 45.90 + 0.25(150) = \$83.40.$$

14.7 (a) $b_0 = 32$, $b_1 = 1.8$

    (b)   (c)

x (°C)	y (°F)
-40	$32 + 1.8(-40) = -40$
0	$32 + 1.8(0) = 32$
20	$32 + 1.8(20) = 68$
100	$32 + 1.8(100) = 212$

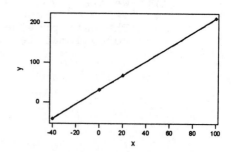

    (d) The visual Fahrenheit temperature estimate corresponding to a Celsius temperature of 28° is about 80°F. The exact temperature is $y = 32 + 1.8(28) = 82.4°F$.

14.9 (a) $b_0 = 45.90$, $b_1 = 0.25$

    (b) The y-intercept $b_0 = 45.90$ gives the y-value at which the straight line $y = 45.90 + 0.25x$ intersects the y-axis. The slope $b_1 = 0.25$

338

indicates that the y-value increases by 0.25 units for every increase in x of one unit.

(c)   The y-intercept $b_0 = 45.90$ is the cost (in dollars) for driving the car zero miles. The slope $b_1 = 0.25$ represents the fact that the cost per mile is \$0.25; it is the amount the total cost increases for each additional mile driven.

14.11 (a) $b_0 = 32$, $b_1 = 1.8$

(b)   The y-intercept $b_0 = 32$ gives the y-value at which the straight line $y = 32 + 1.8x$ intersects the y-axis. The slope $b_1 = 1.8$ indicates that the y-value increases by 1.8 units for every increase in x of one unit.

(c)   The y-intercept $b_0 = 32$ is the Fahrenheit temperature corresponding to 0°C. The slope $b_1 = 1.8$ represents the fact that Fahrenheit temperature increases by 1.8° for every increase of the Celsius temperature of 1°.

14.13 (a) $b_0 = 3$, $b_1 = 4$ **14.15** (a) $b_0 = 6$, $b_1 = -7$

(b)   slopes upward   (b)   slopes downward

(c)   (c)

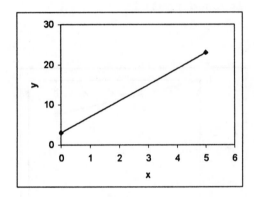   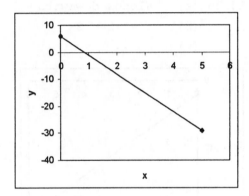

14.17 (a) $b_0 = -2$, $b_1 = 0.5$ **14.19** (a) $b_0 = 2$, $b_1 = 0$

(b)   slopes upward   (b) horizontal

(c)   (c)

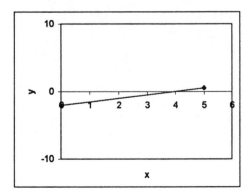   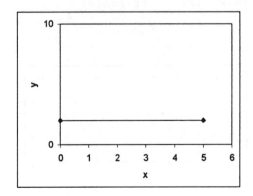

14.21 (a) $b_0 = 0$, $b_1 = 1.5$

      (b)   slopes upward

      (c)

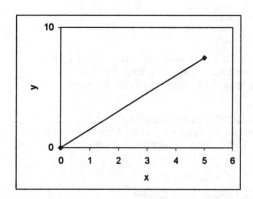

14.23 (a)   slopes upward

      (b)   y = 5 + 2x

      (c)

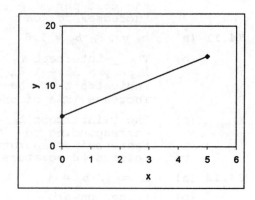

14.25 (a)   slopes downward

      (b) $y = -2 - 3x$

      (c)

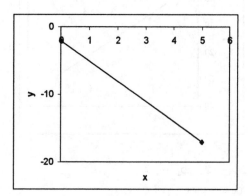

14.27 (a)   slopes downward

      (b) $y = -0.5x$

      (c)

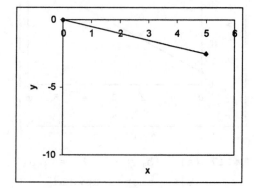

14.29 (a)   horizontal

      (b) $y = 3$

      (c)

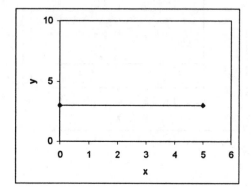

14.31 (a)   If we can express a straight line in the form $y = b_0 + b_1 x$, then there will be one and only one y-value corresponding to each x-value.   However, that is not the case for a vertical straight line; one value of x results in an infinite number of y-values.

(b)   The form of the equation of a vertical straight line is $x = x_0$, where x_0 is the x-coordinate of the vertical straight line.

(c)   For a linear equation, the slope indicates how much the y-value on the straight line increases (or decreases) when the x-value increases by one unit.   We cannot apply this concept to a vertical straight line, since the x-value is *not* permitted to change. Thus, a vertical straight line has no slope.

Exercises 14.2

14.33 (a)   The criterion used to decide on the line that best fits a set of data points is called the least squares criterion.

(b)   The criterion is that the straight line that best fits a set of data points is the one that has the smallest possible sum of the squares of the errors (errors are the differences between and actual and predicted y values).

14.35 (a)   The dependent variable is called the response variable.

(b)   The independent variable is called the predictor variable or the explanatory variable.

14.37 (a)   outlier

(b)   influential observation

14.39 (a)   Line A: $y = 3 - 0.6x$ Line B: $y = 4 - x$

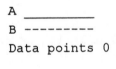

A _____
B - - - - - - - - -
Data points 0

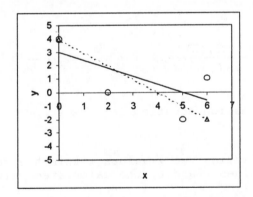

(b)

| Line A: $y = 3 - 0.6x$ | | | | | | Line B: $y = 4 - x$ | | | | |
x	y	$\hat{y}$	e	e^2		x	y	$\hat{y}$	e	e^2
0	4	3.0	1.0	1.00		0	4	4	0	0
2	2	1.8	0.2	0.04		2	2	2	0	0
2	0	1.8	-1.8	3.24		2	0	2	-2	4
5	-2	0.0	-2.0	4.00		5	-2	-1	-1	1
6	1	-0.6	1.6	2.56		6	1	-2	3	9
				10.84						14

(c)   According to the least-squares criterion, Line A fits the set of data points better than Line B.   This is because the sum of squared errors, i.e., Σe^2, is smaller for Line A than for Line B.

14.41 (a)   Formulas for the slope (b_1) and intercept (b_0) of the regression equation are, respectively:

$$b_1 = \frac{\Sigma xy - (\Sigma x)(\Sigma y)/n}{\Sigma x^2 - (\Sigma x)^2/n} \quad \text{and} \quad b_0 = \frac{1}{n}(\Sigma y - b_1 \Sigma x).$$

To compute b_0 and b_1, construct a table of values for x, y, xy, x^2, and their sums:

x	y	xy	x^2
0	4	0	0
2	2	4	4
2	0	0	4
5	-2	-10	25
6	1	6	36
15	5	0	69

Thus: $b_1 = \dfrac{0 - (15)(5)/5}{69 - (15)^2/5} = -0.625$ $b_0 = \dfrac{1}{5}(5 - (-0.625)(15)) = 2.87$

and the regression equation is: $\hat{y} = 2.875 - 0.625x$.

(b)   Begin by selecting two x-values within the range of the x-data. For the x-values 0 and 6, the calculated values for y are, respectively:

$$\hat{y} = 2.875 - 0.625(0) = 2.875$$

$$\hat{y} = 2.875 - 0.625(6) = -0.875 \quad .$$

The regression equation can be graphed by plotting the pairs (0, 2.875) and (6,-0.875) and connecting these points with a straight line. This equation and the original set of five data points are presented as follows:

14.43 (a) Formulas for the slope (b_1) and intercept (b_0) of the regression equation are, respectively:

$$b_1 = \frac{\Sigma xy - (\Sigma x)(\Sigma y)/n}{\Sigma x^2 - (\Sigma x)^2/n} \quad and \quad b_0 = \frac{1}{n}(\Sigma y - b_1 \Sigma x).$$

To compute b_0 and b_1, construct a table of values for x, y, xy, x^2, and their sums:

Thus,

$$b_1 = \frac{122,224 - (761)(1761)/11}{52,729 - (761)^2/11} = 4.83631$$

$$b_0 = \frac{1}{11}(1761 - 4.83631(761)) = -174.494$$

and the regression equation is:

$$\hat{y} = -174.494 + 4.83631x .$$

x	y	xy	x^2
65	175	11,375	4,225
67	133	8,911	4,489
71	185	13,135	5,041
71	163	11,573	5,041
66	126	8,316	4,356
75	198	14,850	5,625
67	153	10,251	4,489
70	163	11,410	4,900
71	159	11,289	5,041
69	151	10,419	4,761
69	155	10,695	4,761
761	1761	122,224	52,729

(b) Begin by selecting two x-values within the range of the x-data. For the x-values 65 and 75, the calculated values for y are, respectively:

$\hat{y} = -174.494 + 4.83631(65) = 139.866$

$\hat{y} = -174.494 + 4.83631(75) = 188.229.$

The regression equation can be graphed by plotting the pairs (65, 139.866) and (75, 188.229) and connecting these points with a straight line. This equation and the original set of 11 data points are presented as follows:

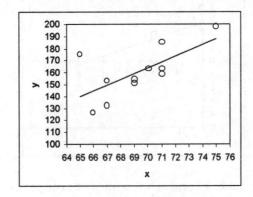

(c)   Weight tends to increase as height increases.

(d)   The weights of 18-24 year-old males increase an estimated 4.84 lb for each increase in height of one inch.

(e)   Weight of 67-inch male: $\hat{y} = -174.494 + 4.83631(67) = 149.539$ lb

Weight of 73-inch male: $\hat{y} = -174.494 + 4.83631(73) = 178.557$ lb

(f)   The predictor variable is height. The response variable is weight.

(g)   The observation (65, 175) is an outlier; (75, 198) is a potential influential observation.

14.45 (a)   Formulas for the slope (b_1) and intercept (b_0) of the regression equation are, respectively:

$$b_1 = \frac{\Sigma xy - (\Sigma x)(\Sigma y)/n}{\Sigma x^2 - (\Sigma x)^2/n} \quad and \quad b_0 = \frac{1}{n}(\Sigma y - b_1 \Sigma x).$$

To compute b_0 and b_1, construct a table of values for x, y, xy, x^2, and their sums:

x	y	xy	x^2
30	186	5,580	900
38	183	6,954	1,444
41	171	7,011	1,681
38	177	6,726	1,444
29	191	5,539	841
39	177	6,903	1,521
46	175	8,050	2,116
41	176	7,216	1,681
42	171	7,182	1,764
24	196	4,704	576
368	1803	65,865	13,968

Thus,
$$b_1 = \frac{65,865 - (368)(1803)/10}{13,968 - (368)^2/10} = -1.1405$$

$$b_0 = \frac{1}{10}(1803 - (-1.1405)(368)) = 222.271$$

and the regression equation is: $\hat{y} = 222.271 - 1.1405x$.

(b) Begin by selecting two x-values within the range of the x-data. For the x-values 29 and 46, the calculated values for y are, respectively:

$$\hat{y} = 222.271 - 1.1405(24) = 194.899$$
$$\hat{y} = 222.271 - 1.1405(46) = 169.808.$$

The regression equation can be graphed by plotting the pairs (24, 194.899) and (46, 169.808) and connecting these points with a straight line. This equation and the original set of 10 data points are presented as follows:

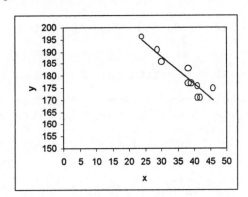

(c) Peak heart rate tends to decrease as age increases.

(d) The peak heart rate an individual can reach during intensive exercise decreases by an estimated 1.14 for each increase in age of one year.

(e) Peak heart rate of 28-year old person: $\hat{y} = 222.271 - 1.1405(28) = 190.34$.

(f) The predictor variable is age. The response variable is peak heart rate.

(g) There are no outliers or potential influential observations.

14.47 (a) Formulas for the slope (b_1) and intercept (b_0) of the regression equation are, respectively:

$$b_1 = \frac{\Sigma xy - (\Sigma x)(\Sigma y)/n}{\Sigma x^2 - (\Sigma x)^2/n} \quad and \quad b_0 = \frac{1}{n}(\Sigma y - b_1 \Sigma x).$$

To compute b_0 and b_1, construct a table of values for x, y, xy, x^2, and their sums:

x	y	xy	x^2
30	55	1,650	900
36	60	2,160	1,296
27	42	1,134	729
20	40	800	400
16	37	592	256
24	26	624	576
19	39	741	361
25	43	1,075	625
197	342	8,776	5,143

Thus,

$$b_1 = \frac{8776 - (197)(342)/8}{5143 - (197)^2/8} = 1.2137$$

$$b_0 = \frac{1}{8}(342 - (1.2137)(197)) = 12.8625$$

and the regression equation is: $\hat{y} = 12.8625 + 1.2137x$.

(b)   Begin by selecting two x-values within the range of the x-data. For the x-values 16 and 36, the calculated values for y are, respectively:

$$\hat{y} = 12.8625 + 1.2137(16) = 32.2817$$
$$\hat{y} = 12.8625 + 1.2137(36) = 56.5557.$$

The regression equation can be graphed by plotting the pairs (16, 32.2817) and (36, 56.5557) and connecting these points with a straight line. This equation and the original set of eight data points are presented as follows:

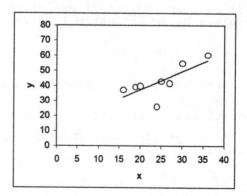

(c)   Annual food expenditure tends to increase as disposable income increases.

(d)   Annual food expenditures increase an estimated $121.37 (1.2137 hundred dollars) for each increase in disposable income of $1000.

(e) Annual food expenditure of a family with a disposable income of $25,000:

$$\hat{y} = 12.8625 + 1.2137(25) = 43.2051 = \$4,320.51.$$

(f) The predictor variable is disposable income. The response variable is annual food expenditure.

(g) The observation (24, 26) is an outlier; there are no potential influential observations.

14.49 The idea behind finding a regression line is based on the assumption that the data points are actually scattered about a straight line. Only the second data set appears to be scattered about a straight line. Thus, it is reasonable to determine a regression line only for the second set of data.

14.51 (a) It is acceptable to use the regression equation to predict the weight of an 18-24-year-old male who is 68 inches tall since that height lies within the range of the heights in the sample data. It is not acceptable (and would be extrapolation) to use the regression equation to predict the weight of an 18-24-year-old male who is 60 inches tall since that height lies outside the range of the heights in the sample data.

(b) It is reasonable to use the regression equation to predict weight for heights between 65 and 75 inches, inclusive.

14.53

x	y	$x-\overline{x}$	$y-\overline{y}$	$(x-\overline{x})(y-\overline{y})$
0	4	-3	3	-9
2	2	-1	1	-1
2	0	-1	-1	1
5	-2	2	-3	-6
6	1	3	0	0
15	5	0	0	-15

$$s_{xy} = \frac{\sum(x-\overline{x})(y-\overline{y})}{n-1}$$

$$= \frac{-15}{4} = -3.75$$

14.55 The equations in Formula (2) are repeated as follows:

$$b_1 = s_{xy}/s_x^2; \quad b_0 = \overline{y} - b_1\overline{x} \quad .$$

The equation for s_{xy} is found in Formula (1). This and the (familiar) equation for s_x^2 are presented as follows:

$$s_{xy} = \frac{\sum (x-\overline{x})(y-\overline{y})}{n-1} \quad ; \quad s_x^2 = \frac{\sum (x-\overline{x})^2}{n-1} \quad .$$

The column manipulations required to calculate s_{xy} for the data in Exercise 14.39 were presented in Exercise 14.53. These are repeated as the first five columns of the following table. The sixth column of the following table presents the calculations for s_x^2. Thus, $s_{xy} = -3.75$ and $s_x^2 = 6$.

x	y	$x-\overline{x}$	$y-\overline{y}$	$(x-\overline{x})(y-\overline{y})$	$(x-\overline{x})^2$
0	4	-3	3	-9	9
2	2	-1	1	-1	1
2	0	-1	-1	1	1
5	-2	2	-3	-6	4
6	1	3	0	0	9
15	5	0	0	-15	24

Substituting these last two calculations into the equation for b_1 presented in Formula (2), we get $b_1 = -3.75/6 = -0.625$. Finally, substituting the calculations for b_1, $\overline{x}$, and $\overline{y}$ into the equation for b_0 presented in Formula (2), we get $b_0 = 1 - (-0.625)(3) = 2.875$. Thus, the regression equation using the equations in Formula (2) is:

$$\hat{y} = b_0 + b_1 x = 2.875 - 0.625x.$$

This is the same result obtained in part (a) of Exercise 14.41.

14.57 (a)   Choose **Graph ▶ Plot...**, select FOODEX for the **Y** variable for Graph **1** and INCOME for the **X** variable, and click **OK**. The result is

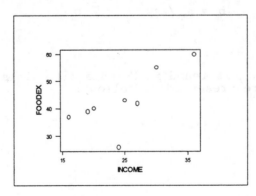

(b) We choose **Stat ▶ Regression ▶ Regression...**, select FOODEX in the **Response** text box, select INCOME in the **Predictors** text box, and click **OK**. The results are

```
The regression equation is
FOODEX = 12.9 + 1.21 INCOME

Predictor     Coef      Stdev  t-ratio          p
Constant     12.86      11.39     1.13      0.302
INCOME      1.2137     0.4493              2.70      0.036

s = 7.676   R-sq = 54.9%      R-sq(adj) = 47.4%

Analysis of Variance

SOURCE              DF         SS         MS         F       p
Regression           1     429.95     429.95      7.30   0.036
Error                6     353.55      58.92
Total                7     783.50
```

Unusual Observations
```
    Obs.      x       y     Fit   Stdev.Fit   Residual   St.Resid
      6    24.0   26.00   41.99        2.73     -15.99      -2.23R
```

R denotes an obs. with a large st. resid.

(c) The potential outlier is (24, 26).

(d) From part (a), the data points appear to be scattered about a straight line; from part (c), there is only a mild outlier and there are no influential observations. Of course, one should investigate the cause of the outlier, and remove it if deemed appropriate.

14.59 (a)

$$\hat{y} = b_0 + b_1 x = 2152.3 + 60.82x.$$

(b) $\hat{y} = 2152.3 + 60.82(17) = 3186.24$ g

(c) The observation (24, 2800) is a potential outlier.

Exercises 14.3

14.61 (a) The coefficient of determination is a descriptive measure of the utility of the regression equation for making predictions. The symbol for the coefficient of determination is r^2.

(b) Two interpretations of the coefficient of determination are:

(i) It represents the percentage reduction obtained in the total squared error by using the regression equation to predict the observed values, instead of simply using the mean $\overline{y}$.

(ii) It represents the percentage of variation in the observed y-values that is explained by the regression.

14.63 (a) The coefficient of determination is $r^2 = SSR/SST = 7626.6/8291.0 = 0.9199$. 91.99% of the variation in the observed values of the response variable is explained by the regression. The fact that r^2 is near 1 indicates that the regression equation is extremely useful for making predictions.

(b)   SSE = SST - SSR = 8291.0 - 7626.6 = 664.4

14.65 To use the defining formulas, we begin with the following table.

x	y	$(y-\overline{y})$	$(y-\overline{y})^2$	$\hat{y}$	$(\hat{y}-\overline{y})$	$(\hat{y}-\overline{y})^2$	$(y-\hat{y})$	$(y-\hat{y})^2$
0	4	3	9	2.875	1.875	3.51563	1.125	1.26563
2	2	1	1	1.625	0.625	0.39063	0.375	0.14063
2	0	−1	1	1.625	0.625	0.39063	−1.625	2.64062
5	−2	−3	9	−0.250	−1.250	1.56250	−1.750	3.06250
6	1	0	0	−0.875	−1.875	3.51563	1.875	3.51563
15	5	0	20		0	9.375	0	10.625

Each $\hat{y}$ value is obtained by substituting the respective value of x into the regression equation $\hat{y} = 2.875 - 0.625x$. (This equation was derived in Exercise 14.41).

(a) $SST = \sum(y-\overline{y})^2 = 20$; $SSR = \sum(\hat{y}-\overline{y})^2 = 9.375$; $SSE = \sum(y-\hat{y})^2 = 10.625$

(b) $r^2 = 1 - SSE/SST = 1 - 10.625/20 = 0.46875$
       $r^2 = SSR/SST = 9.375/20 = 0.46875$

(c)   The percentage of variation in the observed y-values that is explained by the regression is 46.875%.

(d)   Based on the answer in part (c), the regression equation appears to be moderately useful for making predictions.

14.67 To use the shortcut formulas, we begin with the following table. Notice that columns 1-4 are presented in the solution to part (a) of Exercise 14.43, so that only the column for y^2 needs to be developed.

x	y	xy	x^2	y^2
65	175	11,375	4,225	30,625
67	133	8,911	4,489	17,689
71	185	13,135	5,041	34,225
71	163	11,573	5,041	26,569
66	126	8,316	4,356	15,876
75	198	14,850	5,625	39,204
67	153	10,251	4,489	23,409
70	163	11,410	4,900	26,569
71	159	11,289	5,041	25,281
69	151	10,419	4,761	22,801
69	155	10,695	4,761	24,025
761	1761	122,224	52,729	286,273

(a)   Using the last row of the table and Formula 14.2 of the text, we obtain the three sums of squares as follows.

$$SST = S_{yy} = \Sigma y^2 - (\Sigma y)^2/n = 286,273 - (1761)^2/11 = 4352.91$$

$$SSR = \frac{S_{xy}^2}{S_{xx}} = \frac{[\Sigma xy - (\Sigma x)(\Sigma y)/n]^2}{\Sigma x^2 - (\Sigma x)^2/n} = \frac{[122,224 - (761)(1761)/11]^2}{52,729 - (761)^2/11} = 1909.46$$

$$SSE = S_{yy} - \frac{S_{xy}^2}{S_{xx}} = 4352.91 - 1909.47 = 2443.44$$

(b) $r^2 = SSR/SST = 1909.46/4352.91 = 0.4387$

(c)   The percentage of variation in the observed y-values that is explained by the regression is 43.9%. In words, 43.9% of the variation in the weight data is explained by height.

(d)   Based on the answers to parts (b) and (c), the regression equation appears to be moderately useful for making predictions.

14.69 To use the shortcut formulas, we begin with the following table. Notice that columns 1-4 are presented in the solution to part (a) of Exercise 14.45, so that only the column for y^2 needs to be developed.

x	y	xy	x^2	y^2
30	186	5,580	900	34,596
38	183	6,954	1,444	33,489
41	171	7,011	1,681	29,241
38	177	6,726	1,444	31,329
29	191	5,539	841	36,481
39	177	6,903	1,521	31,329
46	175	8,050	2,116	30,625
41	176	7,216	1,681	30,976
42	171	7,182	1,764	29,241
24	196	4,704	576	38,416
368	1803	65,865	13,968	325,723

(a)   Using the last row of the table and Formula 14.2 of the text, we obtain the three sums of squares as follows.

$$SST = S_{yy} = \Sigma y^2 - (\Sigma y)^2/n = 325,723 - (1803)^2/10 = 642.1$$

$$SSR = \frac{S^2_{xy}}{S_{xx}} = \frac{[\Sigma xy - (\Sigma x)(\Sigma y)/n]^2}{\Sigma x^2 - (\Sigma x)^2/n} = \frac{[65,865 - (368)(1803)/10]^2}{13,968 - (368)^2/10} = 553.6$$

$$SSE = S_{yy} - \frac{S^2_{xy}}{S_{xx}} = 642.1 - 553.6 = 88.5$$

(b) $r^2 = SSR/SST = 553.6/642.1 = 0.86217$

(c)   The percentage of variation in the observed y-values that is explained by the regression is 86.2%. In words, 86.2% of the variation in the peak-heart-rate data is explained by age.

(d)   Based on the answers to parts (b) and (c), the regression equation appears to be very useful for making predictions.

14.71 To use the shortcut formulas, we begin with the following table. Notice that columns 1-4 are presented in the solution to part (a) of Exercise 14.47, so that only the column for y^2 needs to be developed.

x	y	xy	x^2	y^2
30	55	1,650	900	3,025
36	60	2,160	1,296	3,600
27	42	1,134	729	1,764
20	40	800	400	1,600
16	37	592	256	1,369
24	26	624	576	676
19	39	741	361	1,521
25	43	1,075	625	1,849
197	342	8,776	5,143	15,404

(a)   Using the last row of the table and Formula 14.2 of the text, we obtain the three sums of squares as follows.

$$SST = S_{yy} = \Sigma y^2 - (\Sigma y)^2/n = 15,404 - (342)^2/8 = 783.5$$

$$SSR = \frac{S^2_{xy}}{S_{xx}} = \frac{[\Sigma xy - (\Sigma x)(\Sigma y)/n]^2}{\Sigma x^2 - (\Sigma x)^2/n} = \frac{[8776 - (197)(342)/8]^2}{5143 - (197)^2/8} = 429.95$$

$$SSE = S_{yy} - \frac{S_{xy}^2}{S_{xx}} = 783.5 - 429.95 = 353.55$$

(b) $r^2 = SSR/SST = 429.95/783.5 = 0.5488$

(c)   The percentage of variation in the observed y-values that is explained by the regression is 54.9%. In words, 54.9% of the variation in food-expenditures is explained by disposable income.

(d)   Based on the answers to parts (b) and (c), the regression equation appears to be moderately useful for making predictions.

14.73 If $r^2 = 0$ for a data set:

(a)   SSE = SST   (b)   SSR = 0

(c)   The regression equation is totally useless for making predictions.

14.74 (a) $r^2 = 1 - SSE/SST = (SST - SSE)/SST$. If the mean were used to predict the observed values of the response variable, the total squared error would be SST. By using the regression line to predict the observed values of the response variable, the sum of squares of the differences between the predicted values and the mean is SSR = SST - SSE. This is a reduction in the error sum of squares. Dividing this quantity by SST (and converting to a percentage) gives us the percentage reduction in the squared error when we use the regression equation instead of the mean to predict the observed values of the response variable.

(b)   From Exercise 14.66, $r^2 = 0.9367$, so the percentage reduction obtained in the total squared error by using the regression equation instead of the mean of observed prices to predict the observed prices is 93.67%.

(c)   From Exercise 14.67, $r^2 = 0.4387$, so the percentage reduction obtained in the total squared error by using the regression equation instead of the mean of observed weights to predict the observed weights is 43.87%.

14.75 We choose **Stat ▶ Regression ▶ Regression...**, select FOODEX in the **Response** text box, select INCOME in the **Predictors** text box, and click **OK**. The results are

The regression equation is
FOODEX = 12.9 + 1.21 INCOME

Predictor	Coef	Stdev	t-ratio	p
Constant	12.86	11.39	1.13	0.302
INCOME	1.2137	0.4493	2.70	0.036

s = 7.676   R-sq = 54.9%   R-sq(adj) = 47.4%

Analysis of Variance

SOURCE	DF	SS	MS	F	p
Regression	1	429.95	429.95	7.30	0.036
Error	6	353.55	58.92		
Total	7	783.50			

From the output:

$r^2 = 0.549$; SST = 783.50; SSR = 429.95; SSE = 353.55 .

14.77 (a) $r^2 = 0.372$

(b)   SSR = 2,505,745; SSE = 4,234,255; SST = 6,740,000

(c)   37.2%

Exercises 14.4

14.79 The linear correlation coefficient is a descriptive measure of the strength of linear relationship between two variables.

14.81 (a)   r   (b)   strong   (c)   zero

14.83 (a)   positively (b)   negatively (c)   uncorrelated

14.85 (a)   Positive. The sign of the slope and of r are always the same.

This can be seen from $r = \dfrac{S_{xy}}{\sqrt{S_{xx}S_{yy}}} = \dfrac{S_{xy}}{S_{xx}} \cdot \dfrac{\sqrt{S_{xx}}}{\sqrt{S_{yy}}} = b_1 \cdot \dfrac{\sqrt{S_{xx}}}{\sqrt{S_{yy}}}$.

Since the ratio of square roots in the last term is always positive, r will have the same sign as b_1.

(b)   The coefficient of determination is $r^2 = 0.846^2 = 0.716$

14.87 To compute the linear correlation coefficient, begin with the following table.

x	y	xy	x^2	y^2
0	4	0	0	16
2	2	4	4	4
2	0	0	4	0
5	-2	-10	25	4
6	1	6	36	1
15	5	0	69	25

The linear correlation coefficient r is computed using the formula in Definition 14.6 of the text:

14.89 To compute the linear correlation coefficient, return to the table presented at the beginning of the solution to Exercise 14.67. Use the last row of this table to perform the calculations in part (a).

(a)   The linear correlation coefficient r is computed using the formula in Definition 14.6 of the text:

$$r = \frac{S_{xy}}{\sqrt{S_{xx}\,S_{yy}}} = \frac{\Sigma xy - (\Sigma x)(\Sigma y)/n}{\sqrt{[\Sigma x^2 - (\Sigma x)^2/n][\Sigma y^2 - (\Sigma y)^2/n]}}$$

$$= \frac{122{,}224 - (761)(1761)/11}{\sqrt{[52{,}729 - (761)^2/11][286{,}273 - (1761)^2/11]}} = 0.662319 \ .$$

(b)   The value of r in part (a) suggests a moderately strong positive linear correlation.

(c)   Data points are clustered moderately closely about the regression line.

(d) $r^2 = (0.662319)^2 = 0.4387$. This matches the coefficient of determination that was calculated in part (b) of Exercise 14.67.

14.91 To compute the linear correlation coefficient, return to the table presented at the beginning of the solution to Exercise 14.69. Use the last row of this table to perform the calculations in part (a).

(a) The linear correlation coefficient r is computed using the formula in Definition 14.6 of the text:

$$r = \frac{S_{xy}}{\sqrt{S_{xx}\,S_{yy}}} = \frac{\Sigma xy - (\Sigma x)(\Sigma y)/n}{\sqrt{[\Sigma x^2 - (\Sigma x)^2/n][\Sigma y^2 - (\Sigma y)^2/n]}}$$

$$= \frac{65,865 - (368)(1803)/10}{\sqrt{[13,968 - (368)^2/10][325,723 - (1803)^2/10]}} = -0.928536 \quad .$$

(b) The value of r in part (a) suggests a strong negative linear correlation.

(c) Data points are clustered closely about the regression line.

(d) $r^2 = (-0.928536)^2 = 0.86217$. This matches the coefficient of determination that was calculated in part (b) of Exercise 14.69.

14.93 To compute the linear correlation coefficient, return to the table presented at the beginning of the solution to Exercise 14.71. Use the last row of this table to perform the calculations in part (a).

(a) The linear correlation coefficient r is computed using the formula in Definition 14.6 of the text:

$$r = \frac{S_{xy}}{\sqrt{S_{xx}\,S_{yy}}} = \frac{\Sigma xy - (\Sigma x)(\Sigma y)/n}{\sqrt{[\Sigma x^2 - (\Sigma x)^2/n][\Sigma y^2 - (\Sigma y)^2/n]}}$$

$$= \frac{8776 - (197)(342)/8}{\sqrt{[5143 - (197)^2/8][15,404 - (342)^2/8]}} = 0.740785 \quad .$$

(b) The value of r in part (a) suggests a moderately strong positive linear correlation.

(c) Data points are clustered relatively closely about the regression line.

(d) $r^2 = (0.740785)^2 = 0.5488$. This matches the coefficient of determination that was calculated in part (b) of Exercise 14.71.

14.95 To compute the linear correlation coefficient, begin with the following table.

x	y	xy	x^2	y^2
-3	9	-27	9	81
-2	4	-8	4	16
-1	1	-1	1	1
0	0	0	0	0
1	1	1	1	1
2	4	8	4	16
3	9	27	9	81
0	28	0	28	196

(a)   The linear correlation coefficient r is computed using the formula in Definition 14.6 of the text:

$$r = \frac{S_{xy}}{\sqrt{S_{xx}\,S_{yy}}} = \frac{\Sigma xy - (\Sigma x)(\Sigma y)/n}{\sqrt{[\Sigma x^2 - (\Sigma x)^2/n][\Sigma y^2 - (\Sigma y)^2/n]}}$$

$$= \frac{0 - (0)(28)/7}{\sqrt{[28 - (0)^2/7][196 - (28)^2/7]}} = 0 \quad .$$

(b)   We cannot conclude from the result in part (a) that x and y are unrelated. We can conclude only that there is no *linear* relationship between x and y.

(c)   Graph for part (e)

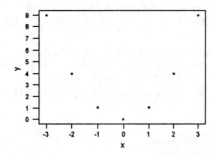   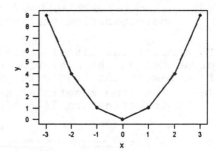

(d)   It is not appropriate to use the linear correlation coefficient as a descriptive measure for the data because the data points are not scattered about a straight line.

(e)   For each data point (x, y), we have $y = x^2$. (See columns 1 and 2 of the previous table.) See the graph above at the right.

14.97 (a)   No. We only know that the linear correlation coefficient is the square root of the coefficient of determination or it is the negative of that square root.

(b)   No. The slope is positive if r is positive and negative if r is negative, but we can not determine the sign of r.

(c)   Yes. $r = -\sqrt{0.716} = -0.846$

(d)   Yes. $r = \sqrt{0.716} = 0.846$

14.99 We choose **Stat ▶ Basic Statistics ▶ Correlation...**, select INCOME and FOODEX in the **Variables** text box, and click **OK**. The result is

Correlation of INCOME and FOODEX = 0.741

REVIEW TEST FOR CHAPTER 13

1.   (a) x   (b) y   (c) b_1 (d) b_0

2.   (a)   It intersects the y-axis when x = 0. Therefore y = 4.

(b)   It intersects the x-axis when y = 0. Therefore x = 4/3.

(c)   slope = -3

    (d)   The y-value decreases by 3 units when x increases by 1 unit.

    (e)   The y-value increases by 6 units when x decreases by 2 units.

3.   (a)   True. The y-intercept is determined by b_0 and that value is independent of b_1, the slope.

    (b)   False. A horizontal line has a slope of zero.

    (c)   True. If the slope is positive, the x-values and y-values increase and decrease together.

4.   Scatterplot or scatter diagram

5.   A regression equation can be used to make predictions of the response variable for specific values of the predictor variable.

6.   (a)   predictor variable or explanatory variable

    (b)   response variable

7.   (a)   smallest

    (b)   regression

    (c)   extrapolation

8.   (a)   An outlier is a data point that lies far from the regression line relative to the other data points.

    (b)   An influential observation is a data point whose removal causes the regression equation to change considerably. Often this is a data point for which the x-value is considerably to the left or right of the rest of the data points.

9.   The coefficient of determination is the percentage of the total variation in the y-values that is explained by the regression equation.

10.   (a)   SST is the total sum of squares and measures the variation in the observed values of the response variable.

    (b)   SSR is the regression sum of squares and measures the variation in the observed values of the response variable that is explained by the regression. It can also be thought of as the variation in the predicted values of the response variable corresponding to the x-values in the data points.

    (c)   SSE is the error sum of squares and measures the variation in the observed values of the response variable that is not explained by the regression.

11.   (a)   linear   (b)   increases   (c)   negative   (d)   zero

12.   True. It is quite possible that both variables are strongly affected by one (or more) other variables (called lurking variables).

13.   (a) $y = 72 - 12x$ (b) $b_0 = 72$, $b_1 = -12$

    (c)   The line slopes downward since $b_1 < 0$.

    (d)   After two years: $y = 72 - 12(2) = \$48$ hundred $= \$4800$

                After five years: $y = 72 - 12(5) = \$12$ hundred $= \$1200$.

(e)

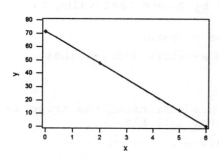

(f)   From the graph, we estimate the value to be about $2500 after 4 years. The actual value is y = 7200 - 1200(4) = $2400.

14.   (a)

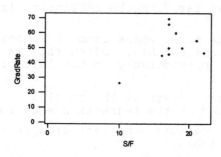

(b)   It is moderately reasonable to find a regression line for the data because the data points appear to be scattered about a straight line. This perception is, however, heavily influenced by the single data point at x = 10.

(c)   The regression equation can be determined by calculating its slope and intercept. Formulas for the slope (b_1) and intercept (b_0) of the regression equation are, respectively:

$$b_1 = \frac{\Sigma xy - (\Sigma x)(\Sigma y)/n}{\Sigma x^2 - (\Sigma x)^2/n} \quad and \quad b_0 = \frac{1}{n}(\Sigma y - b_1 \Sigma x).$$

To compute b_0 and b_1, construct a table of values for x, y, xy, x^2, and their sums. (Note: A column for y^2 is also presented. This is used in Problems 15 and 16.)

x	y	xy	x^2	y^2
16	45	720	256	2,025
20	55	1,100	400	3,025
17	70	1,190	289	4,900
19	50	950	361	2,500
22	47	1,034	484	2,209
17	46	782	289	2,116
17	50	850	289	2,500
17	66	1,122	289	4,356
10	26	260	100	676
18	60	1,080	324	3,600
173	515	9,088	3,081	27,907

Thus,

$$b_1 = \frac{9,088 - (173)(515)/10}{3,081 - (173)^2/10} = 2.02611$$

$$b_0 = \frac{1}{10}(515 - 2.02611(173)) = 16.448$$

and the regression equation is: $\hat{y} = 16.448 + 2.0261x$.

To graph the regression equation, begin by selecting two x-values within the range of the x-data. For the x-values 10 and 22, the calculated values for y are, respectively:

$$\hat{y} = 16.448 + 2.0261(10) = 36.71$$
$$\hat{y} = 16.448 + 2.0261(22) = 61.02 \ .$$

The regression equation can be graphed by plotting the pairs (10, 36.71) and (22, 61.02) and connecting these points with a straight line. This equation and the original set of 10 data points are presented as follows:

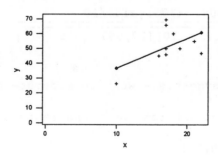

(d)   Graduation rate tends to increase as the student-to-faculty ratio increases.

(e)   Graduation rate increases an estimated 2.026% for each additional 1 unit increase in the student-to-faculty ratio.

(f)   The predicted graduation rate for a university with a student-to-faculty ratio of 17 is 16.448 + 2.0261(17) = 50.89%.

(g)   There is one potential influential observation, (10, 26); no outliers. Looking at the scatterplot in part (c), we suspect that there may be little relationship between the student-to-faculty ratio and the graduation rate if the influential observation is removed from the data.

15.   (a)   To compute SST, SSR, and SSE using the shortcut formulas, begin with the table presented in Problem 14(c). Using the last row of this table and the formula in Definition 14.6 of the text, we obtain the three sums of squares as follows.

$$SST = S_{yy} = \sum y^2 - \left(\sum y\right)^2/n = 27,907 - (515)^2/10 = 1384.50$$

$$SSR = \frac{S_{xy}^2}{S_{xx}} = \frac{\left[\sum xy - \left(\sum x\right)\left(\sum y\right)/n\right]^2}{\sum x^2 - \left(\sum x\right)^2/n} = \frac{[9088 - (173)(515)/10]}{3081 - (173)^2/10} = 361.66$$

$$SSE = S_{yy} - \frac{S_{xy}^2}{S_{xx}} = 1384.5 - 361.66 = 1022.84$$

$$r^2 = SSR/SST = 361.66/1384.5 = 0.261$$

(b)   The percentage reduction obtained in the total squared error by using the regression equation, instead of the sample mean $\bar{y}$, to predict the observed costs is 26.1%.

(c)   The percentage of the variation in the graduation rate that is explained by the student-to-faculty ratio is 26.1%.

(d)   The regression equation appears to be not very useful for making predictions.

16.   (a)   To compute the linear correlation coefficient, begin with the table presented in Problem 14(c). Using the last row of this table and the formula in Definition 14.6 of the text, we get:

$$r = \frac{S_{xy}}{\sqrt{S_{xx}S_{yy}}} = \frac{\sum xy - \left(\sum x\right)\left(\sum y\right)/n}{\sqrt{\left[\sum x^2 - \left(\sum x\right)^2/n\right]\left[\sum y^2 - \left(\sum y\right)^2/n\right]}}$$

$$r = \frac{9088 - (173)(515)/10}{\sqrt{[3,081 - (173)^2/10][27,907 - (515)^2/10]}} = 0.511 \quad .$$

(b)   The value of r suggests a weak to moderate positive linear correlation.

(c)   Data points are clustered about the regression line, but not very closely.

(d) $r^2 = (0.511)^2 = 0.261$

17.   (a)   With the data in two columns named S/F and GRADRATE, we choose
       Graph ▶ Plot..., select GRADRATE for the **Y** variable for **Graph 1**
       and S/F for the **X** variable
       for **Graph 1,** and click **OK.**

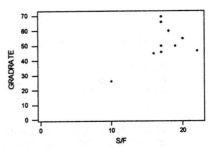

       (b)   We choose **Stat ▶ Regression ▶ Regression...**, select GRADRATE in
              the **Response** text box, select S/F in the **Predictors** text box, and
              click **OK.**   The result is

              The regression equation is
              GRADRATE = 16.4 + 2.03 S/F

Predictor	Coef	StDev	T	P
Constant	16.45	21.15	0.78	0.459
S/F	2.026	1.205	1.68	0.131

              S = 11.31   R-Sq = 26.1%   R-Sq(adj) = 16.9%

              Analysis of Variance

Source	DF	SS	MS	F	P
Regression	1	361.7	361.7	2.83	0.131
Residual Error	8	1022.8	127.9		
Total	9	1384.5			

Unusual Observations

Obs	S/F	GRADRATE	Fit	StDev Fit	Residual	St Resid
9	10.0	26.00	36.71	9.49	-10.71	-1.74 X

X denotes an observation whose X value gives it large influence.

       (c)   From the computer output, r^2 = 26.1%, SST = 1384.5, SSR = 361.7,
              and SSE = 1022.8.

       (d)   There is one potential influential observation, (10, 26).

       (e)   Click on the Data window and move the cursor to the value of 10 in
              the S/F column. Press the DELETE key. Move the cursor to the
              value of 26 in the GRADRATE column and Press the DELETE key. Then
              choose **Stat ▶ Regression ▶ Regression...**, select GRADRATE in the
              Response text box, select S/F in the **Predictors** text box, and
              click **OK.**   The result is

```
The regression equation is
GRADRATE = 72.1 - 0.98 S/F

Predictor        Coef        StDev          T          P
Constant        72.10        32.23       2.24      0.060
S/F             -0.981        1.771      -0.55      0.597

S = 9.518        R-Sq = 4.2%        R-Sq(adj) = 0.0%

Analysis of Variance

Source           DF          SS          MS         F          P
Regression        1        27.79       27.79      0.31      0.597
Residual Error    7       634.21       90.60
Total             8       662.00
```

(f)   The potential influential observation was very influential. For example, the regression equation slope has changed from positive to negative. Even more important is the fact that the coefficient of determination is now only 4.2%, making the regression equation useless for making any predictions at all.

18.   (a)   GRADRATE = 16.4 + 2.03 SFRATIO

      (b)   The coefficient of determination is 26.1%.

      (c)   SSR = 361.7; SSE = 1022.8; SST = 1384.5

      (d)   There is one potential influential observation, (10, 26). No outliers.

19.   We choose **Stat ▶ Basic Statistics ▶ Correlation...**, select S/F and GRADRATE in the **Variables** text box, and click **OK**. The result is

Correlation of S/F and GRADRATE = 0.511, P-Value = 0.131

CHAPTER 15 ANSWERS

Exercises 15.1

15.1 conditional distribution, conditional mean, and conditional standard deviation

15.3 (a) population regression line

(b) σ

(c) normal, $\beta_0 + 6\beta_1$, σ

15.5 The sample regression line is the best estimate of the population regression line.

15.7 residual

15.9 The plot of the residuals against the values of the predictor variable provides the same information as a scatter diagram of the data points. However, it has the advantage of making it easier to spot patterns such as curvature and non-constant standard deviation.

15.11 If the assumptions for regression inferences are satisfied for a model relating an 18-24 year-old male's height to his weight, this means that there are constants β_0, β_1, and σ such that, for each height x, the weights of 18-24 year-old males of that height are normally distributed with mean $\beta_0 + \beta_1 x$ and standard deviation σ.

15.13 If the assumptions for regression inferences are satisfied for a model relating age to peak heart rate, this means that there are constants β_0, β_1, and σ such that, for each age x, the peak heart rates that can be reached during intensive exercise by individuals of that age are normally distributed with mean $\beta_0 + \beta_1 x$ and standard deviation σ.

15.15 If the assumptions for regression inferences are satisfied for a model relating a middle-income family's disposable income to the amount of money that it spends annually on food, this means that there are constants β_0, β_1, and σ such that, for each disposable-income level x, the annual food expenditures made by middle-income families of the same size at that level are normally distributed with mean $\beta_0 + \beta_1 x$ and standard deviation σ.

15.17 To compute the standard error of the estimate, first retrieve the computation for the error sum of squares (SSE) in part (a) of Exercise 14.67 and then apply the formula for s_e in Definition 15.1 of the text.

(a) In part (a) of Exercise 14.67, n = 11 and SSE was computed as 2,443.44. Applying Definition 15.1, the standard error of the estimate is:

$$s_e = \sqrt{\frac{SSE}{n-2}} = \sqrt{\frac{2,443.44}{11-2}} = 16.477.$$

(b) Presuming that the variables height (x) and weight (y) for 18-24 year-old males satisfy Assumptions (1) – (3) for regression inferences, the standard error of the estimate $s_e = 16.477$ lb provides an estimate for the common population standard deviation σ of weights for all 18-24 year-old males of any particular height.

(c)

Height	Residual
x	e
65	35.13
67	-16.54
71	16.12
71	-5.88
66	-18.70
75	9.77
67	3.46
70	-1.05
71	-9.88
69	-8.21
69	-4.21

Residual	Normal score
e	n
-18.70	-1.59
-16.54	-1.06
-9.88	-0.73
-8.21	-0.46
-5.88	-0.22
-4.21	0.00
-1.05	0.22
3.46	0.46
9.77	0.73
16.12	1.06
35.13	1.59

(d)   It appears reasonable to consider the assumptions for regression inferences met for the variables height and weight. However, this is a tough call. There is one potential outlier at (65, 175) which could cast some doubt on the assumptions.

15.19 To compute the standard error of the estimate, first retrieve the computation for the error sum of squares (SSE) in part (a) of Exercise 14.69 and then apply the formula for s_e in Definition 15.1 of the text.

(a)   In part (a) of Exercise 14.69, n = 10 and SSE was computed as 88.5. Applying Definition 15.1, the standard error of the estimate is:

$$s_e = \sqrt{\frac{SSE}{n-2}} = \sqrt{\frac{88.5}{10-2}} = 3.326.$$

(b)   Presuming that the variables age (x) and peak heart rate (y) for individuals satisfy Assumptions (1) - (3) for regression inferences, the standard error of the estimate $s_e = 3.326$ provides an estimate for the common population standard deviation σ of peak heart rates for all individuals of any particular age.

(c)

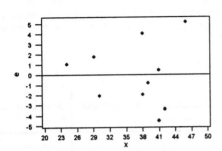

Age	Residual
x	e
30	-2.06
38	4.07
41	-4.51
38	-1.93
29	1.80
39	-0.79
46	5.19
41	0.49
42	-3.37
24	1.10

Residual	Normal score
e	n
-4.51	-1.55
-3.37	-1.00
-2.06	-0.65
-1.93	-0.37
-0.79	-0.12
0.49	0.12
1.10	0.37
1.80	0.65
4.07	1.00
5.19	1.55

(d)   Taking into account the small sample size, we can say that the residuals fall roughly in a horizontal band centered and symmetric about the x-axis.   However, the residual plot casts some doubt on the assumption of equal standard deviations.   We can also say that the normal probability plot for residuals is approximately linear. Therefore, based on the sample data, there are no obvious violations of the assumptions for regression inferences for the variables age and peak heart rate.

15.21 To compute the standard error of the estimate, first retrieve the computation for the error sum of squares (SSE) in part (a) of Exercise 14.71 and then apply the formula for s_e in Definition 15.1 of the text.

(a)   In part (a) of Exercise 14.71, n = 8 and SSE was computed as 353.55. Applying Definition 15.1, the standard error of the estimate is:

$$s_e = \sqrt{\frac{SSE}{n-2}} = \sqrt{\frac{353.55}{8-2}} = 7.6763.$$

(b)   Presuming that the variables disposable income (x) and annual food expenditure (y) for middle-income families of the same size satisfy Assumptions (1) - (3) for regression inferences, the standard error of the estimate s_e = 7.6763 ($767.63) provides an estimate for the common population standard deviation σ of annual food expenditures for all middle-income families of the same size with any particular disposable income.

(c)

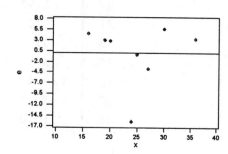

Disposable income	Residual
x	e
30	5.73
36	3.44
27	−3.63
20	2.86
16	4.72
24	−15.99
19	3.08
25	−0.21

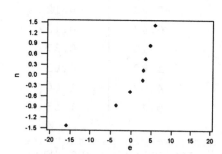

Residual	Normal score
e	n
−15.99	−1.43
−3.63	−0.85
−0.21	−0.47
2.86	−0.15
3.08	0.15
3.44	0.47
4.72	0.85
5.73	1.43

(d)   If the outlier, (24,26), is a legitimate data point, then the assumptions for regression inferences may very well be violated by the variables under consideration; if the outlier is a recording error or can be removed for some other valid reason, then the resulting data reveal no obvious violations of the assumptions for regression inferences (as can be seen by constructing a residual plot and normal probability plot of the residuals for the abridged data).

15.23 (a) The assumption of linearity (Assumption 1) may be violated since the band is not horizontal, as may the assumption of equal standard deviations (Assumption 2) since there is more variation in the residuals for small x than for large x.

(b) It appears that the standard deviation does not remain constant; thus, Assumption 2 is violated.

(c) The graph does not suggest violation of one or more of the assumptions for regression inferences.

(d) The normal probability plot appears to be more curved than linear; thus, the assumption of normality is violated.

15.25 We choose **Stat ▶ Regression ▶ Regression...**, select FOODEX in the **Response** text box, select INCOME in the **Predictors** text box, select **Residuals** from the **Storage** check-box list. Click the **Graphs...** button, select the **Regular** option button from the **Residuals for Plots** list, select the **Normal plot of residuals** check box from the **Residual Plots** list, click in the **Residuals versus the variables** text box and specify INCOME, click **OK**, and click **OK**. The results are

(a) The regression equation is
FOODEX = 12.9 + 1.21 INCOME

Predictor	Coef	Stdev	t-ratio	p
Constant	12.86	11.39	1.13	0.302
INCOME	1.2137	0.4493	2.70	0.036

s = 7.676   R-sq = 54.9%   R-sq(adj) = 47.4%

Analysis of Variance

SOURCE	DF	SS	MS	F	p
Regression	1	429.95	429.95	7.30	0.036
Error	6	353.55	58.92		
Total	7	783.50			

The standard error of the estimate is the first entry in the sixth line of the computer output. It is reported as s = 7.676.

(b) The two graphs which result are shown below.

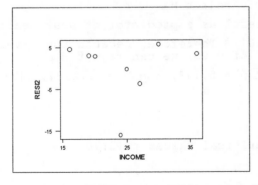 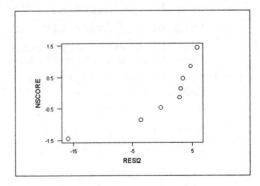

15.27 (a) The standard error of the estimate is the first entry in the sixth line of the computer output. It is reported as s = 382.1.

(b) No; the residual plot falls roughly in a horizontal band centered and symmetric about the x-axis, and the normal probability plot of the residuals is roughly linear.

Exercises 15.2

15.29 normal, β_1

15.31 We can also use the coefficient of determination, r^2, and the linear correlation coefficient, r, as a basis for a test to decide whether a regression equation is useful for prediction.

15.33 From Exercise 14.43, $\Sigma x = 761$, $\Sigma x^2 = 52,729$, and $b_1 = 4.83631$. From Exercise 15.17, $s_e = 16.477$.

Step 1: $H_0: \beta_1 = 0$, $H_a: \beta_1 \neq 0$

Step 2: $\alpha = 0.10$

Step 3:   df = n - 2 = 9;   critical values = ± 1.833

Step 4: $t = \dfrac{b_1}{s_e / \sqrt{\Sigma x^2 - (\Sigma x)^2 / n}} = \dfrac{4.83631}{16.477 / \sqrt{52,729 - (761)^2 / 11}} = 2.65$

Step 5:   Since 2.652 > 1.833, reject H_0.

Step 6:   The data provide sufficient evidence to conclude that the slope of the population regression line is not zero and, hence, that height is useful as a predictor of weight for 18-24 year-old males.

For the P-value approach, $0.02 < P < 0.05$. Therefore, because the P-value is less than the significance level of 0.10, we can reject H_0.

15.35 From Exercise 14.45, $\Sigma x = 368$, $\Sigma x^2 = 13,968$, and $b_1 = -1.1405$. From Exercise 15.19, $s_e = 3.326$.

Step 1: $H_0: \beta_1 = 0$, $H_a: \beta_1 \neq 0$

Step 2: $\alpha = 0.05$

Step 3:   df = n - 2 = 8;   critical values = ± 2.306

Step 4: $t = \dfrac{b_1}{s_e / \sqrt{\Sigma x^2 - (\Sigma x)^2 / n}} = \dfrac{-1.1405}{3.326 / \sqrt{13,968 - (368)^2 / 10}} = -7.07$

Step 5:   Since -7.074 < -2.306, reject H_0.

Step 6:   Evidently, age is useful as a predictor of peak heart rate.

For the P-value approach, $P < 0.01$. Therefore, because the P-value is less than the significance level of 0.05, we can reject H_0.

15.37 From Exercise 14.47, $\Sigma x = 197$, $\Sigma x^2 = 5,143$, and $b_1 = 1.2137$. From Exercise 15.21, $s_e = 7.6763$.

Step 1: $H_0: \beta_1 = 0$, $H_a: \beta_1 \neq 0$

Step 2: $\alpha = 0.01$

Step 3:   df = n - 2 = 6;   critical values = ± 3.707

Step 4: $t = \dfrac{b_1}{s_e / \sqrt{\Sigma x^2 - (\Sigma x)^2 / n}} = \dfrac{1.2137}{7.6763 / \sqrt{5,143 - (197)^2 / 8}} = 2.70$

Step 5:   Since -3.707 < 2.701 < 3.707, do not reject H_0.

Step 6:   The data do not provide sufficient evidence to conclude that disposable income is useful as a predictor of annual food expenditure for middle-income families with a father, mother, and two children.

For the P-value approach, $0.02 < P < 0.05$. Therefore, because the P-value is not less than the significance level of 0.01, we cannot reject H_0.

15.39 From Exercise 14.43, $\Sigma x = 761$, $\Sigma x^2 = 52,729$, and $b_1 = 4.83631$. From Exercise 15.17, $s_e = 16.477$.

(a) Step 1: For a 90% confidence interval, $\alpha = 0.10$. With df = n - 2 = 9, $t_{\alpha/2} = t_{0.05} = 1.833$.

Step 2: The endpoints of the confidence interval for β_1 are

$$b_1 \pm t_{a/2} \cdot s_e / \sqrt{\Sigma x^2 - (\Sigma x)^2 / n}$$

or $4.83631 \pm 1.833 \cdot 16.477 / \sqrt{52,729 - (761)^2 / 11}$

or $4.83631 \pm 1.833 \cdot 1.82363$

or 4.83631 ± 3.34271

or 1.49 to 8.18 *lb*

(b) We can be 90% confident that, for 18-24 year-old males, the increase in mean weight per one inch increase in height is somewhere between 1.49 and 8.18 lb.

15.41 From Exercise 14.45, $\Sigma x = 368$, $\Sigma x^2 = 13,968$, and $b_1 = -1.1405$. From Exercise 15.19, $s_e = 3.326$.

(a) Step 1: For a 95% confidence interval, $\alpha = 0.05$. With df = n - 2 = 8, $t_{\alpha/2} = t_{0.025} = 2.306$.

Step 2: The endpoints of the confidence interval for β_1 are

$$b_1 \pm t_{a/2} \cdot s_e / \sqrt{\Sigma x^2 - (\Sigma x)^2 / n}$$

or $-1.1405 \pm 2.306 \cdot 3.326 / \sqrt{13,968 - (368)^2 / 10}$

or $-1.1405 \pm 2.306 \cdot 0.161222$

or -1.1405 ± 0.3718

or -1.5123 to -0.7687

(b) We can be 95% confident that the decrease in mean peak heart rate per one year increase in age is somewhere between 0.7687 and 1.5123.

15.43 From Exercise 14.47, $\Sigma x = 197$, $\Sigma x^2 = 5,143$, and $b_1 = 1.2137$. From Exercise 15.21, $s_e = 7.6763$.

(a) Step 1: For a 99% confidence interval, $\alpha = 0.01$.

With df = n - 2 = 6, $t_{\alpha/2} = t_{0.005} = 3.707$.

Step 2: The endpoints of the confidence interval for β_1 are

$$b_1 \pm t_{a/2} \cdot s_e / \sqrt{\Sigma x^2 - (\Sigma x)^2 / n}$$

or $1.2137 \pm 3.707 \cdot 7.6763 / \sqrt{5{,}143 - (197)^2 / 8}$

or $1.2137 \pm 3.707 \cdot 0.449318$

or 1.2137 ± 1.6656

or -0.4519 *to* 2.8793 *hundred dollars*

(b)   We can be 99% confident that, for middle-income families with a father, mother, and two children, the change in mean annual food expenditure per \$1,000 increase in family disposable income is somewhere between -\$45.19 and \$287.93.

15.45 We choose **Stat ▶ Regression ▶ Regression...**, select FOODEX in the **Response** text box, select INCOME in the **Predictors** text box, and click **OK**. The results are

The regression equation is
FOODEX = 12.9 + 1.21 INCOME

Predictor	Coef	Stdev	t-ratio	p
Constant	12.86	11.39	1.13	0.302
INCOME	1.2137	0.4493	2.70	0.036

s = 7.676   R-sq = 54.9%   R-sq(adj) = 47.4%

Analysis of Variance

SOURCE	DF	SS	MS	F	p
Regression	1	429.95	429.95	7.30	0.036
Error	6	353.55	58.92		
Total	7	783.50			

The P-value for the slope coefficient is reported in the fifth line of the computer output. Since the P-value of 0.036 is greater than the specified significance level of $\alpha = 0.01$ in Exercise 14.25, we do not reject the null hypothesis.

15.47 (a) $b_1 = 60.82$

(b)   Stdev of $b_1 = 14.68$

(c)   t-ratio = 4.14

(d)   p = 0.000

(e)   Since the P-value of 0.000 is less than the specified significance level of $\alpha = 0.05$, we reject the null hypothesis that the slope of the population regression line is zero. Thus, the data provide sufficient evidence to conclude that estriol level is useful for predicting birth weights.

(f)   Step 1:   For a 95% confidence interval, $\alpha = 0.05$. With n = 31 and df = n - 2 = 29, $t_{\alpha/2} = t_{0.025} = 2.045$.

Step 2:   The endpoints of the confidence interval for β_1 are:

$$b_1 \pm t_{a/2} \cdot s_e / \sqrt{\Sigma x^2 - (\Sigma x)^2/n}$$

or $60.82 \pm 2.045 \cdot 14.68$

or 60.82 ± 30.02

or $30.80 \ to \ 90.84 \ gm$

Thus, we can be 95% confident that the increase in mean birth weight per increase of estriol level by one mg/24 hr is somewhere between 30.80 and 90.84 gm.

(g) In performing the inferences, the four assumptions for regression inferences were made, which are stated in Key Fact 15.1. These assumptions can be checked by obtaining a residual plot and a normal probability plot of the residuals.

Exercises 15.3

15.49 $11,443

15.51 From Exercise 14.43, $\hat{y} = -174.494 + 4.83631x$, $\Sigma x = 761$, and $\Sigma x^2 = 52,729$. From Exercise 15.17, $s_e = 16.477$.

(a) $\hat{y}_p = -174.494 + 4.83631(70) = 164.048$ lb

(b) Step 1:   For a 90% confidence interval, $\alpha = 0.10$.

With df $= n - 2 = 9$, $t_{\alpha/2} = t_{0.05} = 1.833$.

Step 2: $\hat{y}_p = 164.048$

Step 3:   The endpoints of the confidence interval are:

$$\hat{y}_p \pm t_{a/2} \cdot s_e \sqrt{\frac{1}{n} + \frac{(x_p - \Sigma x/n)^2}{\Sigma x^2 - (\Sigma x)^2/n}}$$

or $164.048 \pm 1.833 \cdot 16.477 \sqrt{\dfrac{1}{11} + \dfrac{(70 - 761/11)^2}{52,729 - (761)^2/11}}$

or $164.048 \pm 1.833 \cdot 5.18723$

or 164.048 ± 9.508

or $154.54 \ to \ 173.56$ lb

The interpretation of this interval is as follows: We can be 90% confident that the mean weight of all 18-24 year-old males who are 70 inches tall is somewhere between 154.54 and 173.56 lb.

(c) This is the same as the answer in part (a): $\hat{y}_p = 164.048$ lb.

(d) For a 90% prediction interval, Steps 1 and 2 are the same as Steps 1 and 2, respectively, in part (b). Thus, they are not repeated here. Only Step 3 is presented.

Step 3:   The endpoints of the prediction interval are:

$$\hat{y}_p \pm t_{a/2} \cdot s_e \sqrt{1 + \frac{1}{n} + \frac{(x_p - \Sigma x/n)^2}{\Sigma x^2 - (\Sigma x)^2/n}}$$

or $164.048 \pm 1.833 \cdot 16.477 \sqrt{1 + \frac{1}{11} + \frac{(70 - 761/11)^2}{52,729 - (761)^2/11}}$

or $164.048 \pm 1.833 \cdot 17.2743$

or 164.048 ± 31.664

or 132.38 to 195.71 lb

The interpretation of this interval is as follows: We can be 90% certain that the weight of a randomly selected 18-24 year-old male who is 70 inches tall will be somewhere between 132.38 and 195.71 lb.

(e)

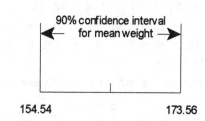

154.54   173.56

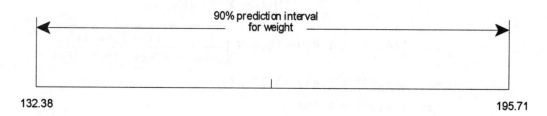

132.38   195.71

(f)   The error in the estimate of the mean weight of all 18-24 year-old males who are 70 inches tall is due only to the fact that the population regression line is being estimated by a sample regression line; whereas, the error in the prediction of the weight of a randomly selected 18-24 year-old male who is 70 inches tall is due to that fact plus the variation in weights of such males.

15.53 From Exercise 14.45, $\hat{y} = 222.271 - 1.1405x$, $\Sigma x = 368$, and $\Sigma x^2 = 13,968$. From Exercise 15.19, $s_e = 3.326$.

(a) $\hat{y}_p = 222.271 - 1.1405(40) = 176.65$

(b)   Step 1:   For a 95% confidence interval, $\alpha = 0.05$.

Step 1: (cont.)   With df $= n - 2 = 8$, $t_{\alpha/2} = t_{0.025} = 2.306$.

Step 2: $\hat{y}_p = 176.65$

Step 3: The endpoints of the confidence interval are:

$$\hat{y}_p \pm t_{a/2} \cdot s_e \sqrt{\frac{1}{n} + \frac{(x_p - \Sigma x/n)^2}{\Sigma x^2 - (\Sigma x)^2/n}}$$

or $176.65 \pm 2.306 \cdot 3.326 \sqrt{\dfrac{1}{10} + \dfrac{(40 - 368/10)^2}{13,968 - (368)^2/10}}$

or $176.65 \pm 2.306 \cdot 1.17148$

or 176.65 ± 2.70

or 173.95 to 179.35

The interpretation of this interval is as follows: We can be 95% confident that the mean peak heart rate of all 40-year-olds is somewhere between 173.95 and 179.35.

(c) This is the same as the answer in part (a): $\hat{y}_p = 176.65$.

(d) For a 95% prediction interval, Steps 1 and 2 are the same as Steps 1 and 2, respectively, in part (b). Thus, they are not repeated here. Only Step 3 is presented.

Step 3: The endpoints of the prediction interval are:

$$\hat{y}_p \pm t_{a/2} \cdot s_e \sqrt{1 + \frac{1}{n} + \frac{(x_p - \Sigma x/n)^2}{\Sigma x^2 - (\Sigma x)^2/n}}$$

or $176.65 \pm 2.306 \cdot 3.326 \sqrt{1 + \dfrac{1}{10} + \dfrac{(40 - 368/10)^2}{13,968 - (368)^2/10}}$

or $176.65 \pm 2.306 \cdot 3.52628$

or 176.65 ± 8.13

or 168.52 to 184.78

The interpretation of this interval is as follows: We can be 95% certain that the peak heart rate of a randomly selected 40-year-old will be somewhere between 168.52 and 184.78.

15.55 From Exercise 14.47, $\hat{y} = 12.8625 + 1.2137x$, $\Sigma x = 197$, and $\Sigma x^2 = 5,143$. From Exercise 15.21, $s_e = 7.6763$.

(a) First, a point estimate for the mean annual food expenditure of all middle-income families that have a disposable income of $25,000 is:

$\hat{y}_p = 12.8625 + 1.2137(25) = 43.2051 = \$4,320.51$

Secondly, the desired confidence interval is derived as follows:

Step 1: For a 99% confidence interval, $\alpha = 0.01$. With df = $n - 2 = 6$, $t_{\alpha/2} = t_{0.005} = 3.707$.

Step 2: $\hat{y}_p = 43.2051 = \$4,320.51$

Step 3: The endpoints of the confidence interval are:

$$\hat{y}_p \pm t_{a/2} \cdot s_e \sqrt{\frac{1}{n} + \frac{(x_p - \Sigma x/n)^2}{\Sigma x^2 - (\Sigma x)^2/n}}$$

or $43.2051 \pm 3.707 \cdot 7.6763 \sqrt{\frac{1}{8} + \frac{(25 - 197/8)^2}{5,143 - (197)^2/8}}$

or $43.2051 \pm 3.707 \cdot 2.71921$

or 43.2051 ± 10.08

or   33.13   to 53.29 hundred dollars

The interpretation of this interval is as follows: We can be 99% confident that the mean annual food expenditure of all middle-income families consisting of a father, mother, and two children that have a disposable income of $25,000 is somewhere between $3,313 and $5,329.

(b)   First, the predicted annual food expenditure of a randomly selected middle-income family having a disposable income of $25,000 is the same as that in part (a): $\hat{y}_p = 43.2051 = $4,320.51.

Secondly, the desired prediction interval is derived as follows:

For a 99% prediction interval, Steps 1 and 2 are the same as Steps 1 and 2, respectively, in part (a). Thus, they are not repeated here. Only Step 3 is presented.

Step 3:   The endpoints of the prediction interval are:

$$\hat{y}_p \pm t_{a/2} \cdot s_e \sqrt{1 + \frac{1}{n} + \frac{(x_p - \Sigma x/n)^2}{\Sigma x^2 - (\Sigma x)^2/n}}$$

or $43.2051 \pm 3.707 \cdot 7.6763 \sqrt{1 + \frac{1}{8} + \frac{(25 - 197/8)^2}{5,143 - (197)^2/8}}$

or $43.2051 \pm 3.707 \cdot 8.14369$

or 43.2051 ± 30.1887

or   13.02 to 73.39 hundred dollars

The interpretation of this interval is as follows: We can be 99% certain that the annual food expenditure of a randomly selected middle-income family consisting of a father, mother, and two children that has a disposable income of $25,000 will be somewhere between $1,302 and $7,339.

15.57 (a)   To obtain both confidence and prediction intervals, we choose **Stat**
▶ **Regression** ▶ **Regression...**, select FOODEX in the **Response** text box, select INCOME in the **Predictors** text box, click on the **Options** button, type 25 in the **Prediction intervals for new observations** text box, type 99 in the **Confidence level** text box, click on the check-boxes for **Confidence limits** and **Prediction limits**, click **OK** and click **OK**. See part (b) for the computer output.

(b)   To obtain the required prediction interval, we do nothing more than what was already done in part (a). The computer output for both part (a) and this part is:

The regression equation is

FOODEX = 12.9 + 1.21 INCOME

Predictor	Coef	Stdev	t-ratio	p
Constant	12.86	11.39	1.13	0.302
INCOME	1.2137	0.4493	2.70	0.036

s = 7.676   R-sq = 54.9%   R-sq(adj) = 47.4%

Analysis of Variance

SOURCE	DF	SS	MS	F	p
Regression	1	429.95	429.95	7.30	0.036
Error	6	353.55	58.92		
Total	7	783.50			

Fit	Stdev.Fit	99.0% C.I.	99.0% P.I.
43.21	2.72	(33.12, 53.29)	(13.01, 73.40)

Both the confidence and prediction intervals are found in the last two lines of the computer output. The 99% confidence interval is 33.12 to 53.29 (i.e., $3,312 to $5,329). The 99% prediction interval is 13.01 to 73.40 (i.e., $1,301 to $7,3140.

15.59 See the last two lines of Printout 15.9 to locate the answers.

(a)   Fit = 3064.6

(b)   95% C.I.:   (2909.2, 3220.1)

(c)   95% P.I.:   (2267.8, 3861.4)

Exercises 15.4

15.61 r

15.63 (a)   uncorrelated

(b)   increases

(c)   negatively

15.65 From Exercise 14.89, r = 0.6623.

Step 1: $H_0: \rho = 0$, $H_a: \rho > 0$

Step 2: $\alpha = 0.05$

Step 3:   df = n - 2 = 9; critical value = 1.833

Step 4: $t = \dfrac{r}{\sqrt{\dfrac{1-r^2}{n-2}}} = \dfrac{0.6623}{\sqrt{\dfrac{1-(0.6623)^2}{11-2}}} = 2.65$

Step 5:   Since 2.652 > 1.833, reject H_0.

Step 6:   The data provide sufficient evidence to conclude that height and weight are positively linearly correlated for 18-24 year-old males.

For the P-value approach, 0.01 < P < 0.025. Therefore, because the P-value is smaller than the significance level of 0.05, we can reject H_0.

15.67 From Exercise 14.91, $r = -0.9285$.

Step 1: $H_0: \rho = 0$, $H_a: \rho < 0$

Step 2: $\alpha = 0.025$

Step 3: $df = n - 2 = 8$; critical value $= -2.306$

Step 4: $$t = \frac{r}{\sqrt{\dfrac{1-r^2}{n-2}}} = \frac{-0.9285}{\sqrt{\dfrac{1-(-0.9285)^2}{10-2}}} = -7.07$$

Step 5: Since $-7.074 < -2.306$, reject H_0.

Step 6: The data provide sufficient evidence to conclude that age and peak heart rate are negatively linearly correlated.

For the P-value approach, $P < 0.005$. Therefore, because the P-value is smaller than the significance level of 0.025, we can reject H_0.

15.69 From Exercise 14.93, $r = 0.7407$.

Step 1: $H_0: \rho = 0$, $H_a: \rho \neq 0$

Step 2: $\alpha = 0.01$

Step 3: $df = n - 2 = 6$; critical values $= \pm 3.707$

Step 4: $$t = \frac{r}{\sqrt{\dfrac{1-r^2}{n-2}}} = \frac{0.7407}{\sqrt{\dfrac{1-(0.7407)^2}{8-2}}} = 2.70$$

Step 5: Since $-3.707 < 2.701 < 3.707$, do not reject H_0.

Step 6: The data do not provide sufficient evidence to conclude that family disposable income and annual food expenditure are linearly correlated for middle-income families with a father, mother, and two children.

For the P-value approach, $0.02 < P < 0.05$. Therefore, because the P-value is not smaller than the significance level of 0.01, we cannot reject H_0.

15.71 With the data in columns called INCOME and FOODEX, choose **Stat ▶ Basic Statistics ▶ Correlation**, select INCOME and FOODEX in the **Variables** text box, make sure that there is a check mark in the Display p-values box, and click **OK**. The result is

Correlation of INCOME and FOODEX = 0.741, P-Value = 0.036

Since the test is two-tailed, the P-value is 0.036. This is larger than the 0.01 significance level, so we can not reject the null hypothesis that $\rho = 0$. The data do not provide sufficient evidence to conclude that disposable income and food expenditures are linearly correlated.

Exercises 15.5

15.73 (a) A normal probability plot is a plot of normal scores against sample data.

(b) An important use of such plots is to help assess the normality of a variable.

(c)   If the sample data come from a normal distribution, the plot should be roughly linear. Therefore, if the plot is roughly linear, we accept as reasonable that the variable is normally distributed; if the plot shows systematic deviations from a straight line, then conclude that the variable is probably not normally distributed.

(d)   The method in (c) is subjective because what constitutes "roughly linear" is a matter of opinion.

15.75 If the population under consideration is normally distributed, then the correlation between the sample data and its normal scores should be near 1. Note that since large normal scores are associated with large data values and vice versa, the correlation between the sample data and its normal scores cannot be negative. Thus, the correlation test for normality is always left-tailed because if the correlation (the test statistic of interest) is smaller than one, the null hypothesis that the population is normally distributed is rejected.

15.77 Step 1: H_0: The population is normally distributed.

Step 2: $\alpha = 0.05$

Step 3:   The critical value is = 0.951.

Step 4:

Exam score	Normal score			
x	w	xw	x^2	w^2
34	−1.87	−63.58	1156	3.4969
39	−1.40	−54.60	1521	1.9600
63	−1.13	−71.19	3969	1.2769
64	−0.92	−58.88	4096	0.8464
67	−0.74	−49.58	4489	0.5476
70	−0.59	−41.30	4900	0.3481
75	−0.45	−33.75	5625	0.2025
76	−0.31	−23.56	5776	0.0961
81	−0.19	−15.39	6561	0.0361
82	−0.06	−4.92	6724	0.0036
84	0.06	5.04	7056	0.0036
85	0.19	16.15	7225	0.0361
86	0.31	26.66	7396	0.0961
88	0.45	39.60	7744	0.2025
89	0.59	52.51	7921	0.3481
90	0.74	66.60	8100	0.5476
90	0.92	82.80	8100	0.8464
96	1.13	108.48	9216	1.2769
96	1.40	134.40	9216	1.9600
100	1.87	187.00	10,000	3.4969
1555	0.00	302.48	126,791	17.6284

$$R_p = \cfrac{\Sigma x\varpi}{\sqrt{[\Sigma x^2 - \cfrac{(\Sigma x)^2}{n}][\Sigma\varpi^2]}}$$

$$= \cfrac{302.48}{\sqrt{[126,791 - \cfrac{(1555^2)}{20}][17.6284]}} = 0.939$$

Step 5:   From Step 4 the value of the test statistic is $R_p = 0.939$, which falls in the rejection region. Hence we reject H_0.

Step 6:   The test results are statistically significant at the 5% level, that is, at the 5% significance level, the data provide sufficient evidence to conclude that final exam scores are not normally distributed.

15.79 Step 1: H_0: The population is normally distributed.

   H_a: The population is not normally distributed.

Step 2: $\alpha = 0.10$

Step 3:   The critical value is = 0.951.

Step 4:

| Miles | Normal score | | | |
x	w	xw	x^2	w^2
3.3	-1.74	-5.742	10.89	3.0276
5.7	-1.24	-7.068	32.49	1.5376
6.6	-0.94	-6.204	43.56	0.8836
7.7	-0.71	-5.467	59.29	0.5041
8.3	-0.51	-4.233	68.89	0.2601
8.6	-0.33	-2.838	73.96	0.1089
8.9	-0.16	-1.424	79.21	0.0256
9.2	0.00	0.000	84.64	0.0256
10.2	0.16	1.632	104.04	0.0256
10.3	0.33	3.399	106.09	0.1089
10.6	0.51	5.406	112.36	0.2601
11.8	0.71	8.378	139.24	0.5041
12.0	0.94	11.280	144.00	0.8836
12.7	1.24	15.748	161.29	1.5376
13.7	1.74	23.838	187.69	3.0276
139.6	0.00	36.705	1407.64	12.695

$$R_p = \cfrac{\Sigma x\varpi}{\sqrt{[\Sigma x^2 - \cfrac{(\Sigma x)^2}{n}][\Sigma\varpi^2]}}$$

$$= \cfrac{36.705}{\sqrt{[1407.64 - \cfrac{(139.6^2)}{15}][12.695]}} = 0.989$$

Step 5:   From Step 4 the value of the test statistic is $R_p = 0.989$, which does not fall in the rejection region. Hence we do not reject H_0.

Step 6:   The test results are not statistically significant at the 10% level, that is, at the 10% significance level, the data do not provide sufficient evidence to conclude that the number of miles cars are driven are not normally distributed.

15.81 We choose **Stat ▶ Basic statistics ▶ Normality test...**, select SCORES

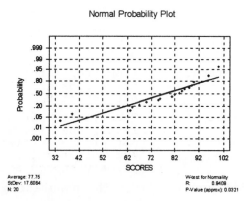

Normal Probability Plot

in the **Variable** text box, select the **Ryan-Joiner** option button from the **Tests for Normality** field, and click **OK**. The result is shown at the right. The P-value shown in the lower right is .0321. Since this is smaller than the 0.05 significance level, we reject the null hypothesis of normality.

Average: 77.75
StDev: 17.6064
N: 20

W-test for Normality
R: 0.9408
P-Value (approx): 0.0321

15.83 Step 1: H_0: The population is normally distributed.

Step 1 (cont.) H_a: The population is not normally distributed.

Step 2: $\alpha = 0.01$

Step 3:   The critical value is = 0.922.

Step 4:

Expenditure x	Normal score w	xw	x^2	w^2
949	−1.82	−1727.18	900,601	3.3124
1059	−1.35	−1429.65	1,121,481	1.8225
1070	−1.06	−1134.20	1,144,900	1.1236
1102	−0.84	−925.68	1,214,404	0.7056
1168	−0.66	−770.88	1,364,224	0.4356
1179	−0.50	−589.50	1,390,041	0.2500
1185	−0.35	−414.75	1,404,225	0.1225
1227	−0.21	−257.67	1,505,529	0.0441
1235	−0.07	−86.45	1,525,225	0.0049
1259	0.07	88.13	1,585,081	0.0049
1327	0.21	278.67	1,760,929	0.0441
1351	0.35	472.85	1,825,201	0.1225
1376	0.50	688.00	1,893,376	0.2500
1393	0.66	919.38	1,940,449	0.4356
1400	0.84	1176.00	1,960,000	0.7056
1452	1.06	1539.12	2,108,304	1.1236
1456	1.35	1965.60	2,119,936	1.8225
1480	1.82	2693.60	2,190,400	3.3124
22,668	0.00	2485.39	28,954,304	15.6425

$$R_p = \frac{\sum x\varpi}{\sqrt{\left[\sum x^2 - \frac{(\sum x)^2}{n}\right]\left[\sum \varpi^2\right]}}$$

$$= \frac{2485.39}{\sqrt{\left[28,954,304 - \frac{(22,668^2)}{18}\right][15.6425]}} = 0.984$$

Step 5:   From Step 4 the value of the test statistic is $R_p = 0.984$, which does not fall in the rejection region. Hence we do not reject H_0.

Step 6:   The test results are not statistically significant at the 1% level, that is, at the 1% significance level, the data do not provide sufficient evidence to conclude that the energy expenditures for households using electricity as their primary energy source is not normally distributed.

15.85 Step 1: H_0: The population is normally distributed.

Ha: The population is not normally distributed.

Step 2: $\alpha = 0.05$

Step 3:   The critical value is = 0.952.

Step 4:

Days x	Normal score w	x^2	xw	w^2
1	-1.89	1	-1.89	3.5721
1	-1.43	1	-1.43	2.0449
3	-1.16	9	-3.48	1.3456
3	-0.95	9	-2.85	0.9025
4	-0.78	16	-3.12	0.6084
4	-0.63	16	-2.52	0.3969
5	-0.49	25	-2.45	0.2401
6	-0.36	36	-2.16	0.1296
6	-0.24	36	-1.44	0.0576
7	-0.12	49	-0.84	0.0144
7	0.00	49	0.00	0.0000
9	0.12	81	1.08	0.0144
9	0.24	81	2.16	0.0576
10	0.36	100	3.60	0.1296
12	0.49	144	5.88	0.2401
12	0.63	144	7.56	0.3969
13	0.78	169	10.14	0.6084
15	0.95	225	14.25	0.9025
18	1.16	324	20.88	1.3456
23	1.43	529	32.89	2.0449
55	1.89	3025	103.95	3.5721
223	0.00	5069	180.21	18.6240

$$R_p = \frac{\sum x\varpi}{\sqrt{\left[\sum x^2 - \frac{(\sum x)^2}{n}\right]\left[\sum \varpi^2\right]}}$$

$$= \frac{180.21}{\sqrt{\left[5069 - \frac{(223^2)}{21}\right][18.624]}} = 0.803$$

Step 5:   From Step 4 the value of the test statistic is $R_p = 0.803$, which falls in the rejection region. Hence we reject H_0.

Step 6:   The test results are statistically significant at the 5% level, that is, at the 5% significance level, the data do provide sufficient evidence to conclude that the lengths of stays by patients in short-term hospitals are not normally distributed.

15.87 (a)   In Minitab, name four columns GP1, GP2, GP3, and GP4. Then choose

Calc ▶ Random data ▶ Normal..., type <u>50</u> in the **Generate rows of data** text box, select GP1, GP2, GP3, and GP4 in the **Store in column(s):** text box, type <u>266</u> in the **Mean** text box, type <u>16</u> in the **Standard deviation** text box, and click **OK**.

(b)   To perform the correlation test on GP1, choose **Stat ▶ Basic**

statistics ▶ Normality test..., select GP1 in the **Variable** text box, click on the **Ryan-Joiner** button and click **OK**. Repeat this process for GP2, GP3, and GP4. The results are shown in the four following graphs.

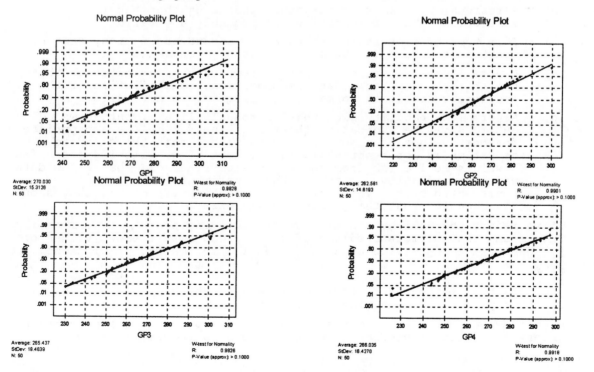

The four values of R_p are 0.9928, 0.9901, 0.9926, and 0.9918. Their P-values are all greater than 0.1000. Thus none of them are significant at the 0.05 significance level.

(c)   Since the samples were generated from a normal distribution and we expect that samples will reflect the properties of the population, the results in (b) are what we should have expected.

REVIEW TEST FOR CHAPTER 14

1.   (a)   conditional

(b)   The four assumptions for regression inferences are:

(1)   *Population regression line:* There is a straight line $y = \beta_0 + \beta_1 x$ such that, for each x-value, the mean of the corresponding population of y-values lies on that straight line.

(2)   *Equal standard deviations:* The standard deviation σ of the population of y-values corresponding to a particular x-value is the same, regardless of the x-value.

(3)   *Normality:* For each x-value, the corresponding population of y-values is normally distributed.

(4)   *Independence:* The observations of the response variable are independent of one another.

2.   (a)   The slope of the sample regression line, b_1

(b)   The y-intercept of the sample regression line, b_0

(c) s_e

3.   We used a plot of the residuals against the values of the predictor variable x and a normal probability plot of the residuals. In the first plot, the residuals should lie in a horizontal band centered on and symmetric about the x-axis. In the second plot, the points should lie roughly in a straight line.

3.   (a)   A residual plot showing curvature indicates that the first assumption, that of linearity, is probably not valid.

(b)   This type of plot indicates that the second assumption, that of constant standard deviation, is probably not valid.

(c)   A normal probability plot with extreme curvature indicates that the third assumption, that the conditional distribution of the response variable is normally distributed, is not valid.

(d)   A normal probability plot that is roughly linear, but shows outliers may be indicating that: the linear model is appropriate for most values of x, but not for all; or, that the standard deviation is not constant for all values of x, allowing for a few y values to lie far from the regression line; or, that a few data values are 'faulty', that is, the experimental conditions represented by the value of the x variable were not as they should have been or that the y value was in error.

5.   If we reject the null hypothesis, we are claiming that β_1 is not zero. This means that different values of x will lead to different values of y. Hence, the regression equation is useful for making predictions.

6. t, r, r^2

7.   No. The best estimate of conditional mean of the response variable and the best prediction of a single future value of y are the same.

8.   A confidence interval estimates the value of a parameter; a prediction interval is used to predict a future value of a random variable.

9. ρ

10.   (a)   If $\rho > 0$, the variables are positively correlated, that is, there is a tendency for one variable to increase as the other one increases.

       (b)   If we know only that $\rho \neq 0$, the two variables are linearly correlated, that is, there is a straight line relationship between them.

       (c)   If $\rho < 0$, the variables are negatively correlated, that is, there is a tendency for one variable to decrease as the other one increases.

11.   If the regression assumptions are satisfied, then there is a linear relationship between the student/faculty ratio and the graduation rate, the conditional standard deviation of the graduation rate is the same for all values of the student/faculty ratio, the conditional distribution of the graduation rate is a normal distribution for all values of the student/faculty ratio, and the values of the graduation rate are independent of each other.

12.   (a)   To determine b_0 and b_1 for the regression line, construct a table of values for x, y, x^2, xy and their sums. A column for y^2 is also presented for later use.

x	y	x^2	xy	y^2
16	45	256	720	2025
20	55	400	1100	3025
17	70	289	1190	4900
19	50	361	950	2500
22	47	484	1034	2209
17	46	289	782	2116
17	50	289	850	2500
17	66	289	1122	4356
10	26	100	260	676
18	60	324	1080	3600
173	515	3081	9088	27907

$S_{xx} = 3081 - 173^2/10 = 88.10$

$S_{xy} = 9088 - (173)(515)/10 = 178.50$

$b_1 = 178.50/88.10 = 2.0261$

$b_0 = 515/10 - 2.0261(173/10) = 16.4484$

$\hat{y}_p = 16.448 + 2.026x$ is the equation of the regression line.

       (b) $S_{yy} = 27907 - 515^2/10 = 1384.50$

       $SSE = S_{yy} - S^2_{xy}/S_{xx} = 1384.50 - 178.50^2/88.10 = 1022.8400$

$$s_e = \sqrt{\frac{SSE}{n-2}} = \sqrt{\frac{1022.8400}{8}} = 11.3073$$

This value of s_e indicates that, roughly speaking, the predicted values of y differ from the observed values of y by about 11.31.

       (c)   Presuming that the variables Student/Faculty ratio and Graduation rate satisfy the assumptions for regression inferences, the standard error of the estimate, $s_e = 11.31$, provides an estimate for the common population standard deviation, σ, of graduation rates of entering freshmen at universities with any particular student/faculty ratio.

13. For each value of x, we compute e = y - ŷ in the following table and plot e against x.

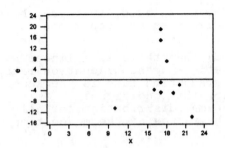

x	e
16	-3.87
20	-1.97
17	19.11
19	-4.94
22	-14.02
17	-4.89
17	-0.89
17	15.11
10	-10.71
18	7.08

Residual e	Normal score n
-14.02	-1.55
-10.71	-1.00
-4.94	-0.65
-4.89	-0.37
-3.87	-0.12
-1.97	0.12
-0.89	0.37
7.08	0.65
15.11	1.00
19.11	1.55

The small sample size makes it difficult to evaluate the plot of e against x. While there are no obvious patterns in the first plot, there are two points (in the lower left and right corners of the plot) which are cause for concern. The normal probability plot is not very linear.

14. (a)   From Review Problem 11, $\sum x = 173$, $\sum x^2 = 3,081$, and $b_1 = 2.0261$.
From Review Problem 12, $s_e = 11.3073$.

Step 1: H_0: $\beta_1 = 0$

H_a: $\beta_1 \neq 0$

Step 2: $\alpha = 0.05$

Step 3:   df = n - 2 = 8; critical values = ± 2.306

Step 4: $t = \dfrac{b_1}{s_e / \sqrt{\sum x^2 - (\sum x)^2 / n}} = \dfrac{2.0261}{11.3073 / \sqrt{3081 - \dfrac{173^2}{10}}} = 1.68$

Step 5:   Since 1.68 < 2.306, do not reject the null hypothesis; at the 5% significance level.

Step 6:   The data do not provide sufficient evidence to conclude that the student-to-faculty ratio is useful as a predictor of graduation rate. For the P-value approach, note that 0.10 < P < 0.20. Since P > α, do not reject the null hypothesis.

(b)   Step 1:   For a 95% confidence interval, $\alpha = 0.05$. With df = n - 2 = 8, $t_{\alpha/2} = t_{0.025} = 2.306$.

Step 2:   The endpoints of the confidence interval for β_1 are

$$b_1 \pm t_{a/2} \cdot s_e / \sqrt{\sum x^2 - (\sum x)^2 / n}$$

or $2.0261 \pm 2.306 \cdot 11.3073 / \sqrt{3,081 - (173)^2 / 10}$

or $2.0261 \pm 2.306 \cdot 1.204678$

or $2.0261 \pm 2.7780 = (-0.7519, 4.8041)$

We can be 95% confident that the change in graduation rate for universities per 1 unit increase in the student/faculty ratio is somewhere between -0.7519 and 4.8041 percent.

15.   From Review Exercise 11, $\hat{y} = 16.4484 + 2.0261x$, $\sum x = 173$, and $\sum x^2 = 3,081$. From Review Exercise 12, $s_e = 11.3073$.

(a) $\hat{y}_p = 16.4484 + 2.0261(17) = 50.89$

(b)   Step 1:   For a 95% confidence interval, $\alpha = 0.05$. With df = n - 2 = 8, $t_{\alpha/2} = t_{0.025} = 2.306$.

Step 2: $\hat{y}_p = 50.89$

Step 3:   The endpoints of the confidence interval are:

$$\hat{y}_p \pm t_{a/2} \cdot s_e \sqrt{\dfrac{1}{n} + \dfrac{(x_p - \sum x / n)^2}{\sum x^2 - (\sum x)^2 / n}}$$

or $50.89 \pm 2.306 \cdot 11.3073 \sqrt{\dfrac{1}{10} + \dfrac{(17 - 173/10)^2}{3,081 - (173)^2 / 10}}$

or $50.89 \pm 2.306 \cdot 3.59390$

or $50.89 \pm 8.29 = (42.60, 59.18)$

The interpretation of this interval is as follows: We can be 95% confident that the graduation rate of universities with a student/faculty ratio of 17 is somewhere between 42.60% and 59.18%.

(c)   This is the same as the answer in part (a): $\hat{y}_p = 50.89$.

(d)   For a 95% prediction interval, Steps 1 and 2 are the same as Steps 1 and 2, respectively, in part (b). Thus, they are not repeated here. Only Step 3 is presented.

Step 3:   The endpoints of the prediction interval are:

$$\hat{Y}_p \pm t_{a/2} \cdot s_e \sqrt{1 + \frac{1}{n} + \frac{(x_p - \Sigma x/n)^2}{\Sigma x^2 - (\Sigma x)^2/n}}$$

or $50.89 \pm 2.306 \cdot 11.3073 \sqrt{1 + \frac{1}{10} + \frac{(17 - 173/10)^2}{3,081 - (173)^2/10}}$

or $50.89 \pm 2.306 \cdot 11.86470$

or 50.89 ± 27.36

or   23.53 to 78.25

The interpretation of this interval is as follows: We can be 95% certain that the graduation rate of a randomly selected university with a student/faculty ratio of 17 will be somewhere between 23.53% and 78.25%.

(e)   The error in the estimate of the mean graduation rate for universities with a student/faculty ratio of 17 is due only to the fact that the population regression line is being estimated by a sample regression line. The error in the prediction of the graduation rate of a randomly chosen university with a student/faculty ratio is due to the estimation error mentioned above plus the variation in graduation rates of universities with a student/faculty ratio of 17.

16.   From Review Problem 11,

$S_{xx} = 88.10$

$S_{xy} = 178.50$

$S_{yy} = 1384.50$

$$r = \frac{S_{xy}}{\sqrt{S_{xx}S_{yy}}} = \frac{178.50}{\sqrt{(88.10)(1384.50)}} = 0.5111$$

Step 1: $H_0: \rho = 0$, $H_a: \rho > 0$

Step 2: $\alpha = 0.025$

Step 3:   df = n − 2 = 8; critical value = 2.306

Step 4: $t = \frac{r}{\sqrt{\frac{1-r^2}{n-2}}} = \frac{0.5111}{\sqrt{\frac{1-(0.5111)^2}{10-2}}} = 1.68$

Step 5:   Since 1.682 < 2.306, do not reject H_0.

Step 6:   The data do not provide sufficient evidence to conclude that graduation rate and student /faculty ratio are positively linearly correlated.

For the P-value approach, 0.05 <P < 0.10. Therefore, because the P-value is larger than the significance level of 0.025, we can not reject H_0.

17. normal scores

18. Step 1: H_0: The population is normally distributed.

                  H_a: The population is not normally distributed.

    Step 2: $\alpha = 0.05$

    Step 3:   The critical value is = 0.938.

    Step 4:

Mileage x	Normal score w	xw	x^2	w^2
25.9	-1.74	-45.066	670.81	3.0276
27.3	-1.24	-33.852	745.29	1.5376
27.3	-0.94	-25.662	745.29	0.8836
27.6	-0.71	-19.596	761.76	0.5041
27.8	-0.51	-14.178	772.84	0.2601
27.8	-0.33	-9.174	772.84	0.1089
28.5	-0.16	-4.560	812.25	0.0256
28.6	0.00	0.000	817.96	0.0000
28.8	0.16	4.608	829.44	0.0256
28.9	0.33	9.537	835.21	0.1089
29.4	0.51	14.994	864.36	0.2601
29.7	0.71	21.087	882.09	0.5041
30.9	0.94	29.046	954.81	0.8836
31.2	1.24	38.688	973.44	1.5376
31.6	1.74	54.984	998.56	3.0276
431.3	0.00	20.856	12,436.95	12.6950

$$R_p = \frac{\sum x\omega}{\sqrt{\left[\sum x^2 - \frac{(\sum x)^2}{n}\right]\left[\sum \omega^2\right]}}$$

$$= \frac{20.856}{\sqrt{\left[12,436.95 - \frac{(431.3^2)}{15}\right][12.6950]}} = 0.981$$

    Step 5:   From Step 4 the value of the test statistic is $R_p = 0.981$, which does not fall in the rejection region. Hence we do not reject H_0.

    Step 6:   The test results are not statistically significant at the 5% level, that is, at the 5% significance level, the data do not provide sufficient evidence to conclude that the gas mileages for this model are not normally distributed.

19. With the data in columns named S/F and GRADRATE, we choose **Stat ▶**

Regression ▶ Regression..., select GRADRATE in the **Response:** text box, select 'S/F' in the **Predictors:** text box, click on the Graphs button, check the **Normal plot of residuals** box, click in the **Residuals versus the variables** text box and select 'S/F', and click **OK**. Looking ahead to Review Problems 21,22 and 24, click on the **Options...** button, type <u>17</u> in the **Prediction intervals for new observations:** text box, type <u>95</u> in the **Confidence level:** text box, check the **Confidence limits** and **Prediction limits** boxes, and click **OK.** Click on the **Storage...** button, check the **Residuals** box, click **OK,** and click **OK.** In addition to the regression output shown in Printout 15.12, the two graphs shown in the text for Review Problem 20 in Printouts 15.10 and 15.11 will be produced. The conclusions are the same as for Review Problem 13.

20. The small sample size makes it difficult to evaluate the plot of e against x. While there are no obvious patterns in the plot, there are two points (in the lower left and right corners of the plot) which are cause for concern about a straight line being the right model. The normal probability plot is not very linear, so the normality of the conditional distribution of the graduation rate is also in question.

21. Following the procedure in Review Problem 19, the regression output is

The regression equation is
GRADRATE = 16.4 + 2.03 S/F

Predictor	Coef	StDev	T	P
Constant	16.45	21.15	0.78	0.459
S/F	2.026	1.205	1.68	0.131

S = 11.31   R-Sq = 26.1%   R-Sq(adj) = 16.9%

Analysis of Variance

Source	DF	SS	MS	F	P
Regression	1	361.7	361.7	2.83	0.131
Residual Error	8	1022.8	127.9		
Total	9	1384.5			

Predicted Values

Fit	StDev Fit	95.0% CI	95.0% PI
50.89	3.59	(42.60, 59.18)	(23.53, 78.25)

(a) The regression equation is shown at the top of the output:

GRADRATE = 16.4 + 2.03 S/F

(b) The standard error of the estimate is shown in the sixth line of output as   S = 11.31.

22. (a) Referring to the output in Review Problem 21, the fifth line of the output shows a t-value of 1.68 with a P-value of 0.131 associated with the regression parameter for the S/F variable. Since .131 > 0.05, the null hypothesis that $\beta_1 = 0$ cannot be rejected. Thus there is not sufficient evidence to conclude that the S/F ratio is useful as a predictor of graduation rate.

(b) The last line of the output gives the 95% confidence interval for the mean graduation rate when S/F = 17 as 42.60 to 59.18 and the prediction interval for a randomly chosen university with an S/F ratio of 17 as 23.53 to 78.25.

23. (a) GRADRATE = 16.4 + 2.03 SFRATIO

(b) s_e = 11.31 (from the sixth line of output)

(c) The slope is 2.026 (from the fifth line of output)

(d) The standard deviation of the slope is 1.205 (from the fifth line)

(e) t = 1.68 (from the fifth line)

(f) P-value = 0.131 (from the fifth line)

(g) Since 0.131 > 0.05, the null hypothesis that $\beta_1 = 0$ cannot be rejected. Thus there is not sufficient evidence to conclude that the S/F ratio is useful as a predictor of graduation rate.

(h) $b_1 \pm t_{\alpha/2} s_{b_1} = 2.026 \pm 2.306(1.205) = 2.026 \pm 2.779 = (-0.753, 4.805)$

(i) The predicted value is 50.89 (from the last line of output)

(j) 42.60 to 59.18 (from the last line of output)

(k) 50.89, same as for part j

(l) 23.53 to 78.25 (from the last line of output)

24. In Review Problem 19, checking the Residuals box after clicking on the **Options...** button resulted in the residuals being stored in a column named RESI1. To perform the correlation test for normality, we now

choose **Stat ▶ Basic statistics ▶ Correlation**, select S/F and GRADRATE in the **Variables:** text box, check the **Display p-values** box, and click **OK**. The results are shown below.

Correlation of S/F and GRADRATE = 0.511, P-Value = 0.131

Since Minitab reports the P-value for a two-tailed test and we wish to test whether the correlation is positive, the actual P-value is 0.131/2 or 0.0655. This is greater than the significance level of 0.025, so we do not reject the null hypothesis that the student/faculty ratio and the graduation rate are uncorrelated. There is insufficient evidence to claim that the variables are positively correlated.

25. We choose **State ▶ Basic statistics ▶**

Normality test..., select MILES in the **Variable** text box, select the **Ryan-Joiner** option button from the **Tests for Normality** field, and click **OK**. The P-value for the normality test is greater than 0.1000. Since this is larger than the 0.025 significance level for the test, we do not reject the null hypothesis of normality. The correlation between MILES and NSCORE = 0.9832.

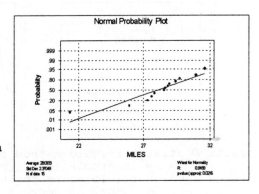

CHAPTER 16 ANSWERS

Exercises 16.1

16.1 We state the degrees of freedom.

16.3 $F_{0.05}$, $F_{0.025}$, F_α

16.5 (a) The first number in parentheses—12—is the number of degrees of freedom for the numerator.

(b) The second number in parentheses—7—is the number of degrees of freedom for the denominator.

16.7 (a) $F_{0.05} = 1.89$ (b) $F_{0.01} = 2.47$ (c) $F_{0.025} = 2.14$

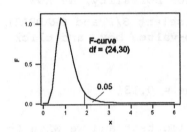 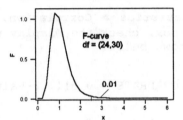 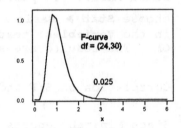

16.9 (a) $F_{0.01} = 2.88$ (b) $F_{0.05} = 2.10$ (c) $F_{0.10} = 1.78$

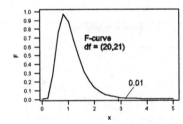 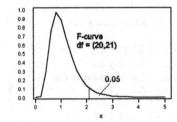 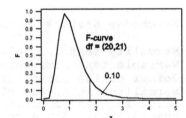

16.11 A straightforward method for finding $F_{0.05}$ for df = (25, 20) using Table X is by interpolation. Specifically, proceed along the top row of the table until locating the numbers 24 and 30, between which is 25, the desired number of degrees of freedom for the numerator. Notice that 25 is not reported in this row but is 1/6 of the way in moving from 24 to 30; i.e., (25 - 24)/(30 - 24) = 1/6.

The outside columns of the table give the degrees of freedom for the denominator, which is 20 in this case. Go down either of the outside columns to the row labeled "20." Then go across that row until under the columns headed "24" and "30." The numbers in the body of the table intersecting these row and column positions are the F-values 2.08 and 2.04. Thus, 24 is matched with 2.08, and 30 is matched with 2.04.

Now, proceed with interpolation. With 25 known to be 1/6 of the distance between 24 and 30, we likewise want to find the F-value that is

390

1/6 of the distance between 2.08 and 2.04. We do this by taking (1/6) · (2.04 - 2.08) = -0.0066. Adding -0.0066 to 2.08 gives us the desired F-value, which is 2.0734, or 2.07.

Thus, an approximation to $F_{0.05}$ for df = (25, 20) using Table X is 2.07. The limitation of this approximation is that it uses a linear technique on values that are related in a nonlinear fashion. However, because the difference between 2.08 and 2.04 is so small in the first place, the approximation errors will likewise be small.

Exercises 16.2

16.13 The pooled-t procedure (Procedure 10.1) is a method for comparing the means of two populations. One-way ANOVA is a procedure for comparing the means of several populations. Thus, one-way ANOVA is a generalization of the pooled-t procedure.

16.15 The reason for the word "variance" in the phrase "analysis of variance" is because the analysis-of-variance procedure for comparing means involves analyzing the *variation* in the sample data.

16.17 (a)   MSTR (or SSTR) is a statistic that measures the variation among the sample means for a one-way ANOVA.

  (b)   MSE (or SSE) is a statistic that measures the variation within the samples for a one-way ANOVA.

  (c)   F = MSTR/MSE is a statistic that compares the variation among the sample means to the variation within the samples.

16.19 It signifies that each piece of sample data is classified in one way, namely, according to the population from which it was sampled.

16.21 No, because the variation among the sample means is not large relative to the variation within the samples.

16.23 If the hypothesis test situation abides by the characteristics outlined in this exercise, we can use the pooled-t test or we can use the one-way ANOVA test discussed in this section.

Exercises 16.3

16.25 A small value of F results when SSTR is small compared to SSE, i.e., when the variation between sample means is small compared to the variation within samples. This describes what should happen when the null hypothesis is true, thus it does not comprise evidence that the null hypothesis is false. Only when the variation between sample means is large compared to the variation within samples, i.e., when F is large, do we have evidence that the null hypothesis is not true.

16.27 SST = SSTR + SSE; this means that the total variation can be partitioned into a component represent variation among the sample means and another component representing variation within the samples.

16.29 Since SSTR/2 = MSTR, we have SSTR = 2(21.652) = 43.304.

Then SST = SSTR + SSE = 43.304 + 84.400 = 127.704.

Since df (Total) = df(Treatment) + df(Error), we have df(Error) = 12.

Then MSE = SSE/12 = 84.400/12 = 7.033

Finally, F = MSTR/MSE = 21.652/7.033 = 3.079. Thus the completed table is

Source	df	SS	MS=S/df	F-statistic
Treatment	2	43.304	21.652	3.079
Error	12	84.400	7.033	
Total	14	127.704		

16.31 The total number of populations being sampled is k = 3. Let the subscripts 1, 2, and 3 refer to A, B, and C, respectively. The total number of pieces of sample data is n = 10. Also,

$n_1 = 2$, $n_2 = 5$, $n_3 = 3$. The following statistics for each sample are: $\bar{x}_1 = 5$, $\bar{x}_2 = 3$, and $\bar{x}_3 = 5$. Also, $s_1 = 5.657$, $s_2 = 1.581$, and $s_3 = 3.000$. Note, the mean of all the sample data is $\bar{x} = 4$. We use this information as follows:

(a) $\text{SST} = \Sigma(x - \bar{x})^2 = (1 - 4)^2 + \ldots + (5 - 4)^2 = 70.0$

(b) $\text{SSTR} = n_1(\bar{x}_1 - \bar{x})^2 + n_2(\bar{x}_2 - \bar{x})^2 + n_3(\bar{x}_3 - \bar{x})^2$

$= 2(5 - 4)^2 + 5(3 - 4)^2 + 3(5 - 4)^2 = 10.0$

(c) $\text{SSE} = (n_1 - 1)s_1^2 + (n_2 - 1)s_2^2 + (n_3 - 1)s_3^2$

$= (2 - 1)(5.657)^2 + (5 - 1)(1.581)^2 + (3 - 1)(3.000)^2 = 60.0$

(d) $\text{SST} = \text{SSTR} + \text{SSE}$ since $70.0 = 10.0 + 60.0$

(e)   Let T_1, T_2, and T_3 refer to the sum of the data values in each of the three samples, respectively. Thus:

$$T_1 = 10 \quad T_2 = 15 \quad T_3 = 15.$$

Also, the sum of all the data values is $\Sigma x = 40$, and their sum of squares is $\Sigma x^2 = 230$.

Consequently:

$$\text{SST} = \Sigma x^2 - \frac{(\Sigma x)^2}{n} = 230 - \frac{(40)^2}{10} = 70 \quad,$$

$$\text{SSTR} = \left(\frac{T_1^2}{n_1} + \frac{T_2^2}{n_2} + \frac{T_3^2}{n_3} \right) - \frac{(\Sigma x)^2}{n} = \left(\frac{10^2}{2} + \frac{15^2}{5} + \frac{15^2}{3} \right) - \frac{(40)^2}{10} = 10 \quad,$$

and

$$\text{SSE} = \text{SST} - \text{SSTR} = 70 - 10 = 60 \ .$$

The two methods of computing yield the same results.

16.33 The total number of populations being sampled is k = 3. Let the subscripts 1, 2, and 3 refer to Hank, Joseph, and Susan, respectively. The total number of pieces of sample data is n = 15. Also,

$n_1 = n_2 = n_3 = 5$. The following statistics for each individual are: $\bar{x}_1 = 9.6$, $\bar{x}_2 = 8.8$, and $\bar{x}_3 = 9.8$. Also, $s_1 = 1.14$, $s_2 = 0.837$, and $s_3 = 0.837$. Note, the mean of all the sample data is $\bar{x} = 9.4$. We use this information as follows:

(i) $\text{SSTR} = n_1(\bar{x}_1 - \bar{x})^2 + n_2(\bar{x}_2 - \bar{x})^2 + n_3(\bar{x}_3 - \bar{x})^2$

$= 5(9.6 - 9.4)^2 + 5(8.8 - 9.4)^2 + 5(9.8 - 9.4)^2 = 2.8$

(ii) $\text{SSE} = (n_1 - 1)s_1^2 + (n_2 - 1)s_2^2 + (n_3 - 1)s_3^2$

$= (5 - 1)(1.14)^2 + (5 - 1)(0.837)^2 + (5 - 1)(0.837)^2 = 10.8$

(iii) $\text{SST} = \Sigma(x - \bar{x})^2 = (8 - 9.4)^2 + \ldots + (9 - 9.4)^2 = 13.6$

(iv) $\text{MSTR} = \dfrac{\text{SSTR}}{k - 1} = \dfrac{2.8}{3 - 1} = 1.4$

(v) MSE $= \dfrac{\text{SSE}}{n - k} = \dfrac{10.8}{15 - 3} = 0.9$

The one-way ANOVA table for the data is

Source	df	SS	MS = SS/df	F-statistic
Treatment	2	2.8	1.4	1.56
Error	12	10.8	0.9	
Total	14	13.6		

With the exception of the F-statistic, every value in the table above has been calculated using (i)-(v). The F-statistic is defined and calculated as

16.35 The total number of populations being sampled is k = 4. Let the subscripts 1, 2, 3, and 4 refer to top tenth, second tenth, second fifth, and third fifth, respectively. The total number of pieces of sample data is n = 20. Also, $n_1 = 4$, $n_2 = 5$, $n_3 = 5$, and $n_4 = 6$. The following statistics for each rank category are: $\bar{x}_1 = 628$, $\bar{x}_2 = 478.8$, $\bar{x}_3 = 518.8$, and $\bar{x}_4 = 397$. Also, $s_1 = 86.733$, $s_2 = 67.385$, $s_3 = 89.547$, and $s_4 = 67.929$. Note, the mean of all the sample data is $\bar{x} = 494.1$. We use this information as follows:

(i)   SSTR $= n_1(\bar{x}_1 - \bar{x})^2 + n_2(\bar{x}_2 - \bar{x})^2 + n_3(\bar{x}_3 - \bar{x})^2 + n_4(\bar{x}_4 - \bar{x})^2$

$= 4(628 - 494.1)^2 + 5(478.8 - 494.1)^2 +$

$5(518.8 - 494.1)^2 + 6(397 - 494.1)^2 = 132508.2$

(ii) SSE $= (n_1 - 1)s_1^2 + (n_2 - 1)s_2^2 + (n_3 - 1)s_3^2 + (n_4 - 1)s_4^2$

$= (4 - 1)(86.733)^2 + (5 - 1)(67.385)^2 +$

$(5 - 1)(89.547)^2 + (6 - 1)(67.929)^2 = 95877.6$

(iii) SST $= \Sigma(x - \bar{x})^2 = (528 - 494.1)^2 + \ldots + (330 - 494.1)^2$

$= 228385.8$

(iv) MSTR $= \dfrac{\text{SSTR}}{k - 1} = \dfrac{132508.2}{4 - 1} = 44169.40$

(v)   MSE $= \dfrac{\text{SSE}}{n - k} = \dfrac{95877.6}{20 - 4} = 5992.35$

The one-way ANOVA table for the data is:

Source	df	SS	MS = SS/df	F-statistic
Treatment	3	132508.2	44169.40	7.37
Error	16	95877.6	5992.35	
Total	19	228385.8		

With the exception of the F-statistic, every value in the table above has been calculated using (i)-(v). The F-statistic is defined and calculated as

$$F = \dfrac{\text{MSTR}}{\text{MSE}} = \dfrac{44169.40}{5992.35} = 7.37.$$

16.37 The total number of populations being sampled is k = 4. Let the subscripts 1, 2, 3, and 4 refer to Brand A, Brand B, Brand C, and Brand D, respectively. The total number of pieces of sample data is n = 20. Also, $n_1 = n_2 = n_3 = n_4 = 5$.

Step 1: H_0: $\mu_1 = \mu_2 = \mu_3 = \mu_4$ (mean lifetimes are equal)

H_a: Not all the means are equal.

Step 2: $\alpha = 0.05$

Step 3:   df = (k - 1, n - k) = (3, 16); critical value = 3.24

Step 4:   Let T_1, T_2, T_3, and T_4 refer to the sum of the data values in each of the four samples, respectively. Thus:

$T_1 = 168$ $T_2 = 154$ $T_3 = 149$ $T_4 = 143$.

Also, the sum of all the data values is $\Sigma x = 614$, and their sum of squares is $\Sigma x^2 = 19410$.

Consequently:

$$SST = \Sigma x^2 - \frac{(\Sigma x)^2}{n} = 19410 - \frac{(614)^2}{20} = 560.2 \ ,$$

$$SSTR = \left(\frac{T_1^2}{n_1} + \frac{T_2^2}{n_2} + \frac{T_3^2}{n_3} + \frac{T_4^2}{n_4} \right) - \frac{(\Sigma x)^2}{n}$$

$$= \left(\frac{168^2}{5} + \frac{154^2}{5} + \frac{149^2}{5} + \frac{143^2}{5} \right) - \frac{(614)^2}{20} = 68.2 \ ,$$

and

$$SSE = SST - SSTR = 560.2 - 68.2 = 492.$$

Step 5:   MSTR = SSTR/(k - 1) = 68.2/(4 - 1) = 22.73

MSE = SSE/(n - k) = 492/(20 - 4) = 30.75

F = MSTR/MSE = 22.73/30.75 = 0.74

The one-way ANOVA table is:

Source	df	SS	MS = SS/df	F-statistic
Treatment	3	68.2	22.73	0.74
Error	16	492.0	30.75	
Total	19	560.2		

Step 6:   From Step 5, F = 0.74. From Step 3, $F_{0.05} = 3.24$. Since 0.74 < 3.24, do not reject H_0.

Step 7:   At the 5% significance level, the data do not provide sufficient evidence to conclude that there is a difference in the mean lifetimes among the four brands of batteries.

For the P-value approach, P(F > 0.74) > 0.10. Therefore, since the P-value is larger than the significance level, do not reject H_0.

16.39 The total number of populations being sampled is k = 5. Let the subscripts 1, 2, 3, 4, and 5 refer to Transp. and Pub. util.; Wholesale trade; Retail trade; Finance, Insurance, Real Estate; and Services, respectively. The total number of pieces of sample data is n = 27. Also, $n_1 = 6$, $n_2 = 5$, $n_3 = 6$, $n_4 = 4$, and $n_5 = 6$.

Step 1: H_0: $\mu_1 = \mu_2 = \mu_3 = \mu_4 = \mu_5$ (mean earnings are equal)

H_a: Not all the means are equal.

Step 2: $\alpha = 0.05$

Step 3: df = (k - 1, n - k) = (4, 22); critical value = 2.82

Step 4: Let T_1, T_2, T_3, T_4, and T_5 refer to the sum of the data values in each of the five samples, respectively. Thus:

$T_1 = 3459$ $T_2 = 2462$ $T_3 = 1380$ $T_4 = 1840$ $T_5 = 2295$

Also, the sum of all the data values is $\Sigma x = 11436$, and their sum of squares is $\Sigma x^2 = 5,290,870$.

Consequently:

$$SST = \Sigma x^2 - \frac{(\Sigma x)^2}{n} = 5,290,870 - \frac{(11436)^2}{27} = 447,088.667 ,$$

$$SSTR = \left(\frac{T_1^2}{n_1} + \frac{T_2^2}{n_2} + \frac{T_3^2}{n_3} + \frac{T_4^2}{n_4} + \frac{T_5^2}{n_5} \right) - \frac{(\Sigma x)^2}{n}$$

$$= \left(\frac{3459^2}{6} + \frac{2462^2}{5} + \frac{1380^2}{6} + \frac{1840^2}{4} + \frac{2295^2}{6} \right) - \frac{(11436)^2}{27} = 404,258.467 ,$$

and

SSE = SST - SSTR = 447,088.667 - 404,258.467 = 42,830.200.

Step 5: MSTR = SSTR/(k - 1) = 404,258.467/(5 - 1) = 101,064.617

MSE = SSE/(n - k) = 42,830.200/(27 - 5) = 1946.827

F = MSTR/MSE = 101,064.617/1946.827 = 51.91

The one-way ANOVA table is:

Source	df	SS	MS = SS/df	F-statistic
Treatment	4	404,258.467	101,064.617	51.91
Error	22	42,830.200	1946.827	
Total	26	447,088.667		

Step 6: From Step 5, F = 51.91. From Step 3, $F_{0.05} = 2.82$. Since 51.91 > 2.82, reject H_0.

Step 7: At the 5% significance level, the data provide sufficient evidence to conclude that there is a difference in the mean weekly earnings among nonsupervisory workers in the five industries.

For the P-value approach, P(F > 41.94) < 0.005. Therefore, since the P-value is smaller than the significance level, reject H_0.

16.41 The total number of populations being sampled is k = 5. Let the subscripts 1, 2, 3, 4, and 5 refer to counterfeiting, drug laws, firearms, forgery, and fraud, respectively. The total number of pieces of sample data is n = 65. Also, $n_1 = 15$, $n_2 = 17$, $n_3 = 12$, $n_4 = 10$, and $n_5 = 11$.

Step 1: H_0: $\mu_1 = \mu_2 = \mu_3 = \mu_4 = \mu_5$ (mean times served are equal)
H_a: Not all the means are equal.

Step 2: $\alpha = 0.01$

Step 3:   df = (k - 1, n - k) = (4, 60); critical value = 3.65

Step 4:   Using the n_j's and $\bar{x}_j$'s given, we can find T_1, T_2, T_3, T_4, and T_5. Let T_1, T_2, T_3, T_4, and T_5 refer to the sum of the data values in each of the five samples, respectively. Thus:
$T_1 = 217.5$ $T_2 = 312.8$ $T_3 = 218.4$ $T_4 = 156$ $T_5 = 126.5$.

Also, the sum of all the data values is $\Sigma x = 1031.2$.

Consequently:

$$SSTR = \left(\frac{T_1^2}{n_1} + \frac{T_2^2}{n_2} + \frac{T_3^2}{n_3} + \frac{T_4^2}{n_4} + \frac{T_5^2}{n_5} \right) - \frac{(\Sigma x)^2}{n}$$

$$= \left(\frac{217.5^2}{15} + \frac{312.8^2}{17} + \frac{218.4^2}{12} + \frac{156^2}{10} + \frac{126.5^2}{11} \right) - \frac{(1031.2)^2}{65}$$

$$= 412.909 \ ,$$

$$SSE = \sum(n_j - 1)s_j^2 = 14(4.5)^2 + 16(3.8)^2 + 11(4.5)^2 + 9(3.6)^2 + 10(4.7)^2$$
$$= 1074.830$$

$$SST = SSTR + SSE = 412.909 + 1074.830 = 1487.739$$

Step 5:   MSTR = SSTR/(k - 1) = 412.909/(5 - 1) = 103.227
MSE = SSE/(n - k) = 1074.830/(65 - 5) = 17.914
F = MSTR/MSE = 103.227/17.914 = 5.76

The one-way ANOVA table is:

Source	df	SS	MS = SS/df	F-statistic
Treatment	4	412.909	103.227	5.76
Error	60	1074.830	17.914	
Total	64	1487.739		

Step 6:   From Step 5, F = 5.76. From Step 3, $F_{0.01} = 3.65$. Since 5.76 > 3.65, reject H_0.

Step 7:   At the 1% significance level, the data provide sufficient evidence to conclude that a difference exists in the mean times served by prisoners in the five offense groups.

For the P-value approach, P(F > 5.76) < 0.005. Therefore, since the P-value is smaller than the significance level, reject H_0.

16.43 From Exercise 16.35, MSE = 5992.35, n = 20, and k = 4. Using this information about MSE, we calculate s = $\sqrt{MSE}$ = $\sqrt{5992.35}$ = 77.41. For a

90% confidence interval with df = n - k = 20 - 4 = 16, $t_{\alpha/2}$ = $t_{0.05}$ = 1.746.

(a)   Also from Exercise 16.35, n_3 = 5 and $\bar{x}_3$ = 518.8. A 90% confidence interval for μ_3 is

$$\bar{x}_3 \pm t_{\alpha/2} \cdot s/\sqrt{n_3}$$

or $518.8 \pm 1.746 \cdot 77.41/\sqrt{5}$

or 518.8 ± 60.44

or   458.36 to 579.24.

(b)   Likewise from Exercise 16.35, n_1 = 4, n_4 = 6, $\bar{x}_1$ = 628, and $\bar{x}_4$ = 397. A 90% confidence interval for $\mu_1 - \mu_4$ is

$$\bar{x}_1 - \bar{x}_4 \pm t_{\alpha/2} \cdot s\sqrt{(1/n_1) + (1/n_4)}$$

or $628 - 397 \pm 1.746 \cdot 77.41\sqrt{(1/4) + (1/6)}$

or 231 ± 87.24

or   143.76 to 318.24.

(c)   The assumptions made in parts (a) and (b) are

(i)   The four random samples are independent.

(ii)   These random samples are taken from four normally distributed populations with means μ_1, μ_2, μ_3, and μ_4, respectively.

(iii) The standard deviations of the four populations are equal.

16.45 Suppose that we define the sample events A and B as follows:

A: the interval constructed around the difference $\mu_1 - \mu_2$

B: the interval constructed around the difference $\mu_1 - \mu_3$.

The 90% confidence interval for each event above is defined as:

P(A) = 0.90 and P(B) = 0.90.

The probability of both A and B occurring simultaneously is written P(A and B). Since A and B are realistically not independent, the general multiplication rule applies; i.e., $P(A \text{ and } B) = P(A) \cdot P(B|A)$.

A difficulty arises in calculating a precise number for P(A and B) because we need additional information about P(B|A), which we do not have. However, P(B|A) is by no means equal to 1.00, which results in the product of P(A) and P(B|A) being less than 0.90 (because 0.90 times a probability less than 1.00 is clearly less than 0.90). Thus, the probability of both A and B occurring simultaneously is not 0.90.

16.47 (a)   With the data in five columns named TPU, WT, RT, FIRE, and SERV,

we choose **Calc ▶ Calculator...**, select TPU in the **Input column** text box and type <u>NTPU</u> in the **Store result in variable** text box, click in the **Expression** text box, select **Normal scores** from the function list, double click on TPU to replace **number** in the NSCOR function, and click **OK**. Then repeat this process to create, respectively, the normal scores columns NWT, NRT, NFIRE, and

NSERV.   To obtain the normal probability plots, we choose **Graph ▶ Plot...,** select NTPU as the **Y** variable and TPU as the **X** variable for **Graph 1,** select NWT as the **Y** variable and WT as the **X** variable for **Graph 2,** select NRT as the **Y** variable and RT as the **X** variable for **Graph 3,** select NFIRE as the **Y** variable and FIRE as the **X** variable for **Graph 4,** select NSERV as the **Y** variable and **SERV** as the X variable for **Graph 5,** and click **OK.** The results are

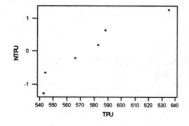
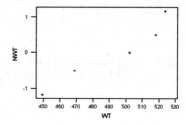
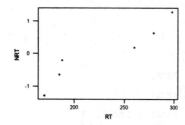
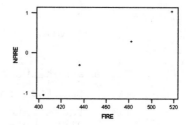
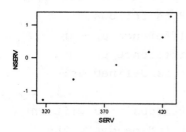

To get the standard deviations, we choose **Stat ▶ Column Statistics...,** select the **Standard deviation** button, select TPU in the **Input variable** text box, and click **OK.**   Repeat this process for the other four variables.   The results are

```
Standard deviation of TPU    =    34.309
Standard deviation of WT     =    32.316
Standard deviation of RT     =    55.343
Standard deviation of FIRE   =    50.200
Standard deviation of SERV   =    44.483
```

(b)   To carry out the residual analysis, we need to have all of the data in a single column which we first name DOLLARS.   In the next column, which we name MEANS, we place the means of the five samples, each one appearing next to each data value in its sample.

The means are respectively

Mean of TPU	=	576.50		Mean of FIRE	=	460.00
Mean of WT	=	492.40		Mean of SERV	=	382.50
Mean of RT	=	230.00				

In the MEANS column, enter 576.50 six times, 492.40 five times, ..., 382.50 six times. One way to enter the data in the DOLLARS column is to choose **Manip ▶ Stack/Unstack ▶ Stack...**, select TPU in the **Stack the following columns** text box, then select WT, RT, FIRE, and SERV; then select DOLLARS in the **Store stacked data in** text box, and click **OK**. Name two more columns RESID and NSCORE R.

To create the values for RESID, we choose **Calc ▶ Calculator...**, select RESID in the **Store result in variable** text box, type DOLLARS - MEANS in the **Expression** text box, and click **OK**. Then choose **Calc ▶ Calculator...**, select NSCORE R in the **Result in** text box, select **Normal scores** from the **function** list, select RESID to replace **number** in the NSCORE function in the **Expression** text box, and click **OK**. The residual analysis consists of plotting the residuals against the means and constructing a normal probability plot. To do this, we choose **Graph ▶ Plot...**, select RESID for the **Y** variable and MEANS for the **X** variable for **Graph 1**, select NSCORE R for the **Y** variable and RESID for the **X** variable for **Graph 2**, and click **OK**. The results are

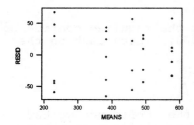
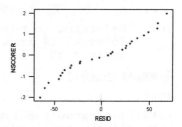

(c)   We choose **Stat ▶ ANOVA ▶ Oneway(Unstacked)...**, select TPU, WT, RT, FIRE, and SERV in the **Responses (in separate columns)** text box, and click **OK**. The result is

Analysis of Variance

Source	DF	SS	MS	F	P
Factor	4	404258	101065	51.91	0.000
Error	22	42830	1947		
Total	26	447089			

```
                                  Individual 95% CIs For Mean
                                  Based on Pooled StDev
Level    N      Mean     StDev    ----+---------+---------+---------+--
TPU      6     576.50    34.31                                (--*--)
WT       5     492.40    32.32                         (--*--)
RT       6     230.00    55.34    (--*--)
FIRE     4     460.00    50.20                    (--*---)
SERV     6     382.50    44.48              (--*--)
                                  ----+---------+---------+---------+--
Pooled StDev =     44.12          240       360       480       600
```

(d)   Based on the graphics in part (a) for the independent samples, the normality assumption appears to be met, and the ratio of the largest and smallest sample standard deviation is less than two, so Assumption 3 is met. Additionally, from the residual analysis in part (b), the normal probability plot of (all) the residuals is roughly linear with no obvious outliers. Also, the plot of the residuals against the sample means shows that the points fall roughly in a horizontal band centered and symmetric about the horizontal axis. Thus, the use of the one-way ANOVA appears reasonable.

16.49 (a)   SSTR = 36859, SSE = 2482910, SST = 2519769

(b)   MSTR = 18430, MSE = 33553

(c)   F = 0.55

(d)   Let μ_1, μ_2, and μ_3 denote the mean costs of owning and operating large, intermediate, and compact cars, respectively. Then the null and alternative hypotheses are:

H_0: $\mu_1 = \mu_2 = \mu_3$

H_a: Not all the means are equal.

(e)   p = 0.580

(f)   Since the P-value of 0.580 is greater than the specified significance level of $\alpha = 0.05$, do not reject H_0. Thus, at the 5% significance level, the data do not provide sufficient evidence to conclude that a difference exists in mean annual insurance premiums among owners of large, intermediate, and compact cars.

(g) $n_1 = 20$, $\overline{x}_1 = \$902.4$, $s_1 = \$152.5$

$n_2 = 30$, $\overline{x}_2 = \$851.7$, $s_2 = \$169.2$

$n_3 = 27$, $\overline{x}_3 = \$890.5$, $s_3 = \$215.8$

(h)   Referring to the third graph, we see that the required confidence interval is from approximately \$822 to \$960. (Each dash on the scale = \$6.)

Exercises 16.4

16.51 zero

16.53 This is because the family confidence level is the confidence we have that all the confidence intervals contain the true differences between the population means. On the other hand, the individual confidence level pertains to the confidence interval for the true difference between (two particular) population means.

16.55 $\nu = n - k$, the degrees of freedom for the denominator in the F-distribution.

16.57 (a)   4.69   (b)   5.98

16.59 Step 1:   Family confidence level = 0.95.

Step 2: $\kappa = 4$ and $\nu = n - k = 20 - 4 = 16$. Consulting Table XIII, we find that $q_{0.05} = 4.05$.

Step 3:   Before obtaining all possible confidence intervals for $\mu_i - \mu_j$, we construct a table giving the sample means and sample sizes.

j	1	2	3	4
$\overline{x}_j$	33.6	30.8	29.8	28.6
n_j	5	5	5	5

In Exercise 16.37, we found that MSE = 30.75. Now we are ready to obtain the required confidence intervals.

The endpoints for $\mu_1 - \mu_2$ are:

$$(33.6 - 30.8) \ \pm \ \frac{4.05}{\sqrt{2}} \cdot \sqrt{30.75} \ \sqrt{\frac{1}{5} + \frac{1}{5}}$$

or

-7.24 *to* 12.84

The endpoints for $\mu_1 - \mu_3$ are:

$$(33.6 - 29.8) \ \pm \ \frac{4.05}{\sqrt{2}} \cdot \sqrt{30.75} \ \sqrt{\frac{1}{5} + \frac{1}{5}}$$

or

-6.24 *to* 13.84

The endpoints for $\mu_1 - \mu_4$ are:

$$(33.6 - 28.6) \ \pm \ \frac{4.05}{\sqrt{2}} \cdot \sqrt{30.75} \ \sqrt{\frac{1}{5} + \frac{1}{5}}$$

or

-5.04 *to* 15.04

The endpoints for $\mu_2 - \mu_3$ are:

$$(30.8 - 29.8) \ \pm \ \frac{4.05}{\sqrt{2}} \cdot \sqrt{30.75} \ \sqrt{\frac{1}{5} + \frac{1}{5}}$$

or

-9.04 *to* 11.04

The endpoints for $\mu_2 - \mu_4$ are:

$$(30.8 - 28.6) \ \pm \ \frac{4.05}{\sqrt{2}} \cdot \sqrt{30.75} \ \sqrt{\frac{1}{5} + \frac{1}{5}}$$

or

-7.84 *to* 12.24

The endpoints for $\mu_3 - \mu_4$ are:

$$(29.8 - 28.6) \ \pm \ \frac{4.05}{\sqrt{2}} \cdot \sqrt{30.75} \ \sqrt{\frac{1}{5} + \frac{1}{5}}$$

or

-8.84 *to* 11.24

Step 4:   Based on the confidence intervals in Step 3, no two population means can be declared different. All of the confidence intervals contain 0.

Step 5:   We summarize the results with the following diagram.

Brand D	Brand C	Brand B	Brand A
(4)	(3)	(2)	(1)
28.6	29.8	30.8	33.6

Interpreting this diagram, we conclude with 95% confidence that no two brand means differ.

16.61 Step 1:   Family confidence level = 0.95.

Step 2: $\kappa = 5$ and $\nu = n - k = 27 - 5 = 22$. Consulting Table XIII, we take the mean of 4.23 and 4.17 to get $q_{0.05} = 4.20$.

Step 3:   Before obtaining all possible confidence intervals for

$\mu_i - \mu_j$, we construct a table giving the sample means and sample sizes.

j	1	2	3	4	5
$\bar{x}_j$	576.5	492.4	230.0	460.0	382.5
n_j	6	5	6	4	6

In Exercise 16.39, we found that MSE = 1946.8. Now we are ready to obtain the required confidence intervals.

The endpoints for $\mu_1 - \mu_2$ are:

$$(576.5 - 492.4) \pm \frac{4.20}{\sqrt{2}} \cdot \sqrt{1946.8} \sqrt{\frac{1}{6} + \frac{1}{5}}$$

or

4.8 *to* 163.4

The endpoints for $\mu_1 - \mu_3$ are:

$$(576.5 - 230.0) \pm \frac{4.20}{\sqrt{2}} \cdot \sqrt{1946.8} \sqrt{\frac{1}{6} + \frac{1}{6}}$$

or

270.8 *to* 422.2

The endpoints for $\mu_1 - \mu_4$ are:

$$(576.5 - 460.0) \pm \frac{4.20}{\sqrt{2}} \cdot \sqrt{1946.8} \sqrt{\frac{1}{6} + \frac{1}{4}}$$

or

31.9 *to* 201.1

The endpoints for $\mu_1 - \mu_5$ are:

$$(576.5 - 382.5) \pm \frac{4.20}{\sqrt{2}} \cdot \sqrt{1946.8} \sqrt{\frac{1}{6} + \frac{1}{6}}$$

or

118.3 *to* 269.7

The endpoints for $\mu_2 - \mu_3$ are:

$$(492.4 - 230.0) \pm \frac{4.20}{\sqrt{2}} \cdot \sqrt{1946.8} \sqrt{\frac{1}{5} + \frac{1}{6}}$$

or
183.1 *to* 341.7

The endpoints for $\mu_2 - \mu_4$ are:

$$(492.4 - 460.0) \pm \frac{4.20}{\sqrt{2}} \cdot \sqrt{1946.8} \sqrt{\frac{1}{5} + \frac{1}{4}}$$

or
-55.5 *to* 120.3

The endpoints for $\mu_2 - \mu_5$ are:

$$(492.4 - 382.5) \pm \frac{4.20}{\sqrt{2}} \cdot \sqrt{1946.8} \sqrt{\frac{1}{5} + \frac{1}{6}}$$

or
30.6 *to* 189.2

The endpoints for $\mu_3 - \mu_4$ are:

$$(230.0 - 460.0) \pm \frac{4.20}{\sqrt{2}} \cdot \sqrt{1946.8} \sqrt{\frac{1}{6} + \frac{1}{4}}$$

or
-314.6 *to* -145.4

The endpoints for $\mu_3 - \mu_5$ are:

$$(230.0 - 382.5) \pm \frac{4.20}{\sqrt{2}} \cdot \sqrt{1946.8} \sqrt{\frac{1}{6} + \frac{1}{6}}$$

or
-228.2 *to* -76.8

The endpoints for $\mu_4 - \mu_5$ are:

$$(460.0 - 382.5) \pm \frac{4.20}{\sqrt{2}} \cdot \sqrt{1946.8} \sqrt{\frac{1}{4} + \frac{1}{6}}$$

or
-7.1 *to* 162.1

Step 4:   Based on the confidence intervals in Step 3, we declare the following means different: μ_1 and μ_2, μ_1 and μ_3, μ_1 and μ_4, μ_1 and μ_5, μ_2 and μ_3, μ_2 and μ_5, μ_3 and μ_4, μ_3 and μ_5. The other two pairs of means are not declared different.

Step 5:   We summarize the results with the following diagram.

Retail Trade	Services	Finance, Insurance, Real Estate	Wholesale Trade	Transp. and Pub. util.
(3)	(5)	(4)	(2)	(1)
230.0	382.5	460.0	492.4	576.5

Interpreting this diagram, we conclude with 95% confidence that the mean weekly earnings of transportation/public-utility workers exceeds those of the other four industries; the mean weekly earnings of retail-trade workers is less than those of the other four industries; the mean weekly earnings of service workers is less than those of wholesale trade workers; no other means can be declared different.

16.63 Step 1:   Family confidence level = 0.99.

Step 2: $\kappa = 5$ and $\nu = n - k = 65 - 5 = 60$. Consulting Table XII, we find that $q_{0.01} = 4.82$.

Step 3:   Before obtaining all possible confidence intervals for $\mu_i - \mu_j$, we construct a table giving the sample means and sample sizes.

j	1	2	3	4	5
$\overline{x}_j$	14.5	18.4	18.2	15.6	11.5
n_j	15	17	12	10	11

In Exercise 16.41, we found that MSE = 17.914. Now we are ready to obtain the required confidence intervals.

The endpoints for $\mu_1 - \mu_2$ are:

$$(14.5 - 18.4) \pm \frac{4.82}{\sqrt{2}} \cdot \sqrt{17.914} \sqrt{\frac{1}{15} + \frac{1}{17}}$$

or
-9.01 *to* 1.21

The endpoints for $\mu_1 - \mu_3$ are:

$$(14.5 - 18.2) \pm \frac{4.82}{\sqrt{2}} \cdot \sqrt{17.914} \sqrt{\frac{1}{15} + \frac{1}{12}}$$

or
-9.29 *to* 1.89

The endpoints for $\mu_1 - \mu_4$ are:

$$(14.5 - 15.6) \pm \frac{4.82}{\sqrt{2}} \cdot \sqrt{17.914} \sqrt{\frac{1}{15} + \frac{1}{10}}$$

or
-6.99 *to* 4.79

The endpoints for $\mu_1 - \mu_5$ are:

$$(14.5 - 11.5) \pm \frac{4.82}{\sqrt{2}} \cdot \sqrt{17.914} \sqrt{\frac{1}{15} + \frac{1}{11}}$$

or
-2.73 *to* 8.73

The endpoints for $\mu_2 - \mu_3$ are:

$$(18.4 - 18.2) \pm \frac{4.82}{\sqrt{2}} \cdot \sqrt{17.914} \sqrt{\frac{1}{17} + \frac{1}{12}}$$

or
-5.24 *to* 5.64

The endpoints for $\mu_2 - \mu_4$ are:

$$(18.4 - 15.6) \pm \frac{4.82}{\sqrt{2}} \cdot \sqrt{17.914} \sqrt{\frac{1}{17} + \frac{1}{10}}$$

or
-2.95 *to* 8.55

The endpoints for $\mu_2 - \mu_5$ are:

$$(18.4 - 11.5) \pm \frac{4.82}{\sqrt{2}} \cdot \sqrt{17.914} \sqrt{\frac{1}{17} + \frac{1}{11}}$$

or
1.32 *to* 12.48

The endpoints for $\mu_3 - \mu_4$ are:

$$(18.2 - 15.6) \pm \frac{4.82}{\sqrt{2}} \cdot \sqrt{17.914} \sqrt{\frac{1}{12} + \frac{1}{10}}$$

or
-3.58 *to* 8.78

The endpoints for $\mu_3 - \mu_5$ are:

$$(18.2 - 11.5) \pm \frac{4.82}{\sqrt{2}} \cdot \sqrt{17.914} \sqrt{\frac{1}{12} + \frac{1}{11}}$$

or
0.68 *to* 12.72

The endpoints for $\mu_4 - \mu_5$ are:

$$(15.6 - 11.5) \pm \frac{4.82}{\sqrt{2}} \cdot \sqrt{17.914} \sqrt{\frac{1}{10} + \frac{1}{11}}$$

or

-2.20 *to* 10.40

Step 4: Based on the confidence intervals in Step 3, we declare the following means different: μ_2 and μ_5, μ_3 and μ_5. All other pairs of means are not declared different.

Step 5: We summarize the results with the following diagram.

Fraud	C-feiting	Forgery	Firearms	Drug laws
(5)	(1)	(4)	(3)	(2)
11.5	14.5	15.6	18.2	18.4

Interpreting this diagram, we conclude with 99% confidence that for prisoners released from federal institutions for the first time, the mean time served for firearms and drug-law offenses exceeds that for fraud offenses; no other means can be declared different.

16.65 The family confidence level is the confidence we have that all the confidence intervals contain the differences between the corresponding population means. We make our simultaneous comparison of the population means based on these confidence intervals; so the family confidence level is the appropriate level for comparing all population means simultaneously.

16.67 With all of the data in a column named DOLLARS and the group number (1, 2, 3, 4, or 5) in a column named INDUSTRY, we choose **Stat ▶ ANOVA ▶ Oneway...**, select DOLLARS in the **Response** text box and INDUSTRY in the **Factor** text box, click on the **Comparisons** button, click on **Tukey's, family error rate** so that an X shows in its check-box, type <u>5</u> in its text box, and click **OK**. The result is [**The first part of the output will be identical to the output given in Exercise 16.47(c).**]

```
    Family error rate = 0.0500
Individual error rate = 0.00707

Critical value = 4.20

Intervals for (column level mean) - (row level mean)
```

	1	2	3	4
2	4.8			
	163.4			
3	270.8	183.1		
	422.2	341.7		
4	31.9	-55.5	-314.6	
	201.1	120.3	-145.4	
5	118.3	30.6	-228.2	-7.1
	269.7	189.2	-76.8	162.1

16.69 Data is *unstacked* when the data in each sample is stored in a different column of Minitab. Data is *stacked* when all of the data is stored in one column of Minitab and the sample or group number for each data value is stored in a second column. Tukey's multiple comparison procedure in Minitab requires that the data be stacked.

16.71 (a) The family confidence level is 0.95. An interpretation would be that we can be 95% confident that all the confidence intervals contain the differences between the corresponding population means.

(b) The individual confidence level is 0.9806. An interpretation would be that we can be 98.06% confident that any particular confidence interval contains the difference between the corresponding population means.

(c) $q_{0.05} = 3.38$

(d) The confidence interval for the difference, $\mu_1 - \mu_3$, between the mean annual insurance premiums of large and compact cars is −$117 to $141.

(e) The confidence interval for the difference, $\mu_3 - \mu_2$, between the mean annual insurance premiums of compact and intermediate cars is −$77 to $155. Note that the confidence limits are the negatives of the limits given in the table for $\mu_2 - \mu_3$.

(f) No pairs of means should be declared different. All of the confidence intervals contain 0.

(g) With 95% confidence, we can state that no significant difference exists between the mean annual insurance premiums of large and intermediate cars, large and compact cars, and finally, intermediate and compact cars.

Exercises 16.5

16.73 The conditions required for using the Kruskal-Wallis test are:

(1) Independent samples
(2) Same-shape populations
(3) All sample sizes are 5 or greater.

16.75 equal

16.77 H has approximately a χ-square distribution with k − 1 degrees of freedom, so for five populations, we use critical values from the right hand side of the χ-square distribution with 4 degrees of freedom.

16.79 The total number of populations being sampled is k = 3. Let the subscripts 1, 2, and 3 refer to 1980, 1990, and 1995, respectively. The total number of pieces of sample data is n = 24. Also, $n_1 = 8$, $n_2 = 7$, and $n_3 = 9$.

Step 1: H_0: $\mu_1 = \mu_2 = \mu_3$ (mean consumption of low-fat milk are equal)
          H_a: Not all the means are equal.

Step 2: $\alpha = 0.01$

Step 3:   df = k − 1 = 2; critical value = 9.210

Step 4:

Sample 1	Rank	Sample 2	Rank	Sample 3	Rank
8.3	1	8.8	3	12.7	10
8.6	2	12.0	9	12.9	12
9.2	4	12.8	11	14.2	16
9.4	5	13.3	13	14.9	18
10.7	6	13.4	14	15.4	20
11.1	7	13.5	15	16.1	21
11.6	8	14.4	17	16.7	22
15.1	19			18.6	23
				18.9	24
	52		82		166

Step 5:

$$H = \frac{12}{n(n+1)} \sum \frac{R_j^2}{n_j} - 3(n+1)$$

$$= \frac{12}{24(24+1)} \left(\frac{52^2}{8} + \frac{82^2}{7} + \frac{166^2}{9} \right) - 3(24+1)$$

$$= 12.207$$

Step 6:   Since 12.207 > 9.210, reject H_0.

Step 7:   At the 1% significance level, the data provide sufficient evidence to conclude that there is a difference in mean consumption of low-fat milk for 1980, 1990, and 1995.

For the P-value approach, $P(\chi^2 > 12.207) < 0.005$. Therefore, because the P-value is smaller than the significance level, reject H_0.

16.81 The total number of populations being sampled is k = 4. Let the subscripts 1, 2, 3, and 4 refer to Northeast, Midwest, South, and West, respectively. The total number of pieces of sample data is n = 32. Also, $n_1 = n_2 = n_3 = n_4 = 8$.

Step 1: $H_0: \eta_1 = \eta_2 = \eta_3 = \eta_4$ (median asking rents are equal)

        H_a: Not all the medians are equal.

Step 2: $\alpha = 0.05$

Step 3:   df = k - 1 = 3; critical value = 7.815

Step 4:

Sample 1	Rank	Sample 2	Rank	Sample 3	Rank	Sample 4	Rank
1293	4	1605	11	642	1	694	2
1581	9	1639	12.5	722	3	1345	5
1781	18	1655	15.5	1354	6	1565	8
2130	23	1691	17	1513	7	1649	14
2149	25	2058	20	1591	10	1655	15.5
2286	27	2115	22	1639	12.5	2068	21
2989	30	2413	28	1982	19	2203	26
3182	31	3361	32	2135	24	2789	29
	167		158		82.5		120.5

Step 5:

$$H = \frac{12}{n(n+1)} \sum \frac{R_j^2}{n_j} - 3(n+1)$$

$$= \frac{12}{32(32+1)} \left(\frac{167^2}{8} + \frac{158^2}{8} + \frac{82.5^2}{8} + \frac{120.5^2}{8} \right) - 3(32+1)$$

$$= 6.369$$

Step 6:   Since 6.369 < 7.815, do not reject H_0.

Step 7:   At the 5% significance level, the data do not provide sufficient evidence to conclude that a difference exists among the median asking rents in the four U.S. regions.

For the P-value approach, $P(\chi^2 > 6.369) > 0.05$. Therefore, because the P-value is larger than the significance level, do not reject H_0.

16.83 (a)   The total number of populations being sampled is k = 5. Let the subscripts 1, 2, 3, 4, and 5 refer to Transp. and Pub. util.; Wholesale trade; Retail trade; Finance, Insurance, Real estate; and Services, respectively. The total number of pieces of sample data is n = 27. Also, $n_1 = 6$, $n_2 = 5$, $n_3 = 6$, $n_4 = 4$, and $n_5 = 6$.

Step 1: H_0: $\mu_1 = \mu_2 = \mu_3 = \mu_4 = \mu_5$ (mean earnings are equal)

   H_a: Not all the means are equal.

Step 2: $\alpha = 0.05$

Step 3:   df = k - 1 = 4; critical value = 9.488

Step 4:

Sample 1	Rank	Sample 2	Rank	Sample 3	Rank	Sample 4	Rank	Sample 5	Rank
543	22	449	15	170	1	404	10	317	7
544	23	469	16	185	2	436	14	343	8
566	24	502	18	188	3	482	17	380	9
583	25	518	19.5	260	4	518	19.5	408	11
588	26	524	21	279	5			420	12
635	27			298	6			427	13
	147		89.5		21		60.5		60

Step 5:

$$H = \frac{12}{n(n+1)} \sum \frac{R_j^{\,2}}{n_j} - 3(n+1)$$

$$= \frac{12}{27(27+1)} \left(\frac{147^2}{6} + \frac{89.5^2}{5} + \frac{21^2}{6} + \frac{60.5^2}{4} + \frac{60^2}{6} \right) - 3(27+1)$$

$$= 23.811$$

Step 6:   Since 23.811 > 9.488, reject H_0.

Step 7:   At the 5% significance level, the data provide sufficient evidence to conclude that a difference exists in the mean weekly earnings among nonsupervisory workers in the five industries.

For the P-value approach, $P(\chi^2 > 23.811) < 0.005$. Because the P-value is less than the significance level, reject H_0.

(b)   It is permissible to perform the Kruskal-Wallis test because normal populations having equal standard deviations have the same shape. It is better to use the one-way ANOVA test because when the assumptions for that test are met, it is more powerful than the Kruskal-Wallis test.

16.85 (a)   Since the populations are nonnormal and do not have the same shape, neither test can be performed.

(b)   Normal distributions with different shapes have different standard deviations, so the ANOVA test cannot be used. Since the shapes are different, the Kruskal-Wallis test cannot be used either.

16.87 Stacked

16.89 With all of the data in one column named GALLONS and the values 1, 2, and 3 representing the three years (1980, 1990, and 1995) in a column

named YEARS, we choose **Stat ▶ Nonparametrics ▶ Kruskal-Wallis...**, select GALLONS in the **Response** text box, select YEARS in the **Factor** text box, and click **OK**. The result is

Kruskal-Wallis Test on GALLONS

YEARS	N	Median	Ave Rank	Z
1	8	10.05	6.5	-2.94
2	7	13.30	11.7	-0.35
3	9	15.40	18.4	3.19
Overall	24		12.5	

$H = 12.21$ $DF = 2$ $P = 0.002$

16.91 (a) $H = 1.29$

(b) H_0: $\mu_1 = \mu_2 = \mu_3$, H_a: Not all the means are equal.

(c)   P-value = 0.525

(d)   At the 5% significance level, the data do not provide sufficient evidence to conclude that a difference exists in mean annual insurance premiums among owners of large, intermediate, and compact cars.

(e) $n_1 = 20$, $M_1 = \$926.20$, $\overline{R}_1 = 43.0$

$n_2 = 30$, $M_2 = \$836.40$, $\overline{R}_2 = 35.8$

$n_3 = 27$, $M_3 = \$859.00$, $\overline{R}_3 = 39.6$

where M_j and $\overline{R}_j$ denote, respectively, the median and the mean rank of the jth sample.

(f)   Neither test rejects the null hypothesis (the P-value is much larger than the significance level of 0.05 for both tests).

REVIEW TEST FOR CHAPTER 16

1.   One-way ANOVA is used to compare means.

2.   (i)   Independent samples

(ii)   Normal populations; Check with normal probability plots, histograms, dotplots.

(iii) All populations have equal standard deviations; this is a reasonable assumption if the ratio of the largest standard deviation to the smallest one is less than 2.

3.   F distribution

4.   There are n = 17 observations and k = 3 samples. The degrees of freedom are (k - 1, n - k) = (2, 14).

5.   (a)   Variation among sample means is measured by the mean square for treatments, MSTR = SSTR/(k - 1).

      (b)   The variation within samples is measured by the error mean square, MSE = SSE/(n - k).

6.   (a)   SST is the total sum of squares. It measures the total variation among all of the sample data. $SST = \sum (x - \bar{x})^2$

             SSTR is the treatment sum of squares. It measures the variation among the sample means. $SSTR = \sum n_i (\bar{x}_i - \bar{x})^2$

             SSE is the error sum of squares. It measures the variation within the samples. $SSE = \sum (x - \bar{x}_i)^2 = \sum (n_i - 1) s_i^2$

      (b)   SST = SSTR + SSE. This means that the total variation in the sample can be broken down into two components, one representing the variation between the sample means and one representing the variation within the samples.

7.   (a)   One purpose of a one-way ANOVA table is to organize and summarize the quantities required for ANOVA.

      (b)

Source	df	SS	MS=SS/df	F-statistic
Treatment	k - 1	SSTR	MSTR=SSTR/(k - 1)	F=MSTR/MSE
Error	n - k	SSE	MSE=SSE/(n - k)	
Total	n - 1	SST		

8.   If the null hypothesis is rejected, a multiple comparison is done to determine which means are different.

9.   The individual confidence level gives the confidence that we have that any particular confidence interval will contain the population quantity being estimated. The family confidence level gives the confidence that we have that all of the confidence intervals will contain all of the population quantities being estimated. The family confidence level is appropriate for multiple comparisons because we are interested in all of the possible comparisons.

10.   Tukey's multiple-comparison procedure is based upon the Studentized Range distribution or q-distribution.

11.   Larger. One has to be less confident about the truth of several statements at once than about the truth of a single statement. For example, if one were 99% confident about a single statement, the confidence that two statements were both true must be smaller since there is no way to have 100% confidence in the second statement. Similarly, each time a statement is added to the list, the overall confidence that all of them are true must decrease.

12.   The parameters for the q-curve are $\kappa = k = 3$, and $\nu = n - k = 17 - 3 = 14$. [k = number of samples and n = total number of observations.]

13.   Kruskal-Wallis test

14. X-square distribution with k − 1 degrees of freedom [k = number of samples]

15. If the null hypothesis of equal population means is true, then the means of the ranks of the k samples should be about equal. If the variation in the means of the ranks for the k samples is too large, then we have evidence against the null hypothesis.

16. Use the Kruskal-Wallis test. The outliers will have a greater effect on the ANOVA than on the Kruskal-Wallis test since an outlier will be replaced by its rank in the Kruskal-Wallis test and the rank of the most distant outlier is either 1 or n regardless of how large or small the actual data value is. In other words, the Kruskal-Wallis test is more robust to outliers than is the ANOVA.

17. (a) 2   (b) 14   (c) 3.74   (d) 6.51   (e) 3.74

18. (a) The total number of populations being sampled is k = 3. Let the subscripts 1, 2, and 3 refer to A, B, and C, respectively. The total number of pieces of sample data is n = 12. Also,

$n_1 = 3$, $n_2 = 5$, $n_3 = 4$. The following statistics for each sample are:

$\bar{x}_1 = 3$, $\bar{x}_2 = 3$, and $\bar{x}_3 = 6$. Also, $s_1 = 2.000$, $s_2 = 2.449$, and $s_3 = 4.243$. Note, the mean of all the sample data is $\bar{x} = 4$. We use this information as follows:

(b) $SST = \Sigma(x - \bar{x})^2 = (1 - 4)^2 + \ldots + (3 - 4)^2 = 110.0$
$SSTR = n_1(\bar{x}_1 - \bar{x})^2 + n_2(\bar{x}_2 - \bar{x})^2 + n_3(\bar{x}_3 - \bar{x})^2$

$= 3(3 - 4)^2 + 5(3 - 4)^2 + 4(6 - 4)^2 = 24.0$

$SSE = (n_1 - 1)s_1^2 + (n_2 - 1)s_2^2 + (n_3 - 1)s_3^2$

$= (3 - 1)(2.000)^2 + (5 - 1)(2.449)^2 + (4 - 1)(4.243)^2 = 86.0$

$SST = SSTR + SSE$ since $110.0 = 24.0 + 86.0$

(c) Let T_1, T_2, and T_3 refer to the sum of the data values in each of the three samples, respectively. Thus:
$$T_1 = 9 \quad T_2 = 15 \quad T_3 = 24.$$
Also, the sum of all the data values is $\Sigma x = 48$, and their sum of squares is $\Sigma x^2 = 302$.
Consequently:

$$SST = \Sigma x^2 - \frac{(\Sigma x)^2}{n} = 302 - \frac{(48)^2}{12} = 110 \ ,$$

$$SSTR = \left(\frac{T_1^2}{n_1} + \frac{T_2^2}{n_2} + \frac{T_3^2}{n_3} \right) - \frac{(\Sigma x)^2}{n} = \left(\frac{9^2}{3} + \frac{15^2}{5} + \frac{24^2}{4} \right) - \frac{(48)^2}{12} = 24 \ ,$$

and
$$SSE = SST - SSTR = 110 - 24 = 86 \ .$$

(d)

Source	df	SS	MS=SS/df	F-statistic
Treatment	2	24	12	1.255
Error	9	86	9.556	
Total	11	110		

19. (a)   MSTR is a measure of the variation between the sample mean losses for highway robberies, gas station robberies, and convenience store robberies.

(b)   MSE is a measure of the variation within the three samples.

(c)   The three assumptions for one-way ANOVA, given in Key Fact 16.2: independent samples, normal populations, and equal standard deviations. Assumption 1 on independent samples is absolutely essential to the one-way ANOVA procedure. Assumption 2 on normality is not too critical as long as the populations are not too far from being normally distributed. Assumption 3 on equal standard deviations is also not that important provided the sample sizes are roughly equal.

20. The total number of populations being sampled is $k = 3$. Let the subscripts 1, 2, and 3 refer to population 1, population 2, and population 3, respectively. The total number of pieces of sample data is $n = 17$. Also, $n_1 = 5$, $n_2 = n_3 = 6$.

Step 1: H_0: $\mu_1 = \mu_2 = \mu_3$ (population means are equal)

H_a: Not all the means are equal.

Step 2: $\alpha = 0.05$

Step 3: $df = (k - 1, n - k) = (2, 14)$; critical value = 3.74

Step 4:   The sums of squares and mean squares have been calculated in Problem 4.

Step 5:   The one-way ANOVA table is:

Source	df	SS	MS = SS/df	F-statistic
Treatment	2	160,601.416	80,300.708	5.34
Error	14	210,540.467	15,038.605	
	16	371,141.882		

The F-statistic is defined and calculated as

$$F = \frac{MSTR}{MSE} = \frac{80,300.708}{15,038.605} = 5.3$$

Step 6:   From Step 5, $F = 5.34$. From Step 3, $F_{0.05} = 3.74$. Since $5.34 > 3.74$, reject H_0.

Step 7:   At the 5% significance level, the data provide sufficient evidence to conclude that a difference in mean losses exists among the three types of robberies.

For the P-value approach, $0.01 < P(F > 5.34) < 0.025$. Therefore, since the P-value is smaller than the significance level, reject H_0.

21. (a) $q_{0.05} = 3.70$ (b)   4.89

22. (a)   Step 1:   Family confidence level = 0.95.

Step 2: $\kappa = 3$ and $\nu = n - k = 17 - 3 = 14$. From Exercise 21(a), we find that $q_{0.05} = 3.70$.

Step 3:   Before obtaining all possible confidence intervals for $\mu_i - \mu_j$, we construct a table giving the sample means and sample sizes.

j	1	2	3
$\overline{x}_j$	635.8	506.8	393.2
n_j	5	6	6

In Problem 19, we found that MSE = 15,039. Now we are ready to obtain the required confidence intervals.

The endpoints for $\mu_1 - \mu_2$ are:

$$(635.8 - 506.8) \pm \frac{3.70}{\sqrt{2}} \cdot \sqrt{15,039} \sqrt{\frac{1}{5} + \frac{1}{6}}$$
$$-65.3 \ to \ 323.3$$

The endpoints for $\mu_1 - \mu_3$ are:

$$(635.8 - 393.2) \pm \frac{3.70}{\sqrt{2}} \cdot \sqrt{15,039} \sqrt{\frac{1}{5} + \frac{1}{6}}$$
$$48.3 \ to \ 436.9$$

The endpoints for $\mu_2 - \mu_3$ are:

$$(506.8 - 393.2) \pm \frac{3.70}{\sqrt{2}} \cdot \sqrt{15,039} \sqrt{\frac{1}{6} + \frac{1}{6}}$$
$$-71.6 \ to \ 298.8$$

Step 4: Based on the confidence intervals in Step 3, we declare means μ_1 and μ_3 different. All other pairs of means are not declared different.

Step 5: We summarize the results with the following diagram.

Convenience store (3)	Gas station (2)	Highway (1)
393.17	506.83	635.80

(b) Interpreting this diagram, we conclude with 95% confidence that the mean loss due to convenience-store robberies is less than that due to highway robberies; no other means can be declared different.

23. (a) The total number of populations being sampled is k = 3. Let the subscripts 1, 2, and 3 refer to highway, gas station, and convenience store, respectively. The total number of pieces of sample data is n = 17. Also, $n_1 = 5$, $n_2 = n_3 = 6$.

Step 1: H_0: $\mu_1 = \mu_2 = \mu_3$ (mean losses are equal)

H_a: Not all the means are equal.

Step 2: $\alpha = 0.05$

Step 3: df = k - 1 = 2; critical value = 5.991

Step 4:

Sample 1	Rank	Sample 2	Rank	Sample 3	Rank
495	7	291	3	234	1
608	12	451	5	246	2
652	15	512	8.5	338	4
680	16	533	10	476	6
744	17	618	13	512	8.5
		636	14	553	11
	67		53.5		32.5

Step 5:

$$H = \frac{12}{n(n+1)} \sum \frac{R_j^2}{n_j} - 3(n+1)$$

$$= \frac{12}{17(17+1)} \left(\frac{67^2}{5} + \frac{53.5^2}{6} + \frac{32.5^2}{6} \right) - 3(17+1)$$

$$= 6.819$$

Step 6:   Since 6.819 > 5.991, reject H_0.

Step 7:   At the 5% significance level, the data provide sufficient evidence to conclude that a difference exists in mean losses among the three types of robberies.

For the P-value approach, $0.025 < P(\chi^2 > 6.819) < 0.05$. Therefore, because the P-value is smaller than the significance level, reject H_0.

(b)   It is permissible to perform the Kruskal-Wallis test because normal populations having equal standard deviations have the same shape. It is better to use the one-way ANOVA test because when the assumptions for that test are met, it is more powerful than the Kruskal-Wallis test.

24. (a)   With the data in three columns named HIGHWAY, GAS, and STORE we choose **Calc ▶ Calculator...**, select HIGHWAY in the **Input column** text box and type <u>NSCOREH</u> in the **Store result in variable** text box, click in the **Expression** text box, select **Normal scores** from the function list, double click on HIGHWAY to replace **number** in the NSCOR function, and click **OK**. Then repeat this process to create, respectively, the normal scores columns NSCOREG and NSCORES. To obtain the normal probability plots, we choose **Graph**

▶ **Plot...**, select NSCOREH as the **Y** variable and HIGHWAY as the **X** variable for **Graph 1**, select NSCOREG as the **Y** variable and GAS as the **X** variable for **Graph 2**, select NSCORES as the **Y** variable and STORE as the **X** variable for **Graph 3**, and click **OK**. The results are

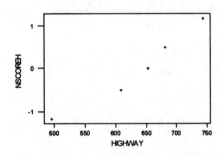

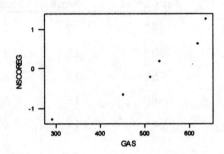

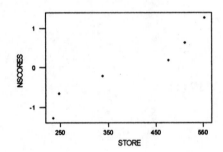

(b)   To get the standard deviations, we choose **Stat ▶ Column Statistics...**, select the **Standard deviation** button, select HIGHWAY in the **Input variable** text box, and click **OK**.   Repeat this process for the other two variables.   The results are

Standard deviation of HIGHWAY = 92.899

Standard deviation of GAS = 126.06

Standard deviation of STORE = 138.97

To carry out the residual analysis, we need to have all of the data in a single column which we first name LOSS.   In the next column, which we name MEAN, we place the means of the three samples, each one appearing next to each data value in its sample. The means are respectively

Mean of HIGHWAY = 635.80

Mean of GAS = 506.83

Mean of STORE = 393.17

In the MEAN column, enter each of the three means above five times, six times, and six times respectively.   One way to enter the data in the LOSS column is to choose **Manip ▶ Stack/Unstack ▶ Stack columns...**, select HIGHWAY in the   **Stack the following**

columns text box, then GAS, and STORE; then select LOSS in the
Store the stacked data in text box, and click **OK**.

Name two more columns RESID and NSCORE R. To create the values
for RESID, we choose **Calc ▶ Calculator...**, select RESID in the
Store result in variable text box, type LOSS - MEAN in the
Expression text box, and click **OK**. Then choose **Calc ▶
Calculator...**, select NSCORE R in the **Store result in variable**
text box, click in the **Expression** text box, select **Normal scores**
from the function list, double click on RESID in the column list
to replace **number** in the NSCOR function, and click **OK**.

The residual analysis consists of plotting the residuals against
the means and constructing a normal probability plot. To do this,
we choose **Graph ▶ Plot...**, select RESID for the **Y** variable and
MEAN for the **X** variable for **Graph 1**, select NSCORE R for the **Y**
variable and RESID for the **X** variable for **Graph 2**, and click **OK**.
The results are

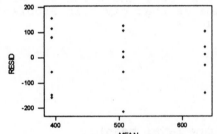 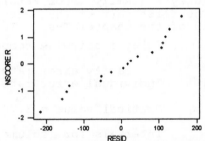

(c)   Based on the graphics in part (a) for the independent samples, the
normality assumption appears to be met, and the ratio of the
largest and smallest sample standard deviation is less than two,
so Assumption 3 is met. Additionally, from the residual analysis
in part (b), the normal probability plot of (all) the residuals is
roughly linear with no obvious outliers. Also, the plot of the
residuals against the sample means shows that the points fall
roughly in a horizontal band centered and symmetric about the
horizontal axis. Thus, the use of the one-way ANOVA appears
reasonable.

25.   We choose **Stat ▶ ANOVA ▶ Oneway(Unstacked)...**, select HIGHWAY, GAS,
and STORE in the **Responses (in separate columns)** text box, and click **OK**.
The result is

```
Analysis of Variance
Source      DF        SS         MS        F        P
Factor       2    160601      80301     5.34    0.019
Error       14    210540      15039
Total       16    371142
```

				Individual 95% CIs For Mean Based on Pooled StDev
Level	N	Mean	StDev	-+---------+---------+---------+-----
HIGHWAY	5	635.8	92.9	(------*-------)
GAS	6	506.8	126.1	(------*------)
STORE	6	393.2	139.0	(------*------)
				-+---------+---------+---------+-----
Pooled StDev =		122.6		300 450 600 750

26. (a)   SSTR = 160,601; SSE = 210,540; SST = 371,142

(b)   MSTR = 80,301; MSE = 15,039

(c)   F = 5.34

(d)   P = 0.019

(e)   Since the P-value is smaller than the significance level, reject H_0. At the 5% significance level, the data provide sufficient evidence to conclude that a difference in mean losses exists among the three types of robberies.

(f) $n_1 = 5$, $\overline{x}_1 = 635.8$, $s_1 = 92.9$

$n_2 = 6$, $\overline{x}_2 = 506.8$, $s_2 = 126.1$

$n_3 = 6$, $\overline{x}_3 = 393.2$, $s_3 = 139.0$

(g)   An individual 95% confidence interval for the mean loss due to convenience-store robberies is approximately $285.00 to $495.00.

27.   With the data in a single column named LOSS and the values 1, 2, and 3 representing HIGHWAY, GAS, and STORE in a column named TYPE, we choose

Stat ▶ ANOVA ▶ Oneway..., select LOSS in the **Response** text box and TYPE in the **Factor** text box, click on the **Comparisons** button, select **Tukey's (family error rate)**, type <u>5</u> in its text box, and click **OK** twice. The first part of the output will be the same as that given in Problem 20 of the Chapter Test. The remaining output is

Tukey's pairwise comparisons

Family error rate = 0.0500
Individual error rate = 0.0203

Critical value = 3.70

Intervals for (column level mean) - (row level mean)

	1	2
2	-65	
	323	
3	48	-72
	437	299

28. (a)   The family confidence level is 0.95.

(b)   The individual confidence level is 0.9797.

(c) $q_{0.05} = 3.70$

(d)   The (simultaneous) confidence interval for the difference, $\mu_2 - \mu_3$, between the mean losses for gas-station and convenience-store robberies is -$72 to $299.

(e)   The (simultaneous) confidence interval for the difference, $\mu_2 - \mu_1$, between the mean losses for gas-station and highway robberies is -$323 to $65.

(f)   The only means that should be declared different are μ_1 and μ_3.

(g)   With 95% confidence, we can state that the mean loss due to convenience-store robberies is less than that due to highway robberies; no other means can be declared different.

29.   We choose **Stat ▶ Nonparametrics ▶ Kruskal-Wallis...**, select LOSS in the **Response** text box and TYPE in the **Factor** text box, and click **OK**.

The result is

LEVEL	NOBS	MEDIAN	AVE. RANK	Z VALUE
1	5	652.0	13.4	2.32
2	6	522.5	8.9	-0.05
3	6	407.0	5.4	-2.16
OVERALL	17		9.0	

H = 6.82 df = 2 p = 0.033
H = 6.83 df = 2 p = 0.033 (adjusted for ties)

30.

(a) H = 6.82 (6.83 when adjusted for ties)

(b) H_0: $\mu_1 = \mu_2 = \mu_3$, H_a: Not all the means are equal.

(c) P-value = 0.033

(d) At the 5% significance level, the data provide sufficient evidence to conclude that a difference in mean losses exists among the three types of robberies.

(e) $n_1 = 5$, $M_1 = \$652.0$, $\overline{R}_1 = 13.4$

$n_2 = 6$, $M_2 = \$522.5$, $\overline{R}_2 = 8.9$

$n_3 = 6$, $M_3 = \$407.0$, $\overline{R}_3 = 5.4$

where M_j and $\overline{R}_j$ denote, respectively, the median and the mean rank of the jth sample.

(f) Both tests reject the null hypothesis, but the one-way ANOVA test provides stronger evidence against the null hypothesis than the Kruskal-Wallis test (P-values are 0.019 and 0.033, respectively).